Ecosystems and Integrated Water Resources Management in South Asia

Ecosystems and Integrated Water Resources Management in South Asia

Editors

E. R. N. Gunawardena
Brij Gopal
Hemesiri Kotagama

Routledge
Taylor & Francis Group
LONDON NEW YORK NEW DELHI

First published 2012 in India
by Routledge
912–915 Tolstoy House, 15–17 Tolstoy Marg, Connaught Place, New Delhi 110 001

Simultaneously published in the UK
by Routledge
2 Park Square, Milton Park, Abingdon, Oxfordshire OX14 4RN

First issued in paperback 2015

Routledge is an imprint of the Taylor & Francis Group, an informa business

Typeset by
Star Compugraphics Private Limited
5, CSC, Near City Apartments
Vasundhara Enclave
Delhi 110 096

British Library Cataloguing-in-Publication Data
A catalogue record of this book is available from the British Library

ISBN 13: 978-1-138-66388-6 (pbk)
ISBN 13: 978-0-415-69305-9 (hbk)

Contents

List of Tables

List of Figures

List of Maps

List of Boxes

Foreword

This is the third volume in a series of books developed by the SaciWATERs, the South Asia Consortium for Interdisciplinary Water Resources Studies (www.saciwaters.org). This series of readers on topical water resources issues in the South Asia region is generated through a project called *Crossing Boundaries: Regional Capacity Building on IWRM and Gender and Water in South Asia*, which seeks to strengthen or establish new Masters education programmes on water resources management that will have a broader scope than the conventional technical focus of water resources education, and incorporate concerns like ecological sustainability, equity and poverty, gender relations, and democratic governance into professional water education. This reconfiguring of professional water education is supported by a series of training, research and networking activities, of which the production of this series of readers is an important one.

The *Crossing Boundaries* project is implemented with financial support from the Government of the Netherlands, which is gratefully acknowledged here. The coordinating South Asian partner is SaciWATERs, based in Hyderabad, India. The South Asian University partners are the Bangladesh Centre for Advanced Studies (BCAS), Dhaka, Bangladesh; Institute of Water and Flood Management (IWFM), Dhaka, Bangladesh; Bangladesh University of Engineering and Technology (BUET), Dhaka, Bangladesh; Centre for Water Resources (CWR), Anna University, Chennai, India; Tata Institute of Social Sciences, Mumbai, India; Nepal Engineering College in Kathmandu, Nepal; and the Postgraduate Institute of Agriculture, University of Peradeniya, Peradeniya, Sri Lanka. The eighth partner is the Irrigation and Water Engineering group at Wageningen University, the Netherlands.

The series is very much the collective product of the project partners, with each volume being edited by two or three editors and comprising contributions from several South Asian countries. The focus on comparisons at the South Asia level and the addressing of regional South Asian water issues is the series' hallmark. The series seeks to provide high-quality collections of 'state-of-the-art' materials on different water resources topics and the policy and research agenda that need to be addressed, which is made accessible to a broad interested audience, and is suitable as resource material in education and other forms of capacity building

This volume attempts to cover issues related to aquatic ecosystems and explains how ecosystems are linked to the concept of integrated water resources management (IWRM). The contributions address three major aspects: (a) concepts related to ecosystems, ecosystem services and their linkages with water; (b) human impacts on ecosystems, particularly aquatic ecosystems, and their assessment; and (c) aquatic ecosystem management, including policy, governance and economics. A key message of the book is that for effective intervention for change it is important to address these different aspects simultaneously. Most of the contributions are based on case studies drawn from South Asian countries except one which examines wetland management in Vietnam. It is hoped that this volume will provide information and insights to inspire new generations of water professionals to think across the boundaries of engineering and hydrological disciplinary territories to secure the future of South Asia's aquatic ecosystems and those who directly and indirectly depend on them for livelihood and quality of life, that is, all of us.

Peter P. Mollinga
London/Voorburg
June 2011

Acknowledgements

The financial support of the Netherlands government through the Crossing Boundaries Project, a regional capacity building project in IWRM, Gender and Water and the coordination activities during the various stages of producing this book by the staff of SaciWATERs, especially Dr Jayati Chourey, are gratefully acknowledged.

All efforts have been made to ensure the accuracy of images. The depiction and use of boundaries shown in maps do not imply official endorsement. Every effort has been made to contact owners of copyright regarding the visual material and illustrations reproduced in this book. Perceived omissions if brought to notice will be rectified in future printing. The publishers apologise if inadvertently any source remains unacknowledged.

1

Ecosystems and Integrated Water Resources Management: The Link and the Need for Integration

Brij Gopal, Hemesiri Kotagama and E. R. N. Gunawardena

Water is the most abundant and almost omnipresent substance on the earth. It forms oceans, often several kilometers deep, that cover nearly three-fourths of the earth's surface — it occurs in enormous polar ice caps, glaciers and snow on mountain peaks; it is buried deep inside the earth in the aquifers; it occurs throughout the atmosphere; it resides in millions of lakes and ponds and flows through an intricate network of innumerable streams and rivers which generally carry it to the oceans. Everywhere on the earth, it sustains life in myriad forms of living organisms — from microscopic bacteria to mighty mammals — and constitutes up to 99 per cent of their body mass. Driven by solar energy, the water remains in a state of continuous circulation, changing its form along its pathways.

This circulation of water, known as the hydrological cycle, unites all the living organisms and their non-living environment (including land, air and energy) into a unique system — the earth's ecosystem — whose functioning sustains the humans and their diverse socio-cultural manifestations. The global ecosystem (called as ecosphere) comprises numerous smaller ecosystems interacting with each other in various ways. Ecosystems (short for ecological systems) are nature's functional units composed of dynamically interacting and interdependent community of plants, animals and micro-organisms, together with their non-living environment. These complex interactions between the organisms and their abiotic environment result in the flow of energy (as the radiant energy transformed by plants into chemical energy gets dissipated as heat energy after passing through the animals as food) and the cycling of elements (the elements utilised by the plants in synthesising organic compounds return to their respective pools after the death and decay of the organisms). Water plays the critical role in carrying various elements with it as well as in energy transfer. During photosynthesis, the water molecule is split to release oxygen and the hydrogen is utilised to reduce carbon dioxide into an organic molecule. The oxidation of the organic molecules during respiration or by microbial activity returns the water molecule to the environment. Organisms in the ecosystems do not produce waste. The excreta of the animals, as well as the dead remains of various organisms, serve as a source of energy and nutrients for numerous organisms responsible for 'scavenging' in different ecosystems. An important characteristic of the 'mature' ecosystems is their resilience to repair themselves and recover from the impacts of various stresses.

Ecosystems differ from each other in the suite of plant and animal communities which comprise them under the given climatic, geomorphic and edaphic conditions (e.g., forests, grasslands, savannas, deserts, peat bogs, etc.). They are generally categorised into terrestrial and aquatic ecosystems on the basis of the land or water being the dominant feature of their physical environment which also determines the kind of organisms that occur there. Thus, rivers are not mere conduits of water from land to the sea, and the lakes are not simply the storages of water but functioning ecosystems. It is not possible to define specific boundaries of an ecosystem though the ponds, lakes and rivers appear to be distinct from the surrounding land. All ecosystems are open systems to the extent that they interact with each other, more actively with the adjacent ecosystems, often exchanging materials, water and energy through physical and biological processes. The terrestrial and aquatic ecosystems interact most intensely as the water moves over land or subsurface to the aquatic ecosystem, and carries with it sediments, nutrients, organic matter and also organisms (or their propagules).

Thus, all natural ecosystems need water for their functioning, and are indeed the primary users of water. They are the direct recipients of all precipitation and are adapted to its natural variability. The natural ecosystems utilise a small amount of water and provide a host of benefits (now called as ecosystem services) that humans cannot obtain from the water themselves. The ecosystems provide food, fiber, fuel, fodder, timber and medicines; protect the soil from erosion; moderate the microclimate; assimilate wastes, and enhance the aesthetics of the landscape. Terrestrial plants absorb a relatively large amount of water that moves into the ground but only a small fraction of this water is converted to biomass and the remaining is transpired as water vapour into the atmosphere. This water, which returns back to the atmosphere, has been called as 'green water' and is important to humans in regulating the climate and microclimate of the region. Plants, animals and micro-organisms living inside water use only a very small amount of water for their growth. Only a few aquatic plants (with their shoots emerging out of water or those floating on the water surface) release water vapours into the atmosphere through transpiration. Animals do not use water for growing their food, or for generating energy. Rather, plants and animals use water to produce food for humans. Organisms, particularly the micro-organisms, living in water utilise wastes discharged by humans and a variety of ecosystem processes utilise solar energy to keep the water clean. The waste assimilation capacity of the flowing rivers cannot be replicated by humans using all the available technologies, even at an enormous cost.

Human use of Water

Humans are also a component of the earth's natural ecosystem. Like all other animals, they draw their water from food and also drink it. The biological requirement of water by humans is estimated at less than 5 liters per day per person. However, ever since humans switched from being a hunter-gatherer to a food cultivator and settled in the floodplain of seasonal rivers, they have diversified their use of water, and increasingly so during the past few decades. The development of human civilisation has rested primarily upon water. Unlike all other animals, humans use water for growing their food, processing this food before consumption, sanitation (bathing, washing and disposal of domestic wastes), manipulating the micro-climate around them (cooling and heating), producing industrial goods and disposal of industrial wastes, generating energy (hydropower), transport of people and materials and for recreation and aesthetics. Interestingly, the human being can utilise only a small amount of water (less than 0.3 per cent of all the freshwater on the earth) that occurs in and moves through the rivers and lakes, for its diverse needs. The use of water in deep aquifers is slowly increasing, and that of the water in the oceans is limited to using it for transportation needs, though coastal waters are being increasingly used for recreation and disposal of domestic and industrial wastes.

The human use of water is complicated by a large number of factors. First and foremost, the earth's water resources are highly unevenly distributed — both spatially and temporally — and at the same time, they vary also in their quality. Second, the human population is also unevenly distributed, and incidentally, the density and growth rate of human populations are generally large in areas relatively poorer in their water resources. The ever-increasing demands of a growing population for better economic development raise the water requirements far beyond its availability. The next important complicating factor is water quality. All human uses of water are not consumptive in nature but invariably affect its 'quality' adversely for most of the other uses. The alterations in the physical (such as temperature and turbidity), chemical (addition of various inorganic or organic substances) and/or biological (such as pathogens) characteristics of water render it unfit for use (wastewater) unless it is treated appropriately to remove these impurities. The discharge of municipal sewage and industrial effluents are causing the most serious problems for treatment of wastewater and availability of water for various uses.

The problem of degradation of the quality of water is however not restricted to the direct disposal of untreated or partly treated wastewater. Human activities such as bathing, washing and recreation, and even transport also significantly contribute to the pollution load in the water bodies. All anthropogenic activities throughout the landscape, near or far away from the water bodies, cause some impact on the quality of water. The intensive use of agrochemicals (fertilisers and pesticides) results in the input of these substances into water bodies through surface and subsurface run-off. Deforestation, fire, overgrazing, mining and other land-based activities promote run-off, erosion and thereby increased inputs of sediments, nutrients, organic matter and toxic substances into water bodies. Urbanisation increases storm run-off that also carries with it a variety of wastes into rivers and lakes. Burning of fossil fuels brings into the atmosphere the oxides of sulphur and nitrogen which enter the water bodies with precipitation, thereby causing a lowering of the water pH (acid rain). The problems of water quality are further linked closely to the quantity of water, and particularly to its spatial and temporal variability in a region. The wastes may get diluted, dispersed, flushed and transformed when large amounts of freshwater are available or else may get concentrated at a place.

Water Management

With increasing dependence on water, first for agriculture and domestic use, and later for industrial development, humans learned also to manage the water resources, particularly in terms of their limited availability and extreme seasonal variability such as that experienced in different parts of South Asia. Scriptures of several cultures make numerous references to the need for managing water properly and protect its quality against all kinds of polluting activities. The rulers were exhorted to construct water storage structures (*taal* or tanks) to meet domestic and irrigation requirements during the dry periods. Canals were also built to carry water to distant areas. However, the post-industrial development period has witnessed the emergence of humans as the major users of water who compete also among themselves for different uses. The traditional community-based water management has been taken over throughout the world by the centralised custodians of water resources (often the ministries of water resources) who administer their development through a storage and distribution network. Laws and policies have been formulated that 'allocate' the available water resources to different users according to perceived priorities. Attempts are also made to transfer water to long distances from so-called 'surplus' areas to 'deficient' areas, across hydrographic basins. These water management practices focus only on the quantitative aspects, often with little coordination between the surface and groundwater resources, and generally ignore water quality considerations. The management of water quality is left to a group of different users (user ministries) such as the ministries of environment, urban development and industry and health. The largest user of water, the Ministry of Agriculture, usually has no say in the management of the resources. The same is true for the water transport and tourism sectors as well.

In recent years, water managers are confronted with a new dimension of environmental consequences which had either not been recognised or simply ignored. As the human demands on water continue to rise steeply, and conflicts arise between demands for agriculture, domestic use, industry and energy generation, the impacts of excessive water abstraction and use started appearing and sometimes, threatening human well-being in several ways. Consequent upon the management's obsession with the quantities of water, the rivers have been increasingly regulated; their flows are stored and diverted by constructing barrages and huge dams. Rivers have stopped flowing as construction of dams on their large stretches upstream have turned them into deep reservoirs, submerging large areas of natural forest and grassland ecosystems, and even agricultural systems. Far longer reaches downstream remain dry for most of the year except when the reservoirs are unable to hold more water in the rainy season. The environmental and socioeconomic impacts of large dams on their upstream areas have been debated for long with practically little remedial action being taken. The rivers are also channelised by constructing embankments, ostensibly for flood control. The elimination of floodplains and their seasonal flooding has decimated the groundwater recharge. The downstream impacts have now become equally obvious. Fisheries are declining, many species of aquatic plants and animals are fast disappearing and exotic invasive species are spreading; the rivers have lost their natural capacity for assimilating wastes and are fast turning into sewers. The groundwater table is falling rapidly. Further, the inflow of run-off into the lakes is diverted or obstructed. While new water storages are being constructed, the earlier ones are getting filled up with sediments eroded from their catchments or with the solid wastes dumped there. Consequent impacts are being felt on the livelihoods of communities dependent upon the water bodies and their resources.

A vicious cycle of excessive water abstraction and use affects both the terrestrial and aquatic ecosystems. Expansion of agricultural lands with increasing supply of water from large reservoirs acts as a double-edged sword to eliminate the natural ecosystems. Intensive agriculture, which relies on excessive use of agrochemicals and irrigation, also results in the modification of the vegetation type and cover and consequently, has significant adverse effects on the partitioning of precipitation into 'green' and 'blue' water, besides promoting soil erosion and nutrient loss which ultimately impinge upon water quality (Falkenmark and Rockstrom 2004).

Unfortunately, the natural ecosystems are still perceived by many as 'water users' which compete with the humans for water, and the water flowing through the rivers is seen as a waste going to the sea. It is in this context that the natural ecosystems have entered into the discussion on water management.

The IWRM Concept

The importance of water for sustaining life and environment was first recognised at the International Conference on Water and the

Environment (ICWE) which was held in Dublin (1992) with the backing of UN agencies and participation of experts from most countries. The conference enunciated four basic principles (now known as Dublin Principles) which recognised freshwater as a 'finite and valuable resource that is essential to sustain life, the environment and development' but at the same time as an 'economic good' and called for a broad participatory approach (including women) to the development and management of water resources.

Later, the Agenda 21 (UNCED 1992) for the first time used the term Integrated Water Resources Management (IWRM) but it was elaborated only by the Global Water Partnership (GWP 2000) which defined it as 'a process which promotes the coordinated development and management of water, land and related resources, in order to maximise the resultant economic and social welfare in an equitable manner without compromising the sustainability of vital ecosystems'. The concept has been discussed in detail in several publications (Jonch-Clausen and Fugl 2001; Biswas 2004; Snellen and Schrevel 2004; Saravanan et al. 2008). In case of water resources, the list of issues that need to be considered for integrated management is too long and complicated. One needs to consider the sources — surface water (glaciers, snow fields, rivers, lakes, wetlands and estuaries) and groundwater; distribution (geographical and seasonal); users (terrestrial and aquatic plants and animals, micro-organisms and humans); human uses (domestic, agriculture, industries, sanitation, hydropower, navigation and recreation); aspects of quantity and quality; the spatio-temporal variability and interactions between all these components; land and water interrelations, upstream–downstream linkages; and numerous drivers of change (laws, policies, institutions and social factors) and the complexities of management (issues and technologies related to water supply, wastewater collection, treatment and disposal; institutional set up, public and private sectors; government and NGOs; economic instruments such as pricing, taxes, subsidies, various social groups, regional, interstate and international issues, costs and benefits, economics of ecosystem services, and many more). Climate change that is now happening brings a very different and complex dimension as it affects everything else as well. The melting of glaciers and polar ice caps, the rising sea level, and above all the increased spatial and temporal variability coupled with increased frequency of extreme events, have all to be factored into 'integrated' management of water resources.

Crossing Boundaries Project and the 'Reader' Series

Despite the complexity of integrating so many issues, and the lack of clarity on how to operationalise it, the term IWRM has become popular with most organisations, governments and individual researchers. The South Asian region shares among its countries a monsoonal climate influenced by the mighty Himalaya, and a long social–cultural history. Most countries share the rivers which arise in the Himalaya and discharge into the Arabian Sea or the Bay of Bengal. The region's countries have many similarities in their biophysical environment, dependence on agriculture, needs for economic development and therefore the problems related to the water resources and their management.

The South Asia Consortium on Interdisciplinary Water Resources Studies (SaciWaters) was formed to build capacity among the water professionals of the region for managing their water resources effectively and in an integrated manner. It formulated the Crossing Boundaries Project with the goal 'to strengthen integrated and gender-sensitive water resources management policy and practice in South Asia by means of a regional, collaborative, partnership-based capacity building programme for active water professionals through higher education, innovation-focussed research and knowledge base development, and networking'.

One of the activities under the project is to produce a 10-volume series, including this book, on topical water resource issues in the South Asia region. The series provides material for an innovative professional education programmes on IWRM for engineers.

Content

This volume, the sixth in the series, attempts to cover the issues related to aquatic ecosystems. It includes 12 selected contributions that were presented and discussed at a workshop, and then modified and elaborated in light of the discussions on the overall theme of water as an integral component of ecosystems. These contributions address three major aspects: (a) the concepts related to ecosystems, ecosystem services and their linkages with water; (b) the human impacts on ecosystems, particularly the aquatic ecosystems, and their assessment; and (c) the management, including policy, governance and economics. Most of the contributions are based on case studies drawn from South Asian countries, except one which examines wetland management in Vietnam.

The first chapter introduces the ecosystem concept and describes the salient characteristics of structure and functioning of inland freshwater ecosystems, namely rivers, lakes and wetlands. It also mentions briefly the groundwater ecosystems and then describes the ecosystem services of these freshwater ecosystems and concludes with a brief mention of anthropogenic impacts on these ecosystems. The chapter also describes how the variability in the hydrological regime (flow in rivers) affects the biodiversity and functioning of freshwater ecosystems.

The importance of flow in the rivers is elaborated in the next chapter which addresses the current issue of environmental flow which is 'the amount of water needed for river and groundwater systems to maintain themselves and their ecological functions, uses and benefit to people'. It describes the concept and the methods used in different countries to determine or prescribe the environmental flow requirements in rivers for different goals and objectives. This chapter then presents a case study from Bangladesh. It points out that water allocation can be optimised through a trade-off between natural and highly managed systems. The next chapter describes in detail the Ganga river ecosystem focusing on its biodiversity, water quality and the management in relation to various human interventions.

The next three chapters deal with the direct and indirect human impacts on freshwater ecosystems. The first of these focuses on the multiple impacts on the river system in Bangladesh and how the flow regulation, embankments and the human interventions affect fisheries and livelihoods. The next chapter examines in detail the science of climate change, and discusses its impacts on freshwater resources, rise of sea level and its consequences, with particular reference to Bangladesh. The global efforts to understand climate change and develop strategies for mitigation and adaptation are also described briefly. The next chapter takes a much broader view of the environmental impacts of water resource development projects and describes how the procedures of Environmental Impact Assessment (EIA) can be applied to manage water resources. A case study from Vietnam focuses on the ecosystem approach to water management and highlights the role of hydrology in the management of the seasonally flooded wetlands in the Mekong delta. It also demonstrates that humans are a part of the ecosystem and that the ecosystems require an adaptive management.

A case study from Tamil Nadu in south India is devoted to ecosystem damages caused by the discharge of industrial effluents into the river. The wastewater effluents of several tanneries over the years have turned the water in the river, in the reservoir on it and further downstream totally unfit for drinking; the crops were damaged and even the fisheries declined, leaving only tilapias which were also contaminated. The damages caused due to pollution in terms of costs of drinking water supplies and loss of crops and fisheries were evaluated in monetary terms. This valuation formed the basis for compensation to be provided to the farmers and the municipality in the river basin area by the industries through the local administration. The study shows that the costs due to damages to even a few obvious components could far exceed the benefits drawn by the industries by neglecting treatment of wastewater.

The final four chapters explore the overarching issues of watershed management, namely; institutions, governance and policies. A case study from Sri Lanka describes the gradual deterioration of urban wetlands near Colombo and the problems of management and institutions. These wetlands served the main function of flood detention and provided resources to the local community. Later parts of these wetlands were converted to paddy fields where urbanisation in the surrounding areas triggered a land use change and pollution with domestic wastes. Large areas of wetlands and paddy fields were reclaimed and brought under habitation. The water quality degraded and the water detention function of the wetland declined greatly. This chapter discusses the gradual change in institutions and policies responsible for the degradation and loss of wetlands and suggests an institutional framework, in consonance with the national wetland policy of the country, for better management of these wetlands.

The subsequent chapter sheds light on the role of forest- and water-based institutions in promoting livelihoods opportunities, equity in benefit-sharing among and across users and maintenance of ecosystem services, its sustainability and their limitations. This chapter also explores the dynamics of natural resource management through understanding of linkages among and between the local level and external institutions. An attempt is made to understand the role resource user groups and their institutions play in the management of natural resources in promoting livelihood opportunities and the constraints and opportunities for integrated management of natural resources through existing institutions.

The next chapter explains the concept of environmental governance, and explores its complexity by providing a contextual illustration of three case studies on Indian wetlands management. It is argued that challenges to environmental governance and its complexity are due to the complex nature of wetland ecosystems with competing uses and resulting multiple livelihood dependencies, and that the complex governance structure due to the historical evolution hinged on a sectoral understanding of natural resources that led to a fragmented institutional structure (policies, laws and organisations).

The final chapter elicits the need of a national policy on financing watershed management considering the case of Sri Lanka. This chapter explains the theoretical ideals of financing watershed management based on economic theory and examines the possibilities of practically achieving such ideals. It is advocated that efforts should be made to acquire finances for watershed management from the environmental/watershed resource users, both domestic and foreign. The acquiring and use of finances for watershed management should be decentralised to provincial government from central government, as much as feasible, in the long run. In the short-run, central government should recognise the level of watershed benefits contributed by provinces in rationalising financial allocations. Finances allocated for watershed management must be clearly earmarked from the central government budget to provincial governments and up to the level of implementing departments.

It has become clear that the ecosystems are being degraded at an alarming rate threatening the very survival of humans. Attempts have been made to understand the drivers of this change and interventions that are being made to contain and reverse this trend through technical, social, institutional, governance and economic means as elaborated in this volume. It is important to ensure that these different aspects are addressed simultaneously in order to make an effective intervention for change. However, much more work is needed to make a lasting impact, including a paradigm shift in perception with regard to 'development', to address problems associated with the degradation of ecosystems. It is hoped that this volume would provide information and thoughts for the new generations of water professionals to take the lead in securing water and associated ecosystems to ensure the survival of humans and associated life forms.

❋

References and Select Bibliography

Biswas, A. K. 2004. 'From Mar del Plata to Kyoto : A Review of Global Water Policy Dialogues', *Global Environmental Change Part A* 14: 81–88.

Falkenmark, Malin and Johan Rockstrom. 2004. *Balancing Water for Humans and Nature: The New Approach in Ecohydrology*. London: Earthscan.

GWP (Global Water Partnership). 2000. 'How to Bring Ecological Services into Integrated Water Resources Management', Report from GWP Seminar, Beijer Occasional Paper series. Stockholm: Beijer Institute and Global Water Partnership.

Jonch-Clausen, T. and J. Fugl. 2001. 'Firming up the Conceptual Basis of Integrated Water Resource Management', *Water Resources Development* 17: 501–10.

Millennium Ecosystem Assessment (MEA). 2005a. *Ecosystems and Human Well-being: Synthesis*. Washington, DC: Island Press.

———. 2005b. *Ecosystems and Human Well-Being: Wetlands and Water Synthesis*. Washington, DC: World Resources Institute.

Postel, S. and S. Carpenter. 1997. 'Freshwater Ecosystem Services', in G. Daily (ed.), *Nature's Services: Societal Dependence on Natural Ecosystems*, pp. 195–214. Washington DC: Island Press.

Saravanan, V. S., G. T. McDonald and P. P. Mollinga. 2008. 'Critical Review of Integrated Water Resources Management: Moving Beyond Polarised Discourse', ZEF Working Paper 29. Bonn: Center for Development Research, University of Bonn.

Snellen, W. B. and A. Schrevel. 2004. *IWRM: for Sustainable Use of Water: 50 years of International Experience with the Concept of Integrated Water Resources Management: Background Document to the FAO/ Netherlands Conference on Water for Food and Ecosystems*. Den Haag: Ministry of Agriculture, Nature and Food Quality; Wageningen: Alterra (Alterra Report 1143).

United Nations Conference on Environment and Development (UNCED). 1992. *Agenda 21: Earth Summit — The United Nations Programme of Action from Rio*. New York: United Nations.

2

Freshwater Ecosystems: Ecological Characteristics and Ecosystem Services

Brij Gopal

Introduction

The Water Resources

Water is the most vital component of the abiotic environment without which life cannot exist. Water occurs on the earth in abundance. Nearly three-fourths of the earth's surface is covered with deep water (oceans and lakes). Widely referred estimates of the global stocks of water show that the earth has nearly 1,386 million km^3 of water of which more than 96.5 per cent resides in the oceans and another 1 per cent is saline or brackish groundwater (Shiklomanov 1993; see also Trenberth et al. 2007). Out of the remaining 2.5 per cent which is freshwater, more than 69.5 per cent is frozen in glaciers, polar ice caps, snow and permafrost, and more than 30 per cent constitutes the groundwater. Less than 0.3 per cent of all freshwater on the earth (i.e., less than 0.0072 per cent of the total water) occurring in the lakes (0.26 per cent), rivers (0.006 per cent) and wetlands (0.03 per cent) is utilisable by humans for their diverse needs (Table 2.1).

However, yet smaller fraction of total water circulates through the hydrological cycle operating between oceans, atmosphere and the earth's surface (Figure 2.1). Water enters the atmosphere principally by evaporation from the surface of oceans, lakes, rivers, land and plants, as well as by transpiration from green plants. Snow and ice at their surfaces can also turn directly into water vapours. Water returns to the earth by condensation in the form of precipitation (snow and rain). The precipitation reaching the ground either infiltrates into the ground, or is stored on the surface, or runs off on the surface until it infiltrates or is stored. The groundwater moves through rocks and sediments by percolation by gravity and pressure, and may get stored in shallow or deep aquifers (confined water bodies). However, the water table may intersect the ground surface so that water is

Table 2.1: **Major Stocks of Water**

	Volume (1,000 km³)	Percentage of total water	Percentage of total freshwater
Salt water			
Oceans	1,338,000.00	96.5400	
Saline/brackish groundwater	12,870.00	0.9300	
Salt water lakes	85.00	0.0060	
Inland waters			
Glaciers, permanent snow cover	24,064.00	1.7400	68.700
Fresh groundwater	10,530.00	0.7600	30.060
Ground ice, permafrost	300.00	0.0220	0.860
Freshwater lakes	91.00	0.0070	0.260
Soil moisture	16.50	0.0010	0.050
Atmospheric water vapour	12.90	0.0010	0.040
Marshes, wetlands	11.50	0.0010	0.030
Rivers	2.12	0.0002	0.006
Incorporated in biota	1.12	0.0001	0.003
Total water	**1,386,000.00**	**100.0000**	
Total freshwater	**35,029.00**		**100.000**

Source: Shiklomanov (1993); Shiklomanov and Rodda (2003).

discharged at the surface as springs or into any other water body (lake, river or ocean).

The hydrological cycle binds together all components of the earth's various ecosystems, drives all natural (physical, chemical and biological) processes and sustains various organisms, plant and animal communities, humans and all life processes and activities. It is estimated that out of an annual evaporation of 577,000 km³ of water from the surface of oceans and land, 119,000 km³ precipitates back on land. Of this amount, 42,700 km³ runs off through the rivers and 2,100 km³ as groundwater to the oceans. Thus, it is the amount of water flowing through the rivers every year that is more important to humans. There is however a far greater volume of water stored in shallow and deep aquifers over the past many millennia. It must be noted that the recharge rate (portion of precipitation that moves into the aquifers every year) is very low, and spatially variable depending upon the permeability of the ground. Yet, the groundwater forms an important source of water for human use. The groundwater accounts for around 30 per cent of drinking water supply in Asia Pacific and South America and as high as 75 per cent in Europe. Likewise, about 30 per cent of total water used in agriculture in India is obtained from groundwater (Kumar et al. 2005) whereas in many countries of the

Figure 2.1: The Water Cycle

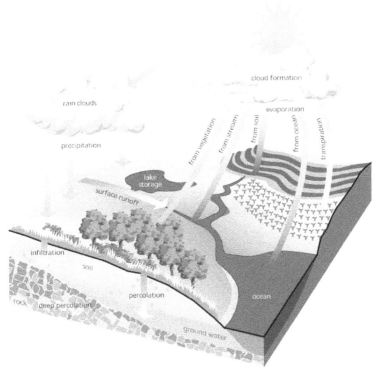

Source: Adapted form FISRWG (1998).

arid regions, the groundwater meets 80–95 per cent of agriculture's irrigation requirement (FAO 2007). It should also be noted that both surface and groundwater are part of the same continuum maintained by the hydrological cycle and interact with each other in various ways.

The forests and other ecosystems affect the hydrological processes such as the partitioning of rainfall into interception, throughfall, stem flow, evaporation, transpiration, groundwater movement and run-off (Figure 2.2), and even influence the rainfall (see Meher Homji 1984). Humans depend upon this freshwater for most of their needs but various anthropogenic activities cause enormous impact on the hydrological cycle that in turn has many feedback effects on humans themselves. Some of these are discussed later in the context of aquatic ecosystems.

Figure 2.2: The Influence of Trees on Hydrological Processes

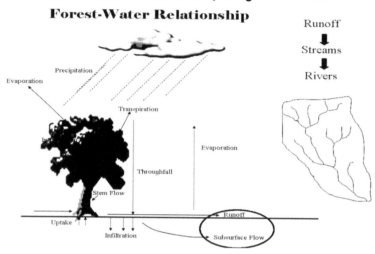

Source: Prepared by the author.

Ecosystems

The term 'ecosystem' was coined by Tansley for 'the basic units of nature 'including not only the organism-complex, but also the whole complex of physical factors forming what we call the environment of the biome — the habitat factors in the widest sense. Though the organism may claim our primary interest, when we are trying to think fundamentally we cannot separate them from their special environment, with which they form one physical system' (1935: 299).

In nature, all living organisms — plants, animals, micro-organisms as well as humans — interact with each other and with their non-living (abiotic) environment comprising water, energy, air and minerals in the earth in a manner that makes them inseparable. The existence of plants depends upon the simple molecules of carbon dioxide and water from which they synthesise complex organic substances utilising solar energy and minerals absorbed from the soil. Some micro-organisms are able to convert gaseous nitrogen to nitrate ions (nitrogen fixation) that become a source of nitrogen required by the plants. All other organisms depend for their food directly or indirectly upon these organic substances produced through photosynthesis by the plants. Animals are linked with each other through the relationship of eating and being eaten (food chain). The dead remains of plants and animals as well their excreta are utilised by a large diversity of animals and micro-organisms which break the complex compounds (decomposition) to release various elements and molecules back into the environment. The solar energy utilised by the plants is returned back as heat energy during respiration and decomposition. Thus, in nature, whereas the chemical elements such as carbon, hydrogen, oxygen, nitrogen and other minerals are cycled through the organisms and the environment (uptake, transfer and release), the energy flows because of transformation in its form — radiant energy converted to chemical energy and then heat energy (Figure 2.3).

Thus, ecosystems are complex dynamic functional units of nature consisting of living organisms (plants, animals and microbes) and all the non-living physical and chemical components of their environment which interact among themselves and with each other to generate cyclic interchange of materials and flow of energy through it. However, definite boundaries cannot be defined for these functional

Figure 2.3: Schematic Representation of Ecosystem Structure with Flow of Energy and Cycling of Nutrients

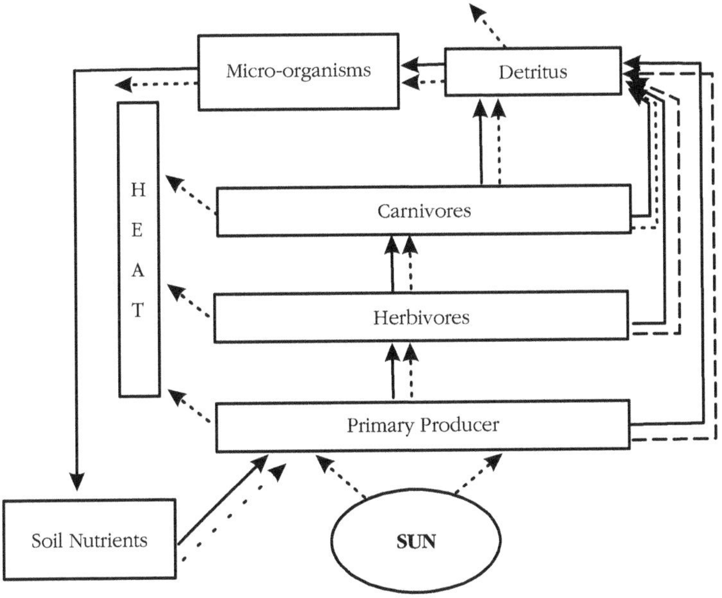

Source: Prepared by the author.
Note: Solid lines represent nutrients and broken lines represent energy.

units. A small pond or a large lake or an ocean, a large river or a small stream, a forest and even the whole earth itself can be considered as an ecosystem. Besides these natural ecosystems, there are human-modified or human-made ecosystems such as an agricultural field, an aquaculture pond or a village. The concept of ecosystem can be extended to a large urban settlement (town, city or metropolis) which depends on continued inputs of large amounts of energy and materials, besides the solar radiation and atmospheric gases, and provides certain goods and services. In most ecosystems, the humans are an important and often dominant component.

Ecosystems are classified in many ways. They are often categorised on the basis of climate (tropical, subtropical, temperate and boreal ecosystems), physiography (coastal, mountain and alpine ecosystems) and dominant vegetation (forest, grassland, desert and agricultural ecosystems). All ecosystems depend on water which is essential for the existence of life and various interactions within the ecosystem. However, ecosystems are usually differentiated into terrestrial and aquatic ecosystems on the basis of the dominant component of the physical environment. Whereas in the terrestrial ecosystems, the land and atmosphere are the dominant components that determine the nature of plants, animals and micro-organisms, in aquatic ecosystems, the abundance of water in and above the substratum significantly alters all other components (energy, gases, minerals) of the physical environment and hence, the kind of organisms inhabiting it. The aquatic ecosystems are further differentiated into freshwater, brackishwater, saline and marine ecosystems (depending upon the salt content).

It must be recognised that no ecosystem is completely independent. All ecosystems depend on the inputs of energy and gases from the earth's common source, the sun and the atmosphere, and on water that circulates through the atmosphere. Animals often move across ecosystems. Thus, all ecosystems are linked with each other even when they are separated geographically by long distances. A landscape comprises a mosaic of ecosystems — terrestrial and aquatic, natural and human-modified as well as human-made. Many similar ecosystems throughout the world — for example forest, desert or freshwater — grouped together are generally known as biomes (forest biome, desert biome, freshwater biome).

In this chapter, we shall discuss the characteristics of only the freshwater ecosystems, with reference to the South Asian region, but inevitably their linkages with the terrestrial and marine ecosystems will be touched upon.

Freshwater Ecosystems

An important and major role of water is in sculpting the earth's surface. The movement of water over land creates different kinds of water-dominated habitats depending upon the nature of the substratum/soil (texture, permeability and chemistry), geomorphology (shape and size of the depressions) and climate. The differences in the water regime cause different plant and animal species to occupy the habitats and result in the diversity of aquatic, mostly freshwater ecosystems. When the water accumulates in a depression and generally remains standing with no or very little movement as in lakes and ponds, these are called lentic ecosystems. On the other hand, when the water flows continuously through a channel as in case of springs, streams and rivers, these are known as lotic ecosystems. Variations in the climate (amount and seasonality of precipitation and temperature regimes) result in the seasonal variations in the amount, duration, frequency and timing of flooding of a particular area. Thus, the rivers, streams and lakes may be seasonal (temporary) or perennial. Shallow lakes in a hot seasonal climate generally dry up during the summer. Many lakes may accumulate salts in different amounts from their underlying geological formations or by concentration through evaporation over very long periods. Thus, the lake water may vary from fresh to brackish and saline.

Seasonally, the increased run-off into the lakes and rivers results in a change in the depth of water over different areas of the lakes and flooding of areas lying adjacent to the rivers (beyond the natural levee bordering the river channels). These marginal areas located between the highest and lowest water levels of lakes are called littoral zones, and the areas flooded seasonally by the rivers are called floodplains (Figure 2.4). These littoral zones and floodplains, together with habitats subject to seasonal or shallow flooding, are now known as wetlands. They are better categorised as aquatic ecosystems because of the dominant influence of water on the organisms inhabiting them. Wetlands are generally sandwiched between the deepwater and terrestrial (upland) habitats and also share some characteristics of both adjoining systems. Marsh and swamp are common terms used for wetlands dominated by herbaceous and woody vegetation respectively.

Figure 2.4: Schematic Diagram of Zonation in Lakes and Change in Temperature with Depth

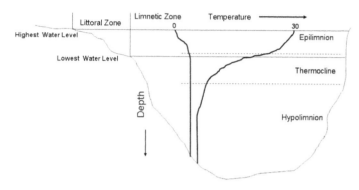

Source: Prepared by the author.
Note: The temperature profile during winter (left) and summer (right) is shown by thick lines.

A few other kinds of aquatic habitats are also categorised as wetlands. Meandering of rivers across their floodplain results in the formation of shallow water bodies (oxbow lakes) which may or may not be periodically connected with the river. These are known in South Asian countries by various names such as *tal, beel, baor, haor, phat,* etc. As the rivers reach the sea, their channels can discharge water only at the time of low tide. At high tide, the sea water enters the rivers and mixes with the freshwater. These lower reaches of the river channels, influenced regularly by the tidal sea water, are called estuaries. The salinity of water in the estuaries varies along a gradient from fresh to marine. In tropical regions, especially where the rivers carry large volumes of fine sediments and deposit them in their delta, there develops a forest community (called mangrove) and its associated animals along the estuaries. The tides also push back the freshwater which often spills over the banks flooding the adjacent areas where a different kind of organisms occur. Further, near the coasts, there are shallow water bodies — the lagoons and backwaters, which are usually connected with the sea through a narrow opening and also receive freshwater from the adjacent upland areas surrounding them. All these habitats and their specialised communities adapted to water-dominated, fresh or marine environment, are recognised as different kinds of wetland ecosystems.

Yet another kind of aquatic ecosystems, which is not readily recognised and appreciated, is the groundwater ecosystem. Infiltration through the soil supplies aquifers with a continual source of water to replace that is pumped out from wells or is discharged naturally (such as at springs). Infiltration is generally a slow processes requiring water to move through a tight maze of sediment which acts as a filter. Together with biological activity, water is cleaned as it moves through the ground. This natural scrubbing is one of the reasons groundwater is so commonly used for drinking.

The movement of water below ground often creates a variety of habitats where the groundwater is a dominant factor influencing and sustaining a specialised community of organisms. These habitats dominate the hyporheic zone — the area below and to the sides of the rivers and other standing water bodies. However, there are often large spaces and channels below ground within the soil and in the unconsolidated substrata through which the water moves swiftly. These areas may be created by natural processes, plant roots or animals, or by dissolution of minerals by water. Habitats which are formed by the dissolution of soluble rocks, including limestone and dolomite, are known as karst. They include caves.

Among the freshwater ecosystems are also several habitats created and managed by humans. Humans impound the surface run-offs and river flows or divert the water to other land areas for agriculture or aquaculture and various other uses. In the process, wetland habitats are created such as reservoirs, fish ponds and paddy fields. Large reservoirs with deep waters still have significant shallow marginal areas (littoral zones) which function as wetlands.

It will be seen from the foregoing account that all kinds of aquatic ecosystems, from lakes at high altitudes to the mangroves, are physically located within the drainage basins of rivers of which they are an integral part (Figure 2.5). The hydrological cycle serves to link them together and with their surrounding terrestrial areas, and provides continuity over time.

Ecological Characteristics

Physicochemical Environment

The water-dominated environment imposes several limitations to the survival and growth of organisms. The light passing through water changes rapidly in its spectral composition as well as intensity. The

Figure 2.5: Schematic Representation of Different Kinds of Aquatic Ecosystems within a River Basin

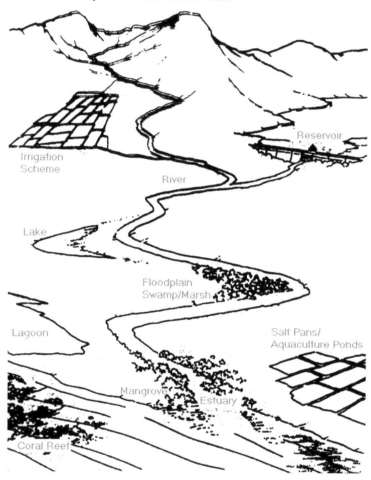

Source: Adapted from an old pamphlet; source unknown.

red part of the light's spectrum is absorbed most by the water and only the shorter wavelengths (blue region) are able to penetrate deeper. Refraction due to high density of the medium, and reflection (and/or absorption) from suspended particulates mean that the amount and quality of light is altered quickly and so much that hardly enough energy is available for photosynthesis by any plant at a depth of about 6–8 m. The incident radiation on the water surface itself is often reduced by the vegetation surrounding the water body.

The availability of essential gases namely, oxygen and carbon dioxide, is severely restricted due to their limited solubility in water. The solubility decreases rapidly with increase in temperature. The presence of organic matter further reduces both penetration of light and availability of oxygen. At higher altitudes, despite a lower temperature, the oxygen solubility declines rapidly with a fall in the atmospheric pressure. Saturation of soils with water eliminates the air from interstitial spaces and hence, reduces the availability of oxygen for roots and soil organisms. The high density of water, relative to that of air, results in steep increase in pressure on the organisms. However, high pressure at greater depths (1 bar for every 10 m depth) helps in increased solubility of oxygen and other gases in the lower layers of water of deep lakes.

Thermal properties of water have a special significance. Water has a low freezing and high boiling point and also, a large amount of latent heat is involved before the water can vaporise or turn into solid ice. Thus it remains liquid over a wide range of temperature. The high specific heat ensures that temperature changes occur rather slowly but at the same time, water bodies act as large reservoirs of heat energy. The highest specific gravity of water is attained at 4°C so that the ice is lighter and floats on the water surface. This ensures that the water below ice does not freeze and allows the organisms to survive under ice.

These thermal properties are also related to the availability of oxygen in deeper layers because it cannot reach there through simple diffusion from the atmosphere. During the summer, the surface layers of the water column become warmer and hence, less dense. The water column below it remains at a low temperature and hence is denser. Thus the water column becomes stratified into two layers, the upper epilimnion and the lower hypolimnion. The relatively thin intermediate zone, through which the temperature falls sharply, is known as thermocline (Figure 2.4). Stratification greatly impedes the exchange of gases and nutrients between the two layers. However, the stratification is broken and the water column in the lake completely overturns when the surface water temperature falls to 4°C making it most dense and causing its sinking to the bottom. The hypolimnic water with a lower density moves up to the surface. In areas which experience seasonal freezing of the lake surface (for example at high altitudes), such mixing of water column occurs twice — shortly before freezing and after the ice melts — when the surface water temperature crosses the 4°C threshold. Such mixing induced by thermal changes allows the oxygen saturated water to reach the deeper layers and the substances dissolved in water from the bottom sediments to become available in the surface layer.

In warmer climates with mild winters, deep lakes remain stratified throughout the year. However, in shallow lakes (2–3 m deep), mixing of the water column occurs due to wind action. Winds blowing at a high velocity over large lakes cause mixing to a greater depth than in smaller water bodies surrounded by vegetation or other physical structures.

In rivers and other flowing waters, the availability of oxygen is not constrained by the lack of diffusion, and the temperature gradients are not large. Both temperature and oxygen availability are influenced by the flow velocity and turbulence which are governed by the slope and roughness of the channel bed. Streams in the hills flow over steep slopes and large boulders create turbulence that keep the water saturated with oxygen. In the plains, the river flow is sluggish and generally without any turbulence and therefore, oxygen deficits may occur.

Though water is a universal solvent and carries various chemical substances during run-off from the drainage basin, their concentrations are usually quite low to support significant growth of the photosynthetic organisms. In lakes, most of the chemical substances that accumulate over time depend upon their hydrology (inflow–outflow of water). However, inputs from anthropogenic sources (fertilisers, detergents, wastewaters, etc.) are major drivers of growth that is often undesirable. Rooted plants occurring in relatively shallow water bodies are able to utilise nutrients from the soil.

Freshwater Biota

Despite many constraints of the aquatic environment, a large proportion of the earth's biodiversity resides in aquatic environments (McAllister et al. 1997; Groombridge and Jenkins 1998). In fact, life originated and evolved in water before migrating to the terrestrial environment. Some plants are also believed to have reverted back to the aquatic environment during their evolution. The diversity of organisms in aquatic ecosystems ranges from micro-organisms (bacteria and fungi) to microscopic algae to large trees and from protozoa to mammals, all of which have a variety of morphological, anatomical and physiological adaptations for survival and growth in water.

Plants

The plant communities (autotrophs or primary producers) of freshwater ecosystems are dominated by algae in open deep waters whereas larger plants (macrophytes, mostly flowering plants) dominate the shallow water areas and wetlands. There are also many bacteria and fungi but very few bryophytes. The algal flora (often referred to as microphytes) comprises chiefly of green algae, blue green algae and diatoms though other groups of algae also occur. It is estimated that in Indian waters, there are about 1,800 species of which about two-thirds belong to the blue-green algae. The algal flora in other countries of South Asia does not differ significantly. However, the diversity of algae is difficult to estimate correctly because of large difference in their taxonomy (see Gopal 1997).

The larger plants — called macrophytes — include many large algae, chiefly charophytes, some bryophytes and mosses, many ferns (pteridophytes) and a large number of families of flowering plants. Common ferns include the members of family Marsileaceae (e.g., *Marsilea*), Salviniaceae (*Salvinia* and *Azolla*), Ceratopteridaceae (*Ceratopteris*) and Siphonales (*Equisetum*). There are at least 1,250 species of flowering plants in different freshwater ecosystems (Cook 1996) of South Asia. These include species which were hitherto not considered as 'aquatic' but may indeed be wetland species. In India alone, the total number of species (including algae) in different freshwater ecosystems constitutes more than 15 per cent of the estimated total floristic diversity.

Animals

Almost all groups of animals are represented in freshwater ecosystems. Among the invertebrates, crustaceans, molluscs and insects are the most abundant. Majority of insects, despite their adults being terrestrial, pass the larval stages of their life-cycle in shallow waters. Crustaceans are represented by three groups, — Cladocera, Rotifera and Copepoda. Numerous species of oligochaetes and leeches, various nematodes and coelenterates, besides the protozoans, occur in water. Among the vertebrates, fishes are wholly aquatic. Amphibians, reptiles and birds have large and significant representation in fresh-waters. Several mammals (dolphins, otters and beavers) are largely aquatic whereas some others such as rhinoceros, hippopotamus and water buffaloes prefer shallow water habitats. A very large proportion of other animals, which are considered to be terrestrial, depend upon the resources of aquatic habitats at some stage of their life-cycle, and in different ways (see Gopal and Junk 2000).

It is estimated that about 16,500 species, i.e., more than 20 per cent of the total fauna in India are aquatic, and majority of them occurs in freshwater. The freshwater ecosystems support about 300 Rotifera, 285 Mollusca, more than 100 Cladocera (Alfred et al. 1998), 100 Ostracoda, more than 300 Copepoda, 742 fishes (Jayaram 1981; Jhingran 1991), and about 1,000 birds (Ali and Ripley 1968–1974).

The organisms in aquatic ecosystems are usually categorised according to their spatial position in the water column. Most common in standing waters are 'plankton' — the organisms which drift freely within the surface layer of water. They include numerous algae, crustaceans (cladocera, rotifera, copepoda) and larval stages of some organisms (such as fish). They are classified into phytoplankton and zooplankton as belonging to plant or animal kingdom respectively. Many bacteria are also planktonic. A variety of algae and invertebrates (nematodes, oligochaetes, insect larvae and crustacean) occur on the bottom sediments of shallow and slow flowing waters. These organisms are collectively known as benthos. The organisms which grow attached to or around various substrates (e.g., stems and leaves of plants) are known as periphyton and include both algae and micro-invertebrates. Other categories are nekton (actively swimming organisms) and pleuston (free-floating on the surface). Similarly, aquatic plants are categorised as free-floating (*Spirodela, Lemna, Salvinia, Azolla, Eichhornia crassipes*), submerged (*Ceratophyllum*), rooted-submerged (*Vallisneria, Hydrilla*), rooted-floating leaved (*Nymphaea, Nelumbo, Trapa*) and emergents (*Typha, Phragmites, Cyperus*, etc.).

Biotic Interactions

The plant and animal communities within an ecosystem interact variously. Food web interactions, which include herbivory, carnivory and detritivory, are major determinants of the community structure. Interspecific competition and allelopathy are of widespread occur-rence (Gopal and Goel 1993). The concentrations and proportions of different nutrients interact with hydrological variables (quantitative) to influence the outcome of competition as well as food web interactions. For example, an increased availability of nutrients results in greater primary production (eutrophication) and a decline in species richness, because a few species that can efficiently utilise the nutrients outcompete others. Also, increased amounts of decomposing organic matter cause oxygen depletion, and the consequent anoxic conditions trigger a series of chemical changes such as denitrification and the reduction of sulphates and carbon compounds to hydrogen sulphide and methane, respectively. Further, plants provide niches for colonisation by various organisms (periphyton and epiphytes) which are in turn consumed by animals such as molluscs and fish.

Hydrology and Freshwater Ecosystems

The aquatic organisms, both plants and animals, differ greatly in their response to water regimes which include gradients of depth, duration, frequency, amplitude and timing of flooding, as well as flow velocity. The organisms' responses also vary depending upon their stage in the life-cycle. Seed germination, seedling establishment, vegetative growth and reproduction of various aquatic plants occur under different water regimes. Seeds of many species require a wetting-drying cycle, and in most aquatic plants a seed germinates on moist soils or in shallow water when light is available. Seedlings of many emergent and woody species often do not tolerate flooding though their adults may survive deep flooding or even complete submergence for varying periods. Accordingly, plants exhibit large variation in their spatial and temporal distribution in different aquatic ecosystems. This is reflected in the well known zonation of plants and associated animals from the wet margins to the deeper areas of a lake or pond. Trees, shrubs and large rooted aquatic plants are confined to relatively shallow standing waters, riparian areas of slow-flowing streams and banks of fast-flowing rivers. Further zonation occurs along the hydrological gradient. Free-floating plants occur mostly in standing waters though they may form significant stands along sheltered margins of rivers and streams. Rarely do submerged species anchored to bottom substrata also occur in swiftly flowing streams. Many species of family Podostemaceae are confined to the river bed below the falls.

All ecosystem processes are directly governed by hydrology (Figure 2.6). As pointed out earlier in this chapter, the depth of water

Figure 2.6: Effect of Different Hydrological Parameters in Biological Processes and Biodiversity

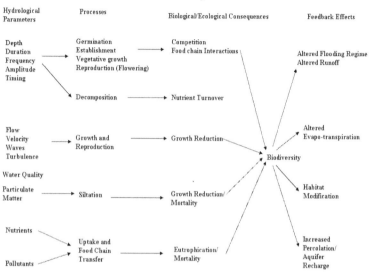

Source: Prepared by the author.

influences the light and oxygen availability as well as temperature and nutrient concentrations. The depletion of oxygen from the water and the underlying sediments retards decomposition of organic matter and this sets in a chain of anaerobic processes. Further, the nutrient, organic matter and organisms are transported into and out of the ecosystem by the flow. In deep water systems, primary production is predominantly carried out by phytoplankton and that too in the surface layers of water where light penetration is adequate for photosynthesis. Hence, energy transfer occurs mainly through the grazing pathway in which zooplankton and fish play a major role. In wetlands (lake littorals, floodplains and other marshes) which are dominated by macrophytes, large herbivores (e.g., rhinos, water buffaloes) remove a greater part of the annual production. In other aquatic ecosystems, primary production is converted to detritus which may decompose and mineralise completely, or may be used by detritivores that in turn support a food chain, or may accumulate as peat. Again hydrology affects these processes through transport and changes in chemistry (oxygen availability, pH, etc.).

Effects of Biota on Hydrology

Plants exert a great influence on the hydrological and topographic features of aquatic systems. They offer resistance to flow, thereby reducing its velocity. They also reduce the capacity of the water body to hold or carry water. This favours the settling of suspended particulates and infiltration of water through soil. The rise in water levels causes flooding of neighbouring areas. The roots and rhizomes help in increasing the hydraulic conductivity of the soil, and hence, promote infiltration. Aquatic vegetation may enhance or reduce the loss of water through evapotranspiration. Besides aiding and abetting siltation, plants may also contribute to the process by producing large amounts of organic matter. The topographic changes, consequently, alter the hydrological regimes which in turn may trigger changes in biotic communities.

Several animals have considerable impacts on hydrology. For example, beavers fell and drag trees to create dams across streams, causing major changes in stream hydrology. Some animals such as rodents and otters make burrows or tunnels on lakeshores and riverbanks, which may lead to increased erosion and flooding. Larger animals such as buffaloes and rhinos affect hydrology by grazing and wallowing in water.

Differences between Rivers, Lakes and Wetlands

The lakes, wetlands and rivers share many common characteristics, and as mentioned earlier in the chapter, are generally interlinked within the river basin. The foregoing account however does not bring out the large differences between them (for details, see Mitsch and Gosselink 2007; Keddy 2000; Wetzel 2000; Calow and Petts 1992, 1994) that arise from the differences in their evolutionary history, geology, climate, watershed characteristics (geomorphology and biota in particular), anthropogenic impacts and size, besides hydrology which is influenced by all these factors. Therefore a brief overview of these three systems is presented below.

Rivers as Ecosystems

Rivers were once considered to be a simple linear feature delimited by the bed and banks of the main channel, and dominated by downstream transfers of water. Research during the past four decades has shown that the rivers are three-dimensional systems dependent upon longitudinal, lateral and vertical transfers of material, energy and biota. A fourth temporal dimension is added to consider their dynamics. It is also well-recognised that their integrity depends upon the interactions of hydrological, geomorphological and biological processes over a range of time scales (Calow and Petts 1992, 1994; Petts and Amoros 1996).

Rivers as a complex ecological system comprise of a diversity of habitats. On a macro-scale, the river consists of a main channel and its fringes (river bed) bounded by the natural levees on both sides. On a meso-scale, the river includes running water, standing water, temporary water, groundwater and terrestrial habitats. River channels contain a mosaic of habitats largely created by flow-substrate inter-actions. Many channels are sinuous, eroding into the outer banks of bend, depositing on the inner banks, actively meandering or passively responding to confining points of resistant geology. Less sinuous channels may modify by braiding. The channel bed is also modified by the flowerosion-deposition regime. Sheltered banks may have backwaters and side channels. Riffles and pools are part of the system. On a micro-scale, further habitat differentiation occurs according to the nature of bed material, vegetation, animals and flow velocity.

The diversity of habitat changes along the river course from headwaters to the mouth. Headwater streams are generally straighter; meanders and migration increase downstream and with age. Mountain streams have fast and turbulent flow, are steep and unstable with the bed composed of rocks or gravel with occasional sandy patches, while channels in the flat plains are slow flowing, have beds composed of sand and silt and meander over large areas. Middle and lower reaches are dominated by the transfer and deposition, respectively, of materials. Other physical characteristics of the river also change with distance from the source. As most rivers arise at high altitudes, the temperature usually increases downstream. In the upper reaches, gaseous exchange, unhindered by stratification and positively encouraged by turbulence, mixing and cascades, keeps the water always saturated with oxygen whereas downstream oxygen deficits increase with greater frequency and duration. Two major zones are generally recognised in hill streams (Illies and Botosaneanu 1961): (a) rhithron — the stretch of the river extending from the source to the point where monthly mean temperature rises to 20°C, and (b) potamon — the region where monthly mean temperature rises above 20°C.

Similarly, the chemical characteristics of water also change, with gradually increasing concentration of various nutrients, as a variety of dissolved and particulate substances and plant litter enter the stream with the run-off. The geological, geochemical and vegetational characteristics of the watershed directly govern the water quality. For example, the water of river Yamuna differs from that of river Ganga due to the nature of their respective catchments.

Biotic Communities

The diversity of habitats within a stream is matched by a similarly high diversity of biotic communities that also vary along the course of the river (Figure 2.7). Most organisms, unless anchored to the channel bed, drift downstream. Fish, birds and mammals can swim against the current. Some fish occur in torrents (Hora 1928). Therefore, species-rich communities are usually confined to shallow, slow moving streams, or to margins where flow velocity is negligible, and shallow pools and riffles abound. The upper reaches with rapid turbulent flow are very poor in plankton but many benthic algae grow attached to boulders. Higher plants are rare except on riverbanks. Aquatic fauna is composed mainly of shredders that feed on freshly fallen leaf litter. They inhabit the river margins with relatively slow flow or sheltered pockets under boulders and gravel. Channel edges have a distinct fauna as compared to mid-channel fauna. Individual boulders and patches of vegetation provide varied habitats for colonisation by different kinds of fauna. Downstream, the leaf litter is fragmented and gradually converted to fine particles. The fauna now comprises of benthic invertebrates feeding on particulate detritus. The increased availability of nutrients allows the development of planktonic communities. Faunal diversity further increases with greater availability of food and niches, and varies with the microhabitat.

Macrophyte communities develop with the reduction in flow velocity and change in the substratum. They vary according to flow rate, substratum, trophic status and general catchment characteristics (Holmes and Newbold 1984). In plains, these communities are quite diverse and exhibit elaborate zonation. Riparian communities comprise of different types of plants and exhibit zonation along a gradient of flooding regimes.

Fishes are a major and economically important component of river biota. Many fishes inhabit the rapids due to their unique physical adaptations. Hooks, spines or suckers, which enable them to fasten

Figure 2.7: The River Continuum — Changes in the Community of Organisms along the River Course

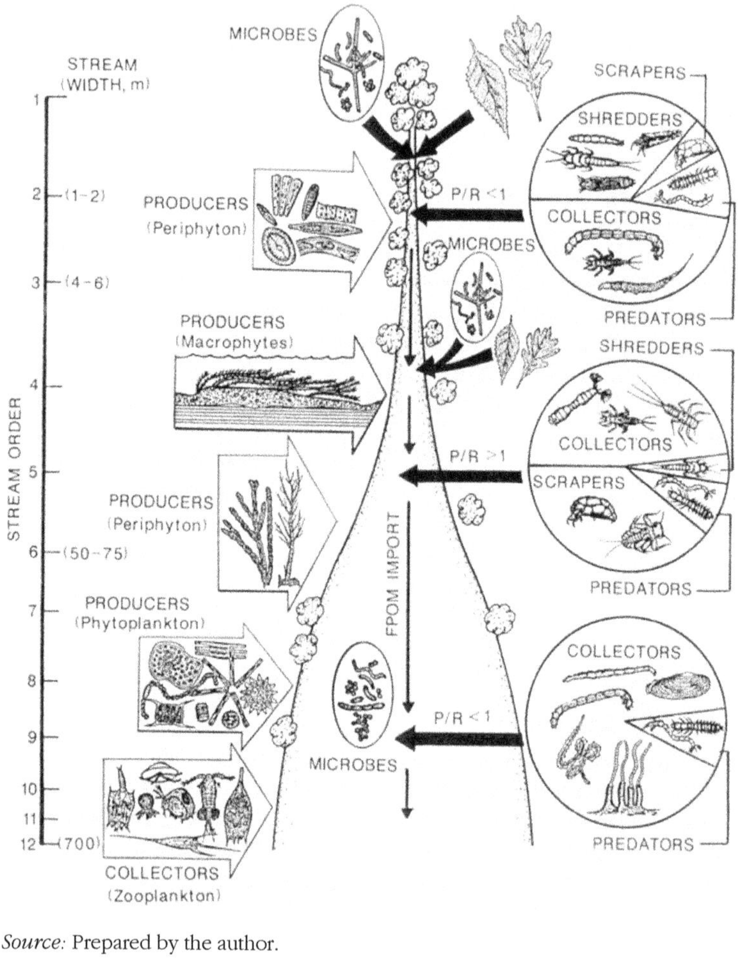

Source: Prepared by the author.

themselves to rocks and vegetation, are common. Small sizes and elongated shapes that allow them to live among rock crevices and rooted vegetation are also common, as are humped shapes that allow for dwelling in the bottom. Pools and riffles in high gradient streams attract different species complexes. Riffles are rich in invertebrate food organisms, and well-adapted to survive in strong currents due to well-aerated waters. Small species or juveniles of larger fish live among the rocks. The quieter waters of pools are inhabited by less energetic swimmers, who take shelter in areas of slack flow, and larger fish, that feed on the drift of organisms dislodged from the riffles. Fish in this area may bury their eggs in gravel pockets or may attach the eggs to rocks or submerged vegetation. Some species scatter their eggs which drift downstream until they reach sufficient size to migrate laterally into floodplain nurseries.

The availability and kind of food are important factors affecting fish communities. In small streams, food items are limited to insects, small amounts of vegetation and to allochthonous material falling into the water from the land. As food is relatively scarce, specialisations to benefit from specific food types are common and resource partitioning is very high (McNeely 1987). In floodplain rivers, food does not appear to be a limiting factor, at least during the flood period. However, a large number of specialised forms exist and predators are common. There are also a large number of generalised feeders who show considerable flexibility in their diet.

Four main behavioural guilds of riverine fishes are recognised (Welcomme 2000):

(i) White fish: large strong migratory fish from many different species which move large distances with river channels between feeding and breeding habitats and are intolerant to low levels of dissolved oxygen. The Indian major carps fall under this category.

(ii) Black fish: These fish move only locally from floodplain water bodies onto the surrounding floodplain when the area is inundated. They return to the pools during the dry season. They are adapted to remain on the floodplains at all times and often have auxiliary respiratory organs that enable them to breathe atmospheric oxygen, or have behaviours which give them access to the well oxygenated surface film. Fish include the clariid catfish of Africa and the anabantids of Africa and Asia.

(iii) Grey fish: an intermediate type between the migratory and floodplain loving species, they migrate over short distances, usually from the floodplain during high water for breeding, to the main river channel where they shelter in marginal vegetation or in deeper pools during the dry season. They are less capable of surviving at extremely low oxygen levels but have elaborate reproductive behaviour which enable them to use the floodplain for breeding. Tilapias, many small characins and cyprinids belong to this group.

(iv) River residents: They inhabit local areas of the main channel, migrate very little, are relatively uncommon and are usually confined to larger rivers. The recently discovered deep water gymnotid fauna of the Amazon river are an example.

Influence of River Biota on Hydrology

In fluvial systems, the biotic influences are greater in the slow-flowing streams and lower stretches of large rivers. The in-stream vegetation impedes flow. The riverbank vegetation influences run-off, subsurface flow and evapotranspiration. It can check or promote erosion and thereby alter hydrological regimes. The role of animals such as beavers has also been pointed out earlier.

Food Web Interactions and Ecosystem Functioning

All ecosystem processes in a river are regulated by its flow regime. The primary production is affected by the availability of nutrients brought in from the catchment. Suspended sediments and organic matter transported from the catchment cause turbidity that reduces photosynthesis. Plant litter from riparian vegetation forms the main source of food for the fauna in upper reaches whereas the fauna in the lower reaches depend to a great extent on the production in floodplains. Stream food webs are thus based on both autochthonous and allochthonous sources which influence the length of food chains, and indirectly, the biodiversity (Figure 2.8). The amount and velocity of flow influence the availability of oxygen, and in turn, the nutrient cycling.

The food chain interactions and biotic communities change continuously along the length of the river from headwaters to its mouth in response to the physical gradients and interactions with adjacent terrestrial communities. These changes led to the river continuum concept (Vannote et al. 1980). The concept has since then been modified in many ways to accommodate a broader range of influences of climate, geology, local geomorphology and tributaries on stream ecology (Minshall et al. 1985; Naiman et al. 1987; Cummins et al. 1989 1995). A related concept is that of nutrient spiraling which considers changes in nutrient cycling in rivers. According to the concept, unidirectional down-gradient flow of streams causes nutrient cycles

Figure 2.8: Processing of Leaf Litter in a Stream and its Food Web

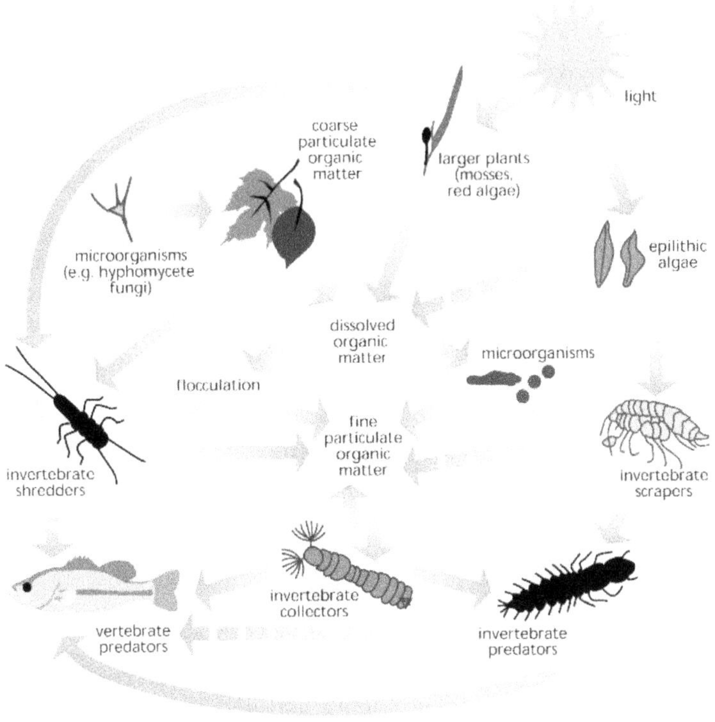

Source: Adapted from FISRWG (1998).

to be open rather than closed, so that there is a gradual downstream displacement of nutrients, the rate of which is controlled primarily by flow.

River-Catchment Interactions

The entire river basin is a landscape unit within which rivers interact with their floodplains, lakes, wetlands and upland terrestrial eco-systems. Rivers transport sediments, nutrients and propagules and distribute them to different parts of the basin. The rivers also exert some influence on the microclimate and the vegetation of their basins. The amount and quality of the run-off from the basin into the rivers is influenced by topographic, edaphic and biological characteristics of the watershed. Further interactions occur with the agency of animals and humans.

Rivers do not remain confined, particularly in the lower reaches (low gradient and/or higher order), to the space delimited by natural levees. High flows that exceed the channel capacity spill over the levees flooding areas on either side — the floodplains. Hydrological processes in the watershed and the rate of downstream discharge determine the depth, duration and frequency of inundation of the floodplain which periodically becomes a part of the river. Flooding forces the exchange of materials and energy between the river and its floodplain. The importance of these exchanges between the river and floodplain has been investigated in great detail in the context of fisheries (Lowe-McConnell 1987; Welcomme 1979). Riverine fishes migrate to the floodplain for spawning. Young larvae and fry grow there feeding on a variety of food (plankton, invertebrates and detritus). Many other animals breed and pass some stages of their life-cycle in different parts of the floodplains. As the floods abate, receding waters carry with them organic matter, propagules and nutrients from floodplains to the river proper. Thus, the structure and function of downstream communities is influenced by not merely the direct upstream–downstream transport processes as envisaged by the continuum concept (Cummins et al. 1995), but more strongly by the river–floodplain interactions as elaborated by the flood pulse concept (Junk et al. 1989).

Various parts of the floodplain are subjected to differential flooding and vary in character between lentic and lotic with time. As most plant species are adapted to a specific hydrological pulse and because different parts of the floodplain experience hydrological pulses of different nature (geomorphic variation and topographic gradient), large biodiversity is obtained in floodplains (ibid.). Nutrient cycling within the floodplains is dominated by flooding from the river, run-off from upland forests, or both, depending upon stream order and season. Vegetation exerts significant biotic control over intrasystem cycling of nutrients, seasonal patterns of growth and decay.

Floodplains influence the rivers in other ways also (Figure 2.9). They lie between the rivers and upland areas, and therefore, water, sediments, and nutrients must pass through them before entering the river. The biological communities in the floodplain control the fate of these substances (Lowrance et al. 1984). Water infiltrates through the soil to the groundwater or moves laterally to the stream. Sediments get trapped and accumulated in the floodplain, causing topographic changes. Organic matter also gets settled and decomposed

Figure 2.9: **Various Processes Occurring in the Floodplains which lie between the Upland and the River that Regulate Water Quality**

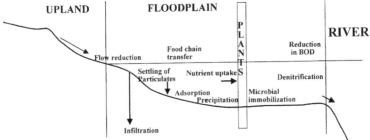

Source: Prepared by the author.

with time and supports many detritus feeding organisms. The nutrients undergo various transformations that reduce their flux to the rivers. Other interactions also occur between the floodplains and uplands, and therefore, floodplains are considered as ecotones (Wissmar and Swanson 1990; Pinay et al. 1990). Many terrestrial animals from up-lands periodically utilise the floodplain resources and numerous insects pass early stages of their life-cycle in the floodplain. Similarly, some aquatic animals (especially waterfowl) depend upon the terrestrial landscape at some stages in their life-cycle. For detailed discussion of these interactions, see Ward (1989).

Lakes as Ecosystems

Lakes are depressions which are formed on the earth's surface in various ways by tectonic, volcanic and glacial activities. A few lakes have also been formed by the impact of meteorites (e.g., lake Lonar in Maharashtra). Fluvial processes (erosion and sedimentation) create many lakes near the rivers. Sometimes lakes are formed by the blocking of surface flow by sand blown and deposited by winds (aeolian process), e.g., lake Sambhar in Rajasthan. Numerous lakes are created by humans to store water by constructing earthen or concrete dams across the flow channels or by excavating earth (man-made lakes or reservoirs). In terms of their age, the lakes formed by fluvial processes and damming of flows (by humans or landslides) are young (from years to several decades) whereas those formed by tectonic and volcanic activities are quite old (geological time scale of thousands to millions of years).

The source of water in the lakes is the precipitation and surface run-off but springs may play an important role. The water may or may not flow out of the lake determining its residence time in the lake. Many glacier-fed lakes at high altitudes are the sources of major rivers. Water is also lost in evaporation and in hot dry climates, water levels drop down considerably by evaporation alone. Some water may also move into the ground depending upon the permeability of the lake bottom.

Lakes differ from the rivers in being invariably depositional systems as all or a part of the sediments, nutrients and organic matter entering them is retained within the lake basin and accumulate over time. Many high altitude lakes in the trans-Himalayan region (Ladakh and Tibet) are understood to have turned brackish or saline due to concentration of salts by evaporation over millennia. Lakes are usually classified into oligotrophic (nutrient poor and least productive) and eutrophic (nutrient rich and highly productive). The process of eutrophication (nutrient enrichment) occurs naturally as the nutrients accumulate in them but is accelerated by the anthropogenic inputs by way of domestic wastewater, agricultural run-off, detergents, etc. Numerous studies have shown that phosphorus is largely responsible for eutrophication (Vollenweider 1968). However, recent studies point out that in tropical regions, nitrogen availability may be the regulating factor.

Lakes differ from the rivers also in their stratification and mixing processes which have already been described earlier in the chapter. Lakes were described as 'microcosms' as early as 1887 by Forbes who considered the lakes to be self-contained systems, 'a world unto themselves'. This view of lakes as closed systems prevailed until mid-twentieth century when overwhelming evidence showed that the lakes are influenced, and even governed, by the natural and anthropogenic processes beyond the shoreline — in their entire watershed. Considerable effort has gone into the demonstration of the importance of organic matter inputs from the littoral zone and the watershed for the structure and functioning of lake ecosystems. In this respect, the lakes are very similar to the rivers.

Wetlands

As described earlier in the chapter, wetlands are areas intermediate in character between deepwater and terrestrial habitats, also transitional in nature and are often located between them. These habitats experience periodic flooding from adjacent deepwater habitats and therefore support plants and animals specifically adapted to such shallow flooding or waterlogging of the substrate. They include lake littorals (marginal areas between highest and lowest water level of the lakes), floodplains (areas lying adjacent to the river channels beyond the natural levees and periodically flooded during high discharge in the river) and other marshy or swampy areas where water gets stagnated due to poor drainage or relatively impervious substrata. Bogs, fens and mangroves are included within the purview of wetlands due to similar ecological characteristics (prolonged or permanent waterlogging). The ancient Sanskrit literature distinguishes between lakes and wetlands which were respectively known as *Sar* (or Sarovar) and *Anup.*

Lakes may be distinguished from wetlands by their greater water depth and the dominance of planktonic communities instead of macrophytes, though the latter may occur along the margins of the lake The lake good webs are therefore dominated by grazing pathway (Figure 2.10) whereas the detritus pathway dominates the wetlands. Lakes have a low primary productivity (of both phytoplankton and macrophytes) and any increase in it is considered undesirable (eutrophication). Wetlands are in general eutrophic systems (except peat bogs). Shallow lakes (generally less than 3 m over most of their area) are usually rich in nutrients (derived from surroundings and their sediments) and have abundant growth of aquatic macrophytes. They support high densities and diversity of fauna, particularly birds, fish and macroinvertebrates, and therefore, have high value for biodiversity conservation. These shallow lakes are rightfully categorised as wetlands.

Ecosystem Services of Freshwater Ecosystems

A variety of processes involving interactions between the physical, chemical and biological components of an ecosystem result in certain ecological functions. For example, photosynthesis by plants results in the production of organic matter. This production of plant material is a basic function of all ecosystems. The cycling of various elements through the processes of uptake, transformation and release after decomposition is another function of the ecosystems. Humans attach a 'value' to several ecosystem functions which are useful to them in various ways. The production function has a value because it provides products which are used as food, fiber, fodder, timber, fuel, etc. Recently, the Millennium Ecosystem Assessment promoted the term 'ecosystem services' for 'the benefits people obtain from ecosystems' (MEA 2005). The MEA grouped various ecosystem services into four types: (a) Provisioning services such as food and water; (b) Regulating services such as regulation of floods, drought, land degradation and disease; (c) Supporting services such as soil formation and nutri-

Figure 2.10: Food Web of a Lake Ecosystem

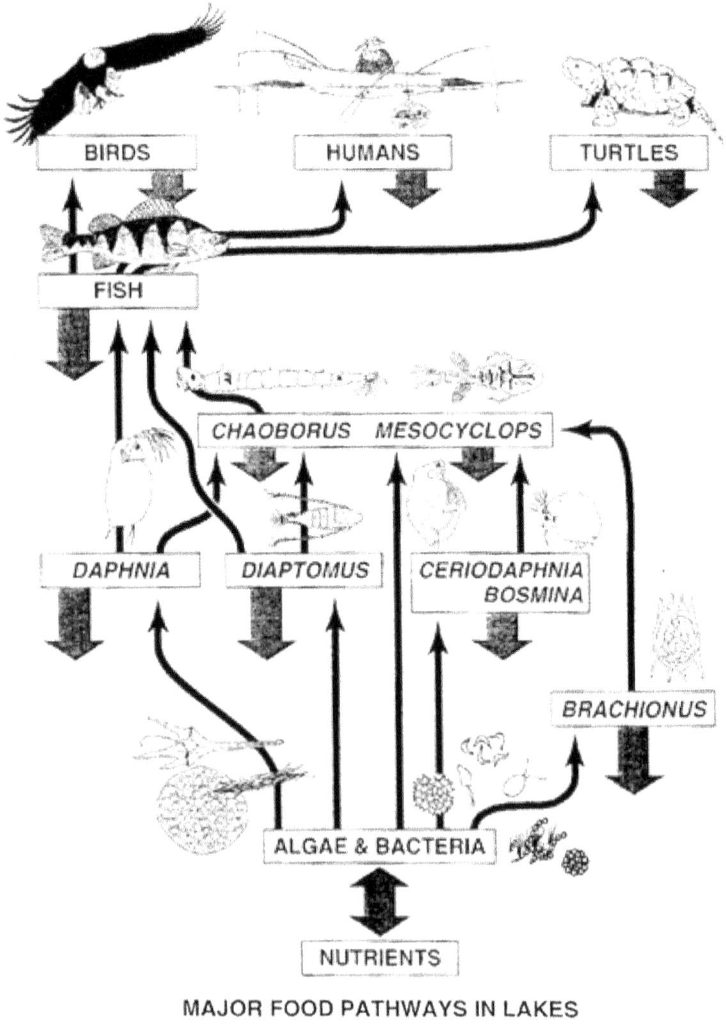

MAJOR FOOD PATHWAYS IN LAKES

Source: Dodson and Hanazato (1995).

ent cycling; and (d) Cultural services such as recreational, spiritual, religious, and other non-material benefits (Table 2.2). An example of an important ecosystem service provided by aquatic ecosystems is the production of fish and prawn which are the source of protein for half of the world's human population. Rivers and wetlands have assimilated the wastes throughout human history until the discharges exceeded their capacity. The gravel and sand transported by rivers are used extensively for construction and contribute to the formation of fertile soils in floodplains. The sediments transported by rivers and the freshwater flows maintain the mangrove forests which protect the inland coastal areas against storms and cyclones, and also support the marine coastal fisheries. Wetlands regulate water quality, moderate the effects of floods and droughts and play an important role in regulating climate. As noted earlier in the chapter, the aquatic ecosystems support a huge biodiversity disproportionate to their areal extent.

The goods and services provided by various aquatic ecosystems, however, differ greatly between them but depend most upon their specific hydrological regimes defined by interacting components such as depth and volume, duration, frequency and season of water flow as well as its quality. The functions and ecosystem services also depend upon specific characteristics of different ecosystems and upon the nature and degree of human interventions. Hence, rivers, lakes and wetlands differ considerably in the nature and magnitude of services provided by them. It must be stressed that we must recognise the differences between these categories of ecosystems instead of clubbing them together under a general term 'wetlands' as is promoted by the Ramsar Convention.

Despite their regular interaction with the main river channel and open deep water, floodplains and lake littorals have distinct ecosystem attributes in the same manner as mangroves and salt marshes lie adjacent to and interact with the open sea. Even wetlands differ among themselves. We must also recognise that all kinds of aquatic ecosystems do not and cannot perform all possible functions, and therefore, do not have similar values. Similarly, no aquatic ecosystem can provide all kinds of services described above and none of the services can be provided by all kinds of wetlands. Shallow lakes, deep lakes and reservoirs differ widely in their potential for various ecosystem services. Small streams in the hills running down steep slopes are different from the large floodplain rivers. Seasonal marshes

Table 2.2: Ecosystem Services Provided by Aquatic Ecosystems

Services	Examples
Provisioning	
Food	Production of fish, wild game, fruits and grains (rice)
Freshwater	Storage and retention of water for domestic, industrial, and agricultural use
Fiber and fuel	Production of timber, fuelwood, peat, fodder
Biochemical	Extraction of medicines and other materials from biota
Genetic materials	Genes for resistance to plant pathogens, ornamental species, etc.
Regulating	
Climate regulation	Source of and sink for greenhouse gases; influence local and regional temperature, precipitation and other climatic processes
Water regulation (hydrological flows)	Groundwater recharge/discharge
Water purification and waste treatment	Retention, recovery, and removal of excess nutrients and other pollutants
Erosion regulation	Retention of soils and sediments
Natural hazard regulation	Flood control, storm protection
Pollination	Habitat for pollinators
Cultural	
Spiritual and inspirational aspects	Source of inspiration; many religions attach spiritual and religious values (sacred lakes, rivers)
Recreational	Opportunities for recreational activities
Aesthetic	Scenic beauty or enhancement of aesthetics of landscape
Educational	Opportunities for formal and informal education and training
Supporting	
Soil formation	Sediment retention and accumulation of organic matter
Nutrient cycling	Storage, recycling, processing and acquisition of nutrients

Source: MEA (2005).

differ from swamp forests and shallow ponds. Lakes, reservoirs and ponds have only insignificant capacity to treat wastewaters whereas rivers, marshes and swamps assimilate large amounts of organic wastes. Large rivers, deep lakes and wetlands without macrophytes have very low primary production and consequently low secondary production, and often have low potential for groundwater recharge. The floodplains and shallow marshes or woody swamps have high production rates and contribute to the productivity of open deep water areas. The potential for carbon sequestration and methane production varies greatly according to the type of vegetation and other biophysical and climatic conditions (see Sahagian and Melack 1998).

Anthropogenic Impacts on Freshwater Ecosystems

Majority of freshwater ecosystems, throughout South Asia as also in other parts of the world, have been severely damaged in different ways by human activities. The human impacts arise from anthropogenic activities not only in the water body or its immediate surroundings but throughout its catchment (watershed). The most important impacts on aquatic ecosystems are caused by the alteration in their hydrological regimes by abstraction and diversion of water for various uses by direct modification of habitats (dams and embankments or landfills) and return of wastewater contaminated with nutrients and a variety of pollutants. The planners and resource managers, with the sole concern for economic development, treated the rivers as mere conduits, lakes as storages of water and the wetlands as wastelands. Water has come to be treated as a commodity worthy of allocation among few selected stakeholders. Thus, rivers have been dammed and channelised and water has been diverted to different uses at far away places. Such withdrawal and diversion of water turns long reaches of rivers almost dry, restrict and/or interfere with the migration of fishes to their seasonal breeding and feeding sites in upstream regions, and eliminates the seasonal connectivity of the river channels with their floodplains which are important habitats for a variety of fishes, birds, reptiles, amphibians and mammals.

Large reaches of the river channels have been converted to sewers by discharge of all kinds of industrial and domestic wastewaters into them. Wetlands have been drained and reclaimed by landfills; floodplains have been eliminated, and their natural water supplies have been cut-off or drastically reduced. The natural resources of these ecosystems have been over-exploited. Many habitats have been extensively modified; for example, mangrove forests have been turned into aquaculture ponds and paddy fields, and numerous freshwater swamps and marshes have been turned into fish ponds. Further, degradation of all aquatic ecosystems has occurred due to the anthropogenic activities in their watersheds. Extensive change in land use, overgrazing, mining and intensive agriculture using agrochemicals accelerate erosion and transport of sediments, nutrients and toxic substances with the run-off to the water bodies (Figure 2.11).

Figure 2.11: Linkages between the Phosphorus Cycle in a Terrestrial and Associated Aquatic Ecosystem

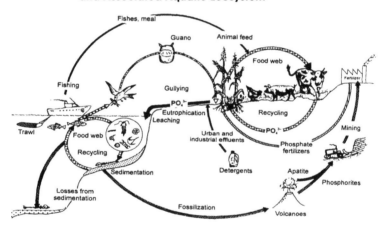

Source: Vyas and Golley (1976).

Introduction of exotic species has created further problems by affecting both quality of water and native biodiversity. It however needs to be pointed out that in South Asia, thousands of water bodies have been created in arid and semi-arid regions for storage of rainwater for irrigation and other domestic uses. Over centuries, many of these irrigation reservoirs have developed characteristics of ecosystems through natural and human assisted (e.g., introduction of plants and animals) processes of succession.

Various values and ecosystem services have been rediscovered and highlighted only in recent years after considerable loss and damage has occurred. However, the management of water resources fails to consider the role of aquatic ecosystems in sustaining the quantity and quality of water. It is not realised by the water resource managers that the quality of water (its physical, chemical and biological composition) determined its use. It is the community of plants, animals and micro-organisms in rivers, lakes and wetlands whose survival and welfare is critical to the human welfare, and therefore, water must first meet their requirements for various life processes. The concept of ecosystem emphasises that various kinds of organisms and their non-living environment function in tandem as one unit. Various freshwater ecosystems function differently but are embedded within the river basin which directly affects them. Therefore, issues of water management have to consider the river basins as one unit.

References and Select Bibliography

Alfred, J. R. B., A. K. Das and A. K. Sanyal. eds. 1998. *Faunal Diversity in India: A Commemorative Volume in the 50th Year of India's Independence*. Kolkata: Zoological Survey of India.

Ali, S. and S. D. Ripley. 1968–1974. *Handbook of Birds of India and Pakistan Together with those of Nepal, Sikkim, Bhutan and Ceylon*, 10 vols. Mumbai: Oxford University Press.

Calow, P. and G. E. Petts. eds. 1992. *Rivers Handbook*, vol. 1. Oxford: Blackwells.

———. 1994. *Rivers Handbook*, vol. 2. Oxford: Blackwells.

Cook, C. D. K. 1996. *Aquatic and Wetland Plant of India*. Oxford: Oxford University Press.

Cummins, K. W., M. A. Wilzbach, D. M. Gates, J. B. Perry and W. B. Taliaferro. 1989. 'Shredders and Riparian Vegetation', *Bio-Science*, 39 (1): 24–30.

Cummins, K. W., C. E. Cushing and G. W. Minshall. 1995. 'Introduction: An Overview of Stream Ecosystems', in C. E. Cushing, K. W. Cummins and G. W. Minshall (eds), *River and Stream Ecosystems of the World*, pp. 1–8. Amsterdam: Elsevier.

Dodson, S. I. and T. Hanazato. 1995. 'Commentary on Effects of Anthropogenic and Natural Organic Chemicals on Development, Swimming Behaviour, and Reproduction of *Daphnia*, a Key Member of Aquatic Ecosystems', *Environmental Health Perspectives* 103, Supplement 4. http://ehp.niehs.nih.gov/members/1995/Suppl-4/dodson-full.html (accessed 31 March 2011).

Federal Interagency Stream Restoration Working Group (FISRWG). 1998. *Stream Corridor Restoration: Principles, Processes, and Practices*. Washington, D.C.: Federal Interagency Stream Restoration Working Group.

Food and Agriculture Organisation of the United Nations (FAO). 2007. 'Agriculture and Water Scarcity: A Programmatic Approach to Water Use Efficiency and Agricultural Productivity', Committee on Agriculture, COAG/2007/7. FAO, Rome.

Forbes, S. A. 1887. 'The Lake as a Microcosm', *Illinois Natural History Survey*, 15 (1925): 537–50.

Gopal, B. 1987. *Water Hyacinth*. Amsterdam: Elsevier.

Gopal, B. 1997. 'Biodiversity in Inland Aquatic Ecosystems in India: An overview', *International Journal of Ecology and Environmental Sciences*, 23: 305–13.

Gopal, B. and W. J. Junk. 2000. 'Biodiversity in Wetlands: An Introduction', in B. Gopal, W. J. Junk and J. A. Davis (eds), *Biodiversity in Wetlands: Assessment, Function and Conservation*, vol. 1, pp. 1–10. Leiden: Backhuys Publishers.

Gopal, B. and U. Goel. 1993. 'Competition and Allelopathy in Aquatic Plant Communities', *Botany Review*, 59 (3): 155–210.

Groombridge, B. and M. Jenkins. eds. 1998. *Freshwater Biodiversity: A Preliminary Global Assessment*, WCMC Biodiversity series 8. Cambridge: World Conservation Press.

Holmes, N. T. H. and C. Newbold. 1984. *River Plant Communities — Reflectors of Water and Substrate Chemistry*, Focus on Nature Conservation 9. Shrewsbury: Nature Conservation Council.

Hora, S. L. 1928. 'Animal Life in Torrential Streams', *Journal of Bombay Natural History Society*, 32 (1): 111–26.

Illies, J. 1955. 'Der biologische Aspekt der limnologischen Fliesswassertypisierung', *Archiv fuer Hydrobiologie*, Supplement 22: 327–46.

Illies, J. and L. Botosaneanu. 1961. 'Preblemes et methodes de la classification et de la zonation ecologique des eaux courrantes, consideres surtout du point de vue faunistique', *Mitteilungen der internationale Vereinigung der theoretische und angewandte Limnologie*, 12: 57.

Jayaram, K. C. 1981. *The Freshwater Fishes of India*. Kolkata: Zoological Survey of India.

Jhingran, V. G. 1991. *Fish and Fisheries of India*, 3rd ed. Delhi: Hindustan Publishing House.

Junk, W. J., P. B. Bayley and R. E. Sparks. 1989. 'The Flood Pulse Concept in River-Floodplain Systems', *Canadian Special Publication of Fisheries and Aquatic Sciences*, 106: 110–27.

Keddy, P. A. 2000. *Wetland Ecology: Principles and Conservation*. Cambridge: Cambridge University Press.

Kumar, R., R. D. Singh and K. D. Sharma. 2005. 'Water Resources of India', *Current Science*, 89 (5): 794–811.

Lowe-McConnell, R. H. 1987. *Ecological Studies in Tropical Fish Communities*. Cambridge: Cambridge University Press.

Lowrance, R., R. Todd, J. Fail, O. Hendrickson, R. Leonard and L. Asmussen. 1984. 'Riparian Forests as Nutrient Filters in Agricultural Watersheds', *Bioscience*, 34 (6): 374–77.

McAllister, D. E., A. L. Hamilton and B. Harvey. 1997. 'Global Freshwater Biodiversity: Striving for the Integrity of Freshwater Ecosystems', *Sea Wind* 11 (3): 1–140.

McNeely, D. L. 1987. 'Niche Relations within an Ozark Stream Cyprinid Assemblage', *Environmental Biology of Fishes*, 18 (3): 195–208.

Meher Homji, V. M. 1984. 'Disasters of Deforestation: Desertification or Deluge!', *International Journal of Ecology and Environmental Sciences*, 10 (1–3): 37–73.

Millennium Ecosystem Assessment (MEA). 2005. *Ecosystems and Human Well-Being: Wetlands and Water Synthesis*. Washington, D.C.: World Resources Institute.

Minshall, G. W., K. W. Cummins, R. C. Petersen, C. E. Cushing, D. A. Bruns, J. R. Sedell and R. L. Vannote. 1985. 'Developments in Stream Ecosystem Theory', *Canadian Journal of Fisheries and Aquatic Sciences*, 42: 1045–55.

Mitsch, W. J. and J. G. Gosselink. 2000. *Wetlands*, 3rd ed. New York: John Wiley.

Naiman, R. J., J. M. Melillo, M. A. Lock, T. E. Ford and S. R. Reice. 1987. 'Longitudinal Patterns of Ecosystem Processes and Community Structure in a Subarctic River Continuum', *Ecology*, 68 (5): 1139–56.

Petts, G. E. and C. Amoros. 1996. *Fluvial Hydrosystems*. London: Chapman and Hall.

Pinay, G., H. Decamps, E. Chauvet and E. Fustec. 1990. 'Functions of Ecotones in Fluvial Systems', in R. J. Naiman and H. Décamps (eds), *The Ecology and Management of Aquatic-Terrestrial Ecotones*, MAB series 4, pp. 141–70. Paris: UNESCO.

Sahagian, D. and J. M. Melack. eds. 1998. *Global Wetland Distribution and Functional Characterisation: Trace Gases and the Hydrologic Cycle*. Stockholm: International Geosphere-Biosphere Programme.

Shiklomanov, I. A. 1993. 'World Fresh Water Resources', in P. Gleick (ed.), *Water in Crisis: A Guide to the World's Fresh Water Resources*, pp. 13–24. Oxford: Oxford University Press.

Shiklomanov, I. A. and J. C. Rodda. 2003. *World Water Resources at the Beginning of the Twenty-First Century*. Cambridge: Cambridge University Press.

Tansley, A. G. 1935. 'The Use and Abuse of Vegetational Concepts and Terms', *Ecology*, 16 (3): 284–307.

Trenberth, K. E., L. Smith, T. Qian, A. Dai and J. Fasullo. 2007. 'Estimates of the Global Water Budget and its Annual Cycle Using Observational and Model Data', *Journal of Hydrometeorology*, 8 (4): 758–69.

Vannote, R. L., G. W. Minshall and K. W. Cummins. 1980. 'The River Continuum Concept', *Canadian Journal of Fisheries and Aquatic Sciences*, 37: 130–37.

Vollenweider, R.A. 1968. *Scientific Fundamentals of the Eutrophication of Lakes and Flowing Waters with Particular Reference to Nitrogen and Phosphorus as Factors in Eutrophication*. Paris: Organisation for Economic Co-operation and Development.

Vyas, A. B. and F. B. Golley. 1976. *Principles of Ecology*. Jaipur: International Scientific Publications.

Ward, J. V. 1989. 'Riverine–Wetland Interactions', in R. R. Sharitz and J. W. Gibbons (eds), *Freshwater Wetlands and Wildlife*, pp. 385–400. Oak Ridge: US Department of Energy.

Welcomme, R. L. 1979. *Fisheries Ecology of Floodplain Rivers*. London: Longman.

———. 2000. 'Fish Biodiversity in Floodplains and their Associated Rivers', in B. Gopal, W. J. Junk and J. A. Davis (eds), *Biodiversity in Wetlands: Assessment, Function and Conservation*, vol. 1, pp. 35–60. Leiden: Backhuys Publishers.

Wetzel, R. G. 2000. *Limnology*, 3rd ed. New York: John Wiley.

Wissmar, R. C. and F. J. Swanson. 1990. 'Landscape Disturbances and Lotic Ecotones', in R. J. Naiman and H. Décamps (eds), *The Ecology and Management of Aquatic-Terrestrial Ecotones*, MAB series 4, pp. 65–90. Paris: UNESCO.

3

Environmental Flow:
Assessments and Applications

Rezaur Rahman

Rivers have been the great nurturer of human civilisation. They have supplied drinking water, provided fish and supported communication. At the same time, they have sustained riparian ecosystems in addition to their own. The provision of flows, including volumes and timings, to maintain downstream aquatic ecosystems and provide services to dependent communities has been recognised in developed countries for more than two decades and is increasingly being adopted in developing countries (World Bank 2009). These services could be categorised into two broad areas, namely human welfare services and ecosystem sustenance services. The human welfare services include clean drinking water, subsistence use, food sources such as fish and invertebrates, opportunities for harvesting wood for fuel, grazing and cropping on riverine corridors and floodplains, flood conveyance, navigation routes, recreational opportunities, cultural, aesthetic and religious benefits. The ecosystem sustenance services include groundwater recharge, biodiversity conservation (including protection of natural habitats, protected areas, and national parks), freshwater saltwater balance and removal of wastes through biogeochemical processes.

An environmental flow is the water regime provided within a river, wetland or coastal zone to maintain ecosystems and their benefits where there is competing water uses and where flows are regulated (IUCN 2003). Environmental flows provide critical contributions to river health, economic development and poverty alleviation. They ensure the continued availability of the many benefits that healthy river and groundwater systems bring to society.

Environmental Flow and IWRM

It is now widely accepted that new thinking on water infrastructure, set within a broader framework of Integrated Water Resource

Management (IWRM), is needed to manage water resources sustainably and equitably (IUCN 2003). Broadly, IWRM considers land-water-environment interactions throughout the entire river basin, in conjunction with surface and groundwater flows, in a more systematic manner. Keeping provision for environmental flow is a key aspect of IWRM.

The environment is linked to IWRM in three fundamental ways (World Bank 2009). First, the aquatic (and related terrestrial) ecosystem provides habitat for fish, invertebrates, and other fauna and flora. The aquatic ecosystem is thus a water-consuming sector just like agriculture, energy, and domestic and industrial supply. Second, the design and operation of hydraulic infrastructure for water supply, sewerage, irrigation, hydropower and flood control often affect ecosystems, both upstream and downstream of the infrastructure, and communities — farming, pastoral and fishing — that are dependent on those ecosystems. Conversely, the reoperation and rehabilitation of existing infrastructure have been used to support the successful restoration of degraded riverine ecosystems. Third, integrated water resources planning and management are facilitated by policies, laws, strategies and plans that are multi-sectoral, based on the allocation of water for all uses; protection of water quality and control of pollution; protection and restoration of lake basins, watersheds, groundwater aquifers, and wetlands; and control and management of invasive species.

Assessment Methods

Environmental flow assessment is a means of describing the potential trade-offs between development gains, such as increased use of water for agriculture or industrial production, and environmental losses, for instance reduced habitats for aquatic plants and animals or reductions in the quality of life of subsistence users of the river. A number of environmental flow assessment methods can be found in the literature (for an excellent inventory, see Tharme 2003). Broadly, the methods can be classified as follows:

- Hydrological methods
- Hydraulic methods
- Habitat simulation methods
- Holistic approaches
- Ecotope method

The methods are described below. Details about these methods can be found in Jowett (1997), IUCN (2003) and BUET-DUT (2006).

Hydrological Methods

Hydrological methods are based on the fact that aquatic organisms need a certain minimum flow for their sustenance. The minimum flow is deduced from the flow data, either observed or synthetically produced, of the river. This method is easy to use but requires historic or estimated flow records of a river. The method is mostly used for long-term planning where detailed assessment is not required. Among the various available methods, the Tennant (1976) is probably the most widely known (Jowett 1997).

The Tennant method is relatively inexpensive, quick and easy to apply. Its development required considerable research and input from experts. The results compare relatively well with those from data-intensive techniques. The approach is based on trends derived from field observations in the United States of the relationship among river conditions, the amount of flow in the river and the resultant fish habitat (BUET-DUT 2006).

The Tennant method assumes that some percentage of the mean flow is needed to maintain a healthy stream environment. Tennant examined cross-section data from 11 streams in Montana, Nebraska and Wyoming. He found that stream width, water velocity and depth increased rapidly from zero flow to 10 per cent of the mean flow, and that the rate of increase declined at flows higher than 10 per cent. At less than 10 per cent of the mean flow, he considered that water velocity and depth were degraded and would provide for 'short-term' survival of aquatic life. He considered that 30 per cent of the average flow would provide satisfactory stream width, depth and velocity for a 'baseflow regime'. Tennant's assessment of the environmental quality of different levels of flow was based on the quality of the physical habitat that they provided. At 10 per cent of average flow, average depth was 0.3 m and velocity 0.25 m/s, and Tennant considered these to be the lower limits for aquatic life. He showed that 30 per cent of average flow or higher provided average depths of 0.45–0.6 m and velocities of 0.45–0.6 m/s and considered these to be in the good to optimum range for aquatic organisms.

The tennant method can be extended to incorporate seasonal variation by specifying monthly minimum flows as a percentage of monthly mean flows. Other historic flow methods recommend flows based on the flow duration curve or an exceedance probability of a low flow, where the level of protection is implicit in the magnitude of the percentage.

Hydrological indices such as those described above are transferable between regions when re-calibrated for a new region. Even then, they do not take account of site-specific conditions. The indices based only on hydrological data are more readily re-calibrated for any region, but have no ecological validity. Thus, the uncertainty to achieve good results is very high. Those indices based on ecological data clearly have more ecological validity, but the ecological data may be costly and time-consuming to collect.

Hydraulic Methods

Hydraulic methods relate various parameters of the hydraulic geometry of stream channels to discharge. The hydraulic geometry is based on surveyed cross-sections, from which parameters such as width, depth, velocity and wetted perimeter are determined. Because of the field and analytical work involved in this, they are more difficult to apply than historic flow methods. Variation in hydraulic geometry with discharge can be established by measurements at different flows, prediction from cross-sectional data and stage-discharge rating curves, Manning's or Chezy's equations, or calculation of water surface profiles.

The most common hydraulic method considers the variation in wetted perimeter with discharge. This method is based on the assumption that fish rearing is related to fish production, which in turn is related to how much of the river bed is wet. The relation of wetland perimeter to cross-section is shown in Figure 3.1. The usual procedure is to choose the break or 'point of diminishing returns' in the stream's wetted perimeter versus discharge relation as a surrogate for minimally accepted habitat. This inflection point represents that flow above which the rates of wetted perimeter gains begin to slow. Because the shape of the channel can influence the results of the analysis, this technique is usually applied to streams with cross-sections that are wide, shallow and relatively rectangular (BUET-DUT 2006). As a rule of thumb, shallow, wide rivers tend to show more sensitivity of their wetted perimeter to changes in flow than do narrow deep rivers.

In some cases, limited field surveys are undertaken while in others the existing stage-discharge curves from river gauging stations are

Figure 3.1: Relation between Wetted Perimeter, Cross-section and Discharge

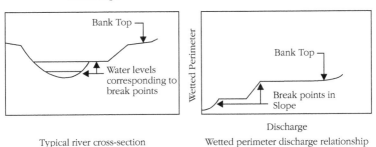

Typical river cross-section   Wetted perimeter discharge relationship

Source: BUET-DUT (2006).

used. This method has been used in the United States and Australia considerably and some researchers have highlighted the problem of trying to identify threshold discharges below which wetted perimeter declines rapidly. Given this limitation, the method is more appropriate to support scenario-based decisionmaking and water allocation negotiations than to determine an ecological threshold.

If flow requirements are based on retaining a percentage of the wetted perimeter at mean flow and there is a linear or near linear relationship between wetted perimeter and flow, the criterion is, in effect, the same as a percentage of the mean flow. Hydraulic methods are not usually used to assess seasonal flow requirements.

Habitat Simulation Methods

Habitat is an encompassing term used to describe the physical surroundings of plants and animals. Some habitat features, such as depth and velocity, are directly related to flow, whereas others describe the river and surroundings. Habitat methods are a natural extension of hydraulic methods. The difference is that the assessment of flow requirements is based on hydraulic conditions that meet specific biological requirements rather than the hydraulic parameters themselves. Hydraulic models predict water depth and velocity throughout a reach. These are then compared with habitat suitability criteria to determine the area of suitable habitat for the target aquatic species. When this is done for a range of flows, it is possible to see how the area of suitable habitat changes with flow.

The hydrologic and hydraulic methods recommend a single instream flow value for a defined period in individual rivers, which is often termed as 'minimum flow'. Such standard-setting recommendations are not very useful in negotiation situations where a compromise is needed between different environmental and socioeconomic uses of the river flow. Habitat methods are particularly suitable in such 'trade-off' situations, where incremental change in habitat can be compared with the benefits of resource use. Habitat/flow relationships can be used to evaluate alternative flow management strategies and are part of the information base used in the process of choosing appropriate flow rules for river management.

Tools that can be used to achieve this result fall into two groups. The first uses statistical analyses to correlate environmental features of a stream with fish population size. An example of this analysis is the Wyoming's Habitat Quality Index (HQI), described by Binns (1982). An HQI is developed by regressing several habitat variables against the standing crop of fish. The second group of tools link open channel hydraulics with known elements of fish behaviour. Examples include In-stream Flow Incremental Methodology (IFIM), developed by the US Fish and Wildlife Services, the main component of which is Physical Habitat Simulation System (PHABSIM). The PHABSIM (Milhous et al. 1984) is a specific model designed to calculate an index to the amount of microhabitat available for different life stages at different flow levels.

Habitat suitability curves are the biological basis of habitat methods. Habitat suitability can be specified as seasonal requirements for different life stages, but this is not limited to aquatic organisms. Depth, velocity and width criteria for bathing, wading, kayaking, canoeing and other recreational pursuits have also been described (see Jowett 1997).

When using habitat methods, there are more ways of determining flow requirements than for either historic flow or hydraulic methods. The relationship between flow and the amount of suitable habitat is usually non-linear (Figure 3.2). Flows can be set so that they maintain optimum levels of fish habitat, retain a percentage of habitat at average or median flow, or set so that they provide a minimum amount of habitat defined either as a minimum percentage of water surface area or as a percentage exceedance value on the habitat duration curve. Flows can also be set at the point of inflection in the habitat/flow relationship. This is possibly the most common method of assessing minimum flow requirements using habitat methods. While there is no percentage or absolute value associated with this level of protection, it is a point of 'diminishing return' where proportionally more habitat is

Figure 3.2: Relationships between Flow and Biological Response for a Hypothetical River

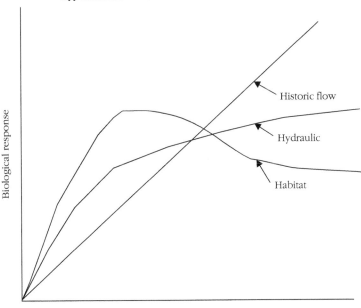

Source: Jowett (1997).
Note: Relationships between flow and biological response for a hypothetical river where biological response is expressed in terms of the measures used in the flow assessment methods; flow for historic flow methods, wetted perimeter for hydraulic methods and weighted usable area for habitat methods.

lost with decreasing flow than is gained with increasing flow. In some rivers, the relationship between flow and habitat for flow-sensitive species is linear, especially in the low flow range. In these cases, flow recommendations using percentage retention or exceedance for instream habitat are, in effect, the same as recommendations of hydraulic and historic flow methods that specify a percentage or exceedance value for flow or wetted perimeter.

Habitat methods are more flexible than either historic flow or hydraulic methods. It is possible to examine the variation of the habitat utilised by many species and life stages throughout the year and to select flows that provide this habitat. However, this means that it is necessary to have a good knowledge of the stream ecosystem and some clear management objectives in order to resolve potential conflicting habitat requirements of different species or life stages.

Holistic Approaches

More and more methods now take a holistic approach that explicitly includes an assessment of the whole ecosystem, such as associated wetlands, groundwater and estuaries. These also account for all species that are sensitive to flow, such as invertebrates, plants and animals, and address all aspects of the hydrological regime including floods, droughts and water quality. A fundamental principle is to maintain natural variability of flows.

Generally, holistic approaches make use of teams of experts and may involve participation of stakeholders, so that the procedure is holistic in terms of interested parties as well as scientific issues. Where methods have the characteristic of being holistic, they clearly have the advantage of covering the whole hydrological-ecological-stakeholder system. The disadvantage is that it is expensive to collect the relevant data.

The family of holistic approach and expert panel based methods has the common feature that they use a team of experts to make judgments on the flow needs of different aquatic biota. The composition of the panel will depend on the specific environmental and social features of the river in question, but typically includes a hydrologist, geomorphologist, aquatic botanist and fish biologist. In many cases, one or more community representative will join the panel. The collective experience of the panel members is used in the absence of reliable, predictive flow-ecology models. By putting these experts on a panel, rather than employing them independently, it is expected that an integrated assessment of flow needs will emerge.

Holistic approaches are essentially ways of organising and using flow-related data and knowledge. One such approach, the Building Block Methodology (BBM) in South Africa (King and Luow 1998) requires early identification of the future desired condition of the river. An environmental flow regime is then constructed on a month by month basis, through separate consideration of different components of the flow regime to achieve and maintain this condition. Each flow component is intended to achieve a particular ecological, geomorphologic or water-quality objective.

Ecotope Method

The ecotope method is a new approach for assessing environmental flow requirements and it enables integration of river and floodplain ecosystems and their functions. Such a method seems useful in low land rivers systems such as in Bangladesh where the floodplain forms an intricate part of the riverine ecosystem (BUET-DUT 2006). The assessment of environmental flows therefore requires an approach that is broader than the instream flow methods used in most cases.

The landscape systems are identified depending on positional factors related to large-scale processes. Ecotope classification is based on conditional factors derived from abiotic processes that determine the appearance of the landscape patterns. The dominant processes are morpho-dynamics (erosion, sediment, substrate, etc.), hydrodynamics (flow, wave regime, etc.) and land use (effects of human intervention). These processes determine the operational factors such as moisture regime, nutrient availability and acidity. Operational factors directly affect the flora and fauna.

Based on remote sensing images and a field survey, an ecotope map of the present situation is defined. Using information from fish and vegetation surveys, ecotope suitability rules are defined. These ecotope suitability rules are used in order to link the ecotopes to the presence of specific species, which are known to be both of ecological and economical importance.

A digital elevation model of the study area combined with hydro-dynamic models enables the generation of spatially defined flood characteristics, such as flood frequency and duration. These results are combined with the ecotope map of the present situation to define ecotope model consisting of specific parameter rules in such a way that ecotopes can be classified into discrete value classes of these parameters.

The procedure described above to develop the ecotope model is thus based on the existing local conditions in the present situation. This model can then be used to define impacts of a changed hydro-dynamic situation that can be simulated by using the hydrodynamic model. The application of the ecotope model generates a map of the predicted future ecotope distribution in the study area. The previously defined ecotope suitability rules are finally used to calculate predicted vegetation and fauna changes including fish abundance. This way, alternative environmental flow requirements can be assessed and the impact on the riverine system can be visualised.

Choice of Appropriate Methods

The purpose of a flow assessment and the intended use of the results should guide the selection of the assessment method. Project-specific flow assessments for large or controversial projects, which are likely to call for considerable negotiation and trade-offs between environment and development issues, require a more comprehensive approach than do flow assessments for coarse-scale planning studies, where a single number might suffice.

Many countries, for example the US and Canada, operate on two-tier system, with the hydrologic method used in a standard-setting problem, such as for level one planning when rapid assessment is required or the value of the fishery is not explicitly considered. Incremental approach — a more rigorous, defensible and detailed method — is used for level two studies enabling quantification of varying instream requirements, incorporating various management options to be assessed (BUET-DUT 2006).

Within either category, the flow assessment method eventually chosen will depend on technical considerations, such as the quality and availability of data on the study river, the location and extent of the study area, expertise available, the prevailing time and financial constraints and the level of confidence required in the final output. Eco-hydrology is a relatively new scientific field and so there is only limited understanding and very few models of species responses to varying hydrologic conditions. Most of the data and understanding required for interactive approaches have to be acquired on a site-by-site basis, considerably adding to the time, funding and expertise required for a flow assessment.

Releasing Environmental Flow through Existing Water Infrastructure

Historically, river water has been used for maximising economic growth (e.g., agriculture, industry) and social welfare (e.g., drinking water supply, flood control). Natural flow regimes of the rivers have been modified by direct impoundments through reservoirs and dams or by diverting water by barrages and pumps. As a result, river flow has reduced and hydrological cycle has been modified. These modifications have adversely affected not only the aquatic ecosystems but also the subsistence use of the poor people.

Table 3.1 (IUCN 2003) shows various types of 'soft' and 'hard' infrastructure used in water management, together with the associated strategies and measures that serve to improve environmental flows. The physical ability to modify releases from existing dams depends on

Table 3.1: Representative 'Soft' and 'Hard' Infrastructure Development and Management Strategies to Improve Environmental Flows

Water management	Representative strategies and measures (To improve environmental flows)		
Function	*Infrastructure/activity*	*Strategy/objective*	*Possible measures*
River water storage, abstraction and flow regulation	Dams, weirs, and river diversions of all scales	Improve the quantity, timing and quality of downstream releases Reduce the quantity of abstractions/diversion flows (via demand management)	Change design standards for new facilities Modify existing reservoir operating strategies *Where feasible:* Retrofit outlet works of existing dams Decommission dams to restore flows
Groundwater abstraction and recharge	Tubewells, groundwater recharge systems, retention/recharge basins, community-scale rainwater harvesting, etc.	Reduce unsustainable abstractions lowering groundwater tables Improve flows (availability) to groundwater-dependent ecosystems Improve infiltration of storm and flood water to groundwater sources Improve groundwater quality	Modify abstraction rates (through pricing, fees and demand-side measures) Introduce infrastructure for storm and flood water retention/groundwater recharge at different scales Introduce sustainable groundwater/aquifer management Introduce/modify infrastructure for conjunctive water use

(Table 3.1 Continued)

(*Table 3.1 Continued*)

Water management	*Representative strategies and measures (To improve environmental flows)*		
Function	*Infrastructure/activity*	*Strategy/objective*	*Possible measures*
Transport, bulk delivery and distribution to consumptive offstream uses	Canals, aqueducts, primary and tertiary distribution canals, pipelines, etc.	Reduce unnecessary losses in distribution systems to take pressure off supply Improve the efficiency of delivery systems	Repair leaks in municipal water distribution systems and infrastructure Line irrigation canals
End-use demand management	Water-efficient end-use devices, water conservation and water management	Reduce abstractions from surface and groundwater abstractions Recycle and reuse water where feasible	Utilise water-efficient end-use devices Increase water metering and control (piped and groundwater) Implement policy measures promoting conservation (e.g., progressive tariffs) Employ technologies and systems for water reuse
Water quality management	Water treatment facilities, drainage systems, land use systems, agrochemical systems	Improve water treatment Control/reduce urban, agricultural and industrial pollutants entering the watercourses Restore wetlands, environmental flows for natural purification	Expand and rehabilitate water treatment infrastructure and facilities Design water treatment facilities for new water quality standards Eliminate/modify infrastructure (e.g., holding or settling ponds) and practices that contaminate groundwater

Catchment and watershed management	Land management systems and farming practices, erosion control, forest and vegetation cover management, etc.	Improve water retention capacities of catchments and reduce uncontrolled run-off Reduce erosion and sediment flow into rivers Improve soil stability	Implement/reinforce catchment management measures where feasible, e.g., adapting: – forest and vegetation cover management; – agriculture land use practices; and local water harvesting technologies
Non-conventional supply	Recycling, desalination of brackish water and seawater, conjunctive water management, traditional water harvesting systems, etc.	Add non-conventional supply to centralise water system and networks Add local supply options Improve integrated management of water sources	Introduce/reinforce infrastructure where feasible, e.g., introduce: – desalination; – conjunctive surface-groundwater management; and – local rural/urban use of rain water harvesting

Source: IUCN (2003).

the type of dam, the provisions for releasing water through the dam, and the state of repair of the key water control outlets and structures.

Reregulation weirs are sometimes constructed downstream of a dam when there are large fluctuations in daily releases from the peaking operation of hydropower units. These weirs can range from a few hundred meters to a few kilometers downstream. They are generally designed to pool water during peak discharge periods to prevent large surges, and release it more regularly.

Improving downstream releases can be a simple matter of lifting a sluice gate, turning a valve to open bottom outlets, or increasing flows through power turbines. New dams can be designed with physical provisions for adjusting releases and accommodating future changes in values for managing the river at limited costs. When it is not physically feasible to adjust releases from older existing dams, retrofit is required.

Lessons from Environmental Flow Implementation Case Studies

World Bank (2009) presents the results of in-depth analysis of 17 selected case studies to identify the lessons from incorporating environmental flows into water resources policy, basin and catchment plans, new infrastructure projects and the rehabilitation and reoperation of existing infrastructure. The assessment criteria included factors that influenced the case study's success as well as the institutional drivers that initiated and supported the introduction of environmental flows.

Inclusion of Environmental Flows in Water Resources Policies

An analysis of five policy case studies found that the inclusion of environmental flows in policy should provide for the following:

1. Legal standing for environmental water allocations.
2. Inclusion of environmental water provisions in basin water resources plans.
3. Assessment of all relevant parts of the water cycle when undertaking Environmental Flow Assessments (EFA).
4. A method or methods for setting environmental objectives in basin plans.

5. Attention to both recovery of over-allocated systems and protection of unstressed systems.
6. Clear requirements for stakeholder involvement.
7. An independent authority to audit implementation.
8. A mechanism for turning value-laden terms into operational procedures.

Inclusion of Environmental Flows in Basin and Catchment Plans

Several lessons emerged from the analysis of four basin and catchment water resources plans:

1. Recognition of environmental flows in water resources policy and legislation provides important backing for including environmental flows in basin or catchment plans.
2. There is a need to demonstrate the benefits from environmental water allocations after plans are implemented.
3. The term 'environmental flows' can be counterproductive if not explained at an early stage.
4. Participatory methods need to be tailored to suit stakeholder capacity.
5. A range of Environmental Flows Assessment (EFA) techniques is needed to suit different circumstances.
6. Ecological monitoring is essential to provide information for adaptive management.

Inclusion of Environmental Flows in Infrastructure Projects

Four new dams and restoration projects were reviewed for lessons in assessing and implementing environmental flows:

1. Engineering improvements usually have to be combined with reoperations to provide the volume of water needed for major ecosystem restoration.
2. Inclusion of environmental flows in water resources policy simplifies the application of EFAs at the project level.
3. Environmental outcomes need to be linked closely to social and economic outcomes.
4. EFAs should be conducted for all components of the hydrological cycle.

5. Traditionally trained water resources professionals can find it difficult to grasp environmental flow concepts.
6. Water resources plans provide benchmarks for water allocations during project assessments.
7. Active monitoring is needed to enforce flow allocation decisions and undertake adaptive management.
8. It is important to present information in terms that are comprehensible to decision makers.
9. Economic studies can support arguments for downstream water allocations.
10. EFAs are yet to be fully mainstreamed into EIAs.
11. The cost of conducting EFAs constitutes a small fraction of project costs.
12. EIAs have not always or adequately identified issues associated with downstream water provisions.

A Case Study of Assessment of Environmental Flow Requirement in the Coastal Region of Bangladesh*

Introduction

The concept of environmental flow is new to Bangladesh. To operationalise this concept, two protocols have been developed by the IUCN for Bangladesh: one for assessment of environmental flow requirement and another for nationwide establishment of environmental flow. The protocol for environmental flow assessment is based mainly on expert assessment considering the dearth of data regarding environmental flow.

The protocol was tested in Bakkahli river in southeast part of Bangladesh. The river originates in the Arakan hills and falls into the Bay of Bengal. Water is stored in the river for dry season irrigation using rubber dams that are inflated in December–January, and are deflated in March–April. Low-lift pumps are used to irrigate surrounding agricultural lands. The Local Government Engineering Department (LGED) constructs these rubber dams and after construction, hands over their operation and management to the locally formed Water Management Cooperative Association (WMCA).

Most of the rubber dams have replaced previous earthen dams that were constructed by the local people to store water during the dry season. Although dam deployment deters navigation through the channel, the primary impact has been on upstream and downstream directed free movement of fish during this period, adversely affecting the spawning and breeding of different fish species.

Most of the water sector structures like the rubber dams in Bangladesh have been constructed for agricultural purposes. User rights are already established in these structures on behalf of the farmers. During the case study, it was apparent that farmers, being mostly marginal and small, are not very willing to sacrifice this water right to ensure environmental flow. Although farmers prioritise on increasing agricultural productivity, they recognise the need for fish, which is a principal component of traditional Bengali diet. Therefore, fish has been recommended as an indicator species for environmental flow assessment. Flow requirement of two locally popular fish species were assessed. The species are Golda (a large freshwater shrimp variety) and Hilsa (Indian Salmon). During this exercise, assessment of flow required for movement of fish during a specified time of the dry season was a major component. The following sections describe the flow assessment procedure and important flow-related issues in the rubber dam projects in general.

Panel of Experts

After initial planning meetings among specialists from different fields, a team of experts was selected to carry out the task of environmental flow assessment. The team included a water resources expert, an environmental flow analyst, a fisheries expert and a fish biologist. The expert panel met on several occasions to exchange information, discuss important issues and decide on the strategies of action, and finally to assess the environmental flow requirement.

Compilation of Available Information

Information on hydrology, dam construction, irrigation coverage and water supply was available from various sources. Information regarding fish habitat, catch composition and fish biology was available from secondary sources via reports and personal communication.

Considering agriculture to be the backbone of the local economy, the main trade-off of flow was between irrigation water requirement and dry season flow requirement for fish movement to increase the fish production. The initially compiled information also indicated that the main incentive for dry season flow release would be the consequent increase in Golda production in the river.

Survey of the Study Area

Survey of the study area was conducted through two visits of the expert panel along with other specialists. During the survey, the team verified information collected from secondary sources, identified potential conflicts in use of water or flow release and assessed the willingness of the WMCA for dry season flow release in exchange for a potential increase in Golda production in the river. Several interviews were conducted to crosscheck and supplement field information. The interviewed personnel included the chairman and members of WMCA, irrigation scheme managers, LGED executive engineer and socioeconomist. Focus group discussions were carried out to determine local people's perception about the relationship between dry season flow release and increase in Golda production.

The WMCA is apparently more interested in culture fisheries in the reservoir than releasing dry season flow for Golda production in the river. Some of the members proposed to raise the embankments with excavated materials of the silted up channel bottom to increase irrigation coverage and culture fisheries. They expressed their willingness to include fisheries within their management. However, they also indicated that they would release excess rainwater and lower the dam if sufficient irrigation water is available. They would also participate in hydrological and fish species monitoring if a future development project is planned.

Local people appeared to be not very aware or concerned about the decline in Golda production because of the operation of the dam. LGED personnel were willing to cooperate with the WMCA for technical and advisory support if they wanted to release dry season flow for Golda production. Water level gauges are installed beside the dam that can be used for water level monitoring.

Assessment of Environmental Flow Requirement

Bakkhali dam conserves about 80 Mm^3 (based on 25 per cent dependable flow) while the average volume of water lifted from the Bakkhali river for irrigation during the dry season is about 14.9 Mm^3. However, it is also indicated that for average and 80 per cent dependable flow, Bakkhali dam conserves 62.8 and 30.2 Mm^3, respectively, between December and April. Based on this flow availability, the surplus water volume available for release at different levels of conservation of the dam is shown in Table 3.2.

Table 3.2: **Volume of Water Available for Release through Bakkhali Dam**

	Flow availability		
	25% dependable flow	Average flow	80% dependable flow
Conservation capacity (Mm³)	80.0	62.8	30.2
Surplus volume (Mm³)	65.1	47.9	15.3

Source: DOE (2005).

There is also a scope for saving water by improving existing irrigation management. A significant volume of water can be conserved in the Bakkhali project by reducing the depth of standing water used in the paddy fields. Large volume of water is lost by seepage and evaporation in the unlined drainage channels of the Bakkhali project. Seepage through the channel bottom may be minimised by lining them. The surplus volume indicated in Table 3.2 and additional volume conserved by improving management practices can meet the environmental flow requirement without compromising for water required for irrigation.

Flow release requirement was assessed for Golda and Hilsa species. An overall flow release requirement was also assessed for all fish movement. The flow is to be released during night for 6 to 8 hours per day, 7 days during new and full moon phases. The downstream pool depth should be about 1.0 m. Other requirements for flow release are given in Table 3.3.

Table 3.3: **Flow Release Requirements at Bakkhali Dam for Fish Movement**

Fish species	Flow release period	Number of days flow released	Downstream velocity (m/sec)
Golda	March 1 to April 30	28	0.30 to 0.68
Hilsa	January 1 to March 31	42	0.6 to 0.8
Overall	February 1 to April 15	35	0.6 to 0.8

Source: DOE (2005).

Required flow release volumes are calculated based on the criteria set forth above, and are given in Tables 3.4, 3.5 and 3.6. The flow velocity at the dam is calculated based on the required downstream velocity for fish movement. To meet the criteria for fish movement, a minimum of 15.6 and 45.2 Mm³ of water is to be released for Golda and Hilsa, respectively. Considering both the species, the minimum overall flow release volume is 37.2 Mm³. These requirements are comparable to the surplus volumes shown in Table 3.2, and may be negotiated for release through the dam.

Table 3.4: Required Flow Release Volume through Bakkhali Dam for Golda Movement

	Volume of water to be released (Mm³)	
Flow velocity (m/sec)	*Flow release (hr/day)*	
	6.0	8.0
0.30	15.6	20.8
0.68	34.9	46.6

Source: DOE (2005).

Table 3.5: Required Flow Release Volume through Bakkhali Dam for Hilsa Movement

	Volume of water to be released (Mm³)	
Flow velocity (m/sec)	*Flow release (hr/day)*	
	6.0	8.0
0.6	45.2	60.3
0.8	60.4	80.5

Source: DOE (2005).

Table 3.6: Required Flow Release Volume through Bakkhali Dam for Overall Fish Movement

	Volume of water to be released (Mm³)	
Flow velocity (m/sec)	*Flow release (hr/day)*	
	6.0	8.0
0.6	37.2	49.6
0.8	49.6	66.2

Source: DOE (2005).

Summary of the Case Study

In the Bakkhali river rubber dam project, the main trade-off of flow is between irrigation water requirement and dry season flow requirement for fish movement and a consequent increase in fish population. Field investigation indicates that the main incentive for dry season flow release would be the increase in Golda production in the river.

This preliminary assessment shows that surplus water is available in the study area which can be used as environmental flow. There is also scope for saving water by improving existing irrigation management. The surplus volume and additional volume conserved by improving management practices can meet the environmental flow requirement without compromising the requirement of water for irrigation.

Fish has been recommended as the indicator species for assessment of environmental flow. Flow release requirement has been assessed for the overall fish population, particularly for Golda and Hilsa species. These requirements are found to be comparable to the surplus volumes and may be negotiated for release through the dam.

In an existing water infrastructure, where water right is already established, it is very important to establish dialogue with the beneficiaries at the very onset in order to identify opportunities, which would convince them about benefits of releasing environmental flow. In this exercise, sustenance of locally popular fish species provided such an entry point. The next task was to explore operation and management practices, modification and improvement of which would allow fulfillment of the environmental flow requirement as well as the existing rights.

Note

* This case study is based on works carried out by Rezaur Rahman and M. Shah Alam Khan for the project entitled 'Minimum Environmental Flow Requirement for Ecosystem and its Functioning' of IUCN — Bangladesh. A brief about the project including Tables 3.3, 3.4, 3.5 and 3.6 appears in DOE (2005).

References and Select Bibliography

Binns, N. A. 1982. *Habitat Quality Index Procedures Manual.* Cheyenne: Wyoming Game and Fish Department.

BUET-DUT. 2006. *Management of Rivers for Instream Requirements and Ecological Protection.* Dhaka: BUET-DUT Linkage Project Phase III, BUET and TUDelft.

Department of Environment (DOE). 2005. *Bangladesh National Programme of Action for Protection of the Coastal and Marine Environment from Land-based Activities.* Dhaka: Department of Environment, Ministry of Environment and Forests, Government of the People's Republic of Bangladesh.

International Union for Conservation of Nature and Natural Resources (IUCN). 2003. *Flow — The Essentials of Environmental Flows.* Gland: IUCN.

Jowett, I. G. 1997. 'Instream Flow Methods: A Comparison of Approaches', *Regulated Rivers: Research and Management,* 13 (2): 115–27.

King, J. and D. Louw. 1998. 'Instream Flow Assessments for Regulated Rivers in South Africa Using the Building Block Methodology', *Aquatic Ecosystem Health and Management,* 1 (2): 109–24.

Milhous, R. T., D. L. Wegner and T. Waddle. 1984. *User's Guide to the Physical Habitat Simulation System.* Washington, D.C.: U.S. Dept. of the Interior, Fish and Wildlife Service.

Tennant, D. L. 1976. 'Instream Flow Regimens for Fish, Wildlife, Recreation, and Related Environmental Resources', in J. F. Orsborn and C. H. Allman (eds), *Proceedings of the Symposium and Speciality Conference on Instream Flow Needs II,* pp. 359–73. Bethesda: American Fisheries Society.

Tharme, R. E. 2003. 'A Global Perspective on Environmental Flow Assessment: Emerging Trends in the Development and Application of Environmental Flow Methodologies for Rivers', *Rivers Research and Applications,* 19 (5–6): 397–441.

World Bank. 2009. *Environmental Flows in Water Resources Policies, Plans, and Projects — Findings and Recommendations.* Washington D.C.: World Bank.

4

Human Interventions on Water Ecosystem and Implications for Fisheries Resources in Bangladesh

M. Monirul Qader Mirza, Mokhlesur Rahman and Anisul Islam

Bangladesh is often termed as the land of *Nodi* (rivers) and *Nala* (rivulets). The country is criss-crossed by a vast network of rivers, tributaries and canals (Figure 4.1). It is the largest delta in the world formed by the alluvial sediments carried down by the three rivers — Ganga, Brahmaputra and Meghna (GBM) — and their numerous tributaries and distributaries through the geologic times. The sediments are generally generated in the Himalaya. Vast agricultural fields in the basins of these rivers are also a significant source of sediment. It is estimated that on average, the Ganges and Brahmaputra transport 316 and 721 million tonnes of sediment, respectively, every year into Bangladesh (Islam et al. 1999). These high sediment loads in the rivers demonstrate the high rate of denudation that is occurring in their drainage basins, mostly located outside Bangladesh. The transportation of sediment varies seasonally, with the highest amounts being carried during the monsoon season. Loss of capacity through sedimentation has implications for open-water fishery resource in Bangladesh. Hydrology of the river basins is a major factor for sustaining fisheries production, maintaining biodiversity and determining fishing practices. Hydrological regimes of the river change annually. During the monsoon season, huge volume of water from the vast watershed drains into the Bay of Bengal through a vast network of rivers in Bangladesh. Water quality also varies with the seasonal water availability.

The people of Bangladesh are generally known as the *Mache-Bhate Bangalee* (the fish and rice eating Bengali), which is particularly appropriate for communities living on a vast floodplain. Rice is the staple food in Bangladesh and capture fisheries and aquaculture

Map 4.1: River Systems in Bangladesh

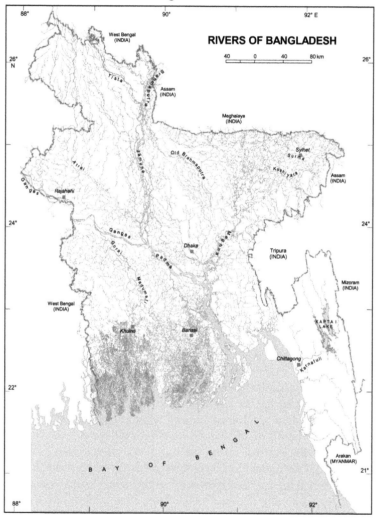

Source: http://www.banglapedia.org/httpdocs/Maps/MR_0207B.GIF (accessed on 9 May 2011).

serve as the major source of protein (World Bank 2007) although its contribution in recent times has declined due mainly to falling per capita availability and reduced buying capacity of the poor. In 2008–2009, the fisheries sector contributes 3.54 per cent to the Gross

Domestic Product (GDP) and 19.75 per cent to the agricultural GDP (BBS, 2011). In Bangladesh, fish is second only to rice as a source of food and each rural family eats fish 3.5 days per week (Minkin et al., 1993). The rural families consume more than 50 species of fish during the course of a year (ibid.). About 9 per cent of the population and over one million full-time commercial fishermen and another 11 million part-time ones depend on fisheries for their livelihood, and over 70 per cent of households are involved in subsistence fishing (World Bank 2007).

Open-water or capture fishery in Bangladesh is inflicted by a string of socioeconomic and environmental issues. The population of the country has more than doubled since its independence in 1971. It is one of the most populated countries in the world with more than 1,000 people per sq. km. land area. For the growing number of people who need an average 170 kg rice per capita per year, the country places very high importance on intensive agriculture. For this purpose, measures were taken to create flood-free environment to facilitate agriculture in floodplains and to expand irrigation from surface and groundwater sources, increase use of agro-chemicals and expand the area under rice cultivation by encroaching upon natural wetlands. By and large, the wetland ecosystems in Bangladesh have lost connections with larger water bodies (rivers and canals) due to gradual siltation and landfill or drainage for agriculture and homestead use. It is estimated that more than 50 per cent of seasonal and perennial wetlands have been affected by growing unplanned agricultural and urban land use (USAID 2008). Further, industrial pollution deteriorates water quality of the fish habitats. Cross-border withdrawal of water from the major rivers has resulted in reduced water supply affecting life-cycle of riverine fish populations. Alien Invasive Species (AIS) and commercial aquaculture are also contributing to the deterioration of open-water fishery in Bangladesh.

The main objective of this chapter is to assess the impact of human interventions in water ecosystems on fisheries in Bangladesh. The chapter discusses hydrology, habitats and fish biological functions, analyses human interventions in Bangladesh, and examines their implications for fish production, livelihood support, protein availability and long-term sustainability. It also highlights coastal and marine fisheries. Finally, the chapter also discusses actions which need to be taken at the planning stages and ways to restore wetland habitats to facilitate open water fisheries.

Hydrology, Habitats and Fish Biological Functions

Hydrology and water resource of Bangladesh are highly dominated by the monsoon rainfall which lasts from June to September. The extent of monsoon floodwater varies over years due to reasons such as the intensity and volume of monsoon rains and the amount and timing of snow melting in the Himalaya. Rainfall is highly skewed as more than 80 per cent of 2,320 mm annual rainfall occurs over only four monsoon months. Within Bangladesh, there is a high regional variation in rainfall. The highest rainfall of 2,830 mm occurs in the north-east of the country while the south-west registers only 1,110 mm. In terms of river basins, the highest rainfall occurs in the Meghna basin and the lowest in the Ganga basin (Mirza and Paul 1992). Note that the annual run-off that is generated within the boundary of Bangladesh is small as it the lower riparian of the GBM systems of rivers. With only 9 per cent of the GBM basin area within its boundary, Bangladesh drains out over 91 per cent of the annual run-off generated outside of its border. River flows strongly follow the precipitation patterns of the basins. For example, the average combined flow of the Ganga (post-Farakka) and Brahmaputra rivers typically varies between less than 5,000 m^3/s in the driest period (March–April) to 80,000–140,000 m^3/s in late August to early September (WARPO 2000).

The life-cycle of fish is integrally connected with the annual hydrological regimes (Figure 4.2). They determine the spatial and temporal aspects of wetland habitats in terms of hectare-months and accordingly changes in the fisheries production is experienced over the years, both in terms of its quality and quantity.

Therefore, habitats alone cannot determine the fishery; rather all these are highly dependent and influenced by the annual hydrological regimes. It is experienced that in the low flooding year, many habitats in some areas at relatively higher elevation could not adequately support the fish to perform their biological functions in full, although the habitats are comparatively in better shape. On the other hand, in the high flood year higher fish production is experienced due to increased spatial and temporal habitat area in terms of hectare-months.

The life-cycles of fishes in the typical floodplains such as those in Bangladesh are adjusted to the annual hydrological regimes and fish accordingly move out to various habitats at different times of the year in order to complete their life-cycles. Therefore, any change in

Figure 4.1: Life-cycle of Floodplain Fish Species in Relation to Flooding Cycle

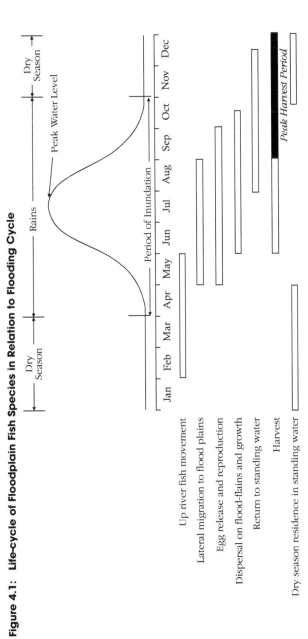

Source: MPO (1987), modified by Mirza and Ericksen (1996).

hydrology affects the habitats (quality and quantity) and eventually fisheries production. The biological features of fish with changing hydrological regime over the season are presented in Table 4.1.

However wetland habitats, including microhabitats, are also very important and essential for the sustenance of the fisheries production and biodiversity. Fish and other aquatic biota require suitable habitats for performing their various biological functions and unless these habitat requirements are met, fish cannot perform their biological functions. For examples, Indian Major Carps (*rohu, catla, mrigal and kalbasu*) require flooded basins in the upper riparian areas of major rivers with current and rainfall for spawning. These species would not spawn if they do not get suitable habitats. It is seen that the adults of these species remain in abundance in tertiary rivers in the lower riparian areas or in beels but they do not spawn there. In culture ponds, eggs develop in the gonads of major carps in the early monsoon but are absorbed at the late monsoon as they do get suitable spawning habitats in ponds.

The annual hydrological regime, which is the major player in the inland capture fishery, is exclusively an external factor and thus very little can be done to fix it towards a favourable environment for fisheries each year. However, opportunities exist to deal with the habitats where some interventions (protection, restoration and enhancement) can be undertaken to make them favourable for the inland fisheries production and maintaining of biodiversity.

Human Interventions in Water Ecosystem and Fish Habitats

Inland fish catch in Bangladesh is divided into two groups: open-water or capture fisheries and culture fisheries. Open-water fisheries refer to life-cycle of fish in natural environment of the floodplain while culture fisheries are commercially managed closed-water fisheries. In recent years, inland fish catch shows an increasing trend mainly due to the inclusion of commercial capture fisheries in the floodplains in the category of 'Flood Land' (WARPO 2000). Recently, the World Bank (2007) reported a very large increase in fish production from all forms of pond aquaculture and decline in most capture fisheries. The following major human interventions are attributed to the decline of open water fisheries in Bangladesh (World Bank 2007; Craig et al. 2004; Parveen and Faisal 2003; Sultana and Thompson 1997; Mirza and Ericksen 1996).

Table 4.1: Hydrological Seasons, Fish Habitat Features and Fish Biological Functions

Seasons/ months	Habitat quantity	Habitat quality	Fish biological functions	Fishing practices	Remarks
1. Dry season (December– February)	Habitats become isolated, lowest spatial water area, lowest depth, many wetlands become dry	Highest pollution concentration, low temperature, bad water quality, many wetlands can not support fish	All biological functions of fish reduced, fish take refuge in dry season rivers, river pools, and perennial *beels*	Fishing intensity peaked, perennial wetlands pumped out to dry for fishing, most wetlands are fished out	Fish become vulnerable or exposed to natural and fishing mortality
2. Pre/early-monsoon (March– May)	Water area started to expand spatially due to early rains	Water quality improves with rising water volume and temperatures	Biological functions initiate, perform longitudinal (long distance) and lateral (short distance) migrations for spawning and feeding	Less fishing intensity. Fishing in beel edges with less harmful gears but fishing in khals cause destruction due to catching on migrating fish	Sensitive time for fish migration, fish productivity and diversity depends on successful migration

(Table 4.1 Continued)

(*Table 4.1 Continued*)

Seasons/ months	Habitat quantity	Habitat quality	Fish biological functions	Fishing practices	Remarks
3. Monsoon (June–August)	Spatial and depth parameters of wetland reached the highest, huge area under water, all sorts of floodplain habitats get integrated	Best water quality with rich fish food, huge area for spawning, feeding and growth	All biological functions of fish (migration, spawning, nursing, feeding, growth, etc.) peaked	Fishing peaked in flooded areas with various gears, open access fishing with few exceptions, use of harmful gears like *kona jal, current jal* peaked	Best time for the entire fish community for natural replenishment, however, harmful gears cause damage to fishery
4. Late/post monsoon (September–November)	Spatial water area rapidly decreased, different wetland habitats become isolated	Water quality still remain good,	Fish migrate to their dry season refuge areas like perennial beels and rivers	Fishing peaked in khals during their return migration, beel, river fishing continued	Sustained natural fish production depends on successful return migration to their refuge areas.

Source: Rahman et al. (2005).

Infrastructural Interventions

Because of the geographical location, hydro-morphological characteristics and flatness of the country, Bangladesh is highly vulnerable to flood hazard. On average, 21 per cent of the country is flooded annually and in extreme cases, flooded area exceeds 70 per cent of the country (Mirza 2002). In order to reduce flood-related losses (agriculture, infrastructure, industrial outputs, services, etc.), a planned mitigation programme was launched in early 1960, in response to the devastating floods of the 1950s. The first water master plan was formulated and approved in 1964 which encompassed 58 large Flood Control and Drainage (FCD) and Flood Control, Drainage and Irrigation (FCDI) projects. These projects aimed at two major objectives: to protect communities from occasional flooding and create a flood-free environment for agriculture. Some of the FCD/FCDI projects, especially located in the northeastern part of the country, allow inundation in the monsoon, for protecting the standing *boro* (irrigated rice crop in Bangladesh, transplanted in winter, from December to February) crops from flash floods in the pre-monsoon period.

Between 1964 and 1984, over 0.81 million hectares of floodplain were permanently eliminated from fish production (MPO 1987). Until 2001, FCD/FCDI facilities have been provided to 5 million ha which is 50 per cent of the total vulnerable area (WARPO 2001). The flood control programme is a continued process and is estimated to expand to 5.74 million ha by 2010 (Mirza and Ericksen 1996; Craig et al. 2004). Major physical infrastructures built by the Bangladesh Water Development Board (BWDB) are listed in Table 4.2, and the flood control areas are shown in Figure 4.3.

Pollution

Pollution in Bangladesh's rivers, canals, wetlands and ponds caused by industrial and urban effluents has reached alarming levels. The most polluted rivers are the Buriganga, Sitalakhya, Turag, Karnafuli and Rupsha. Fish population is affected by pollution at a time when demand for fish is increasing at a faster rate than before. Water pollution occurs from both 'point' and 'non-point' sources. Point source pollution occurs by the discharge of wastewater from industries such as textiles, tanneries, pulp and paper mills, fertiliser, industrial chemical production and refineries (WARPO 2000; UNEP 2002). A complex mixture of hazardous chemicals, both organic and inorganic,

Table 4.2: Major Physical FCD/FCDI Infrastructures

Infrastructure	Unit
Embankment	9,143 km
Coastal	4,195 km
Others (non-coastal)	4,948 km
Irrigation canals	5,072 km
Drainage channels	3,514 km
Number of hydraulic structures	12,447
Number of pump house	98
Number of barrages	4
Number of river closures	1,241
Number of bridges and culverts	5,228

Source: WAPRO (2001).

is discharged into the water bodies usually without treatment. Recent statistics of the polluting industries in Bangladesh is not available. However, JICA (1999) reported the existence of 1,176 polluting industries which were categorised into nine major types: chemical including pharmaceutical, paper and pulp, sugar, food and tobacco, leather, industrial dyes, petroleum, metals and power generation. Most of the older industries are not equipped with treatment facilities. Their effluents generally have high Biological Oxygen Demand (BOD), very high Chemical Oxygen Demand (COD) and contain high levels of sodium sulphate, ethanoic acid, reactive dyes and detergents (Chadwick and Clemett 2003). JICA (1999) reported that 71,200 tons of BOD is discharged annually in Bangladesh of which the highest 61 tons/day comes from the fish and food processing industries. Fish kills often occur in polluted water bodies.

Chemical fertilisers and pesticides are other sources of water pollution and harmful to fish population. Since the introduction of high yielding crop varieties in late 1960s, the use of fertilisers and pesticides has increased to about two million tons (UNEP 2002). Urea, Triple Super Phosphate (TSP), Muriate of Potash (MP) and Gypsum are the four major chemical fertilisers used in Bangladesh.

Pesticide consumption is also increasing in Bangladesh in relation to the acreage of irrigated agriculture (Heijnen 2001). In the 1990s, pesticide use increased by 20 per cent. Pesticides in all forms — granular, liquid and powder — were applied in the agriculture sector and also in health programmes. Bangladesh entered into the pesticide market in 1956 with only 3 tons of product and 500 hand sprayers. Four and a half decades later, their use has increased to 16,200 MT annually (Table 4.3).

Map 4.2: Flood Control Schemes in Bangladesh

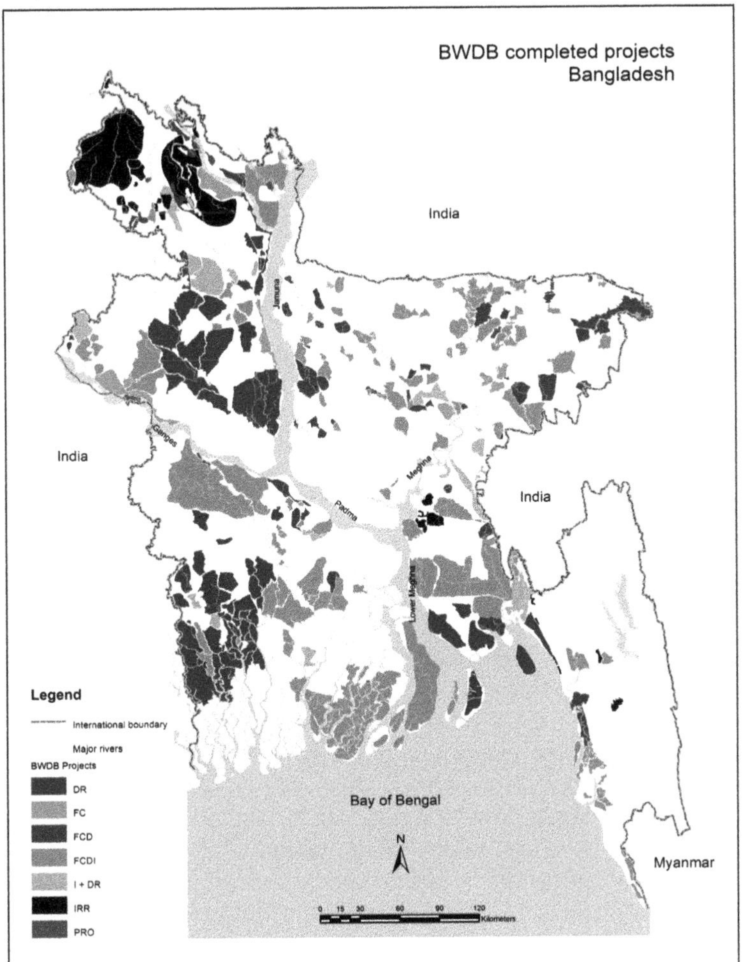
BWDB completed projects
Bangladesh
India
India
India
Jamuna
Ganges
Meghna
Padma
Lower Meghna
Bay of Bengal
Myanmar
N
Legend
International boundary
Major rivers
BWDB Projects
DR
FC
FCD
FCDI
I + DR
IRR
PRO
0 15 30 60 90 120
Kilometers

Source: CEGIS (2011).

The World Bank has recently reported that more than 47 per cent of farmers in Bangladesh use more pesticides than needed to protect their crops. This conclusion was based on a survey of 820 *boro* (winter rice), potato, bean, eggplant, cabbage, sugarcane and mango growers. Through rain and return irrigation flow to water bodies, harmful pesticides can kill fish or can cause harm to their life-cycle. However, there is no documented information on long-term

Table 4.3: Use of Pesticides in Bangladesh

Product type	Forms of product	1993	1994	1995	1996	2001
Insecticide	Granular	5,819	6,006	6,997	8,717	–
	Liquid	1,003	992	1,307	1,340	–
	Powder	84	76	90	109	–
Fungicide	Miticide	20	27	25	20	–
		565	542	585	798	
Herbicide		111	139	140	150	–
Rodenticide		57	66	76	92	–
Total		7,659	7,848	9,220	11,226	16,200

Source: Heijnen (2001); UNEP (2002).

impacts of agricultural pesticides on fish in Bangladesh. A study in Vietnam concludes that pesticides enter ponds with the water flow and feeding of paddy field by-products can contribute to a chronic stress in fish and thereby make them more vulnerable to diseases (Steinbronn et al. 2005).

Water Withdrawal for Dry Season Irrigation

Dry season agriculture in Bangladesh depends on intensive irrigation. Transplanted high yielding rice and wheat are the major crops that require regular irrigation. Until the 1970s, surface water was the major source of irrigation but gradually groundwater has taken over the lead. However, even now, more than one-fifth of the irrigated area of Bangladesh is dependent on surface water irrigation (Table 4.4). Gravity canals, low-lift pumps (LLPs) and traditional pumping methods are used in surface water irrigation.

Dry season irrigation from water bodies is a compromise between cereal crops and fish — the former is more important in Bangladesh. High demand for irrigation water in the dry months is particularly

Table 4.4: Summary of Irrigation Modes in Bangladesh

Mode of irrigation	Area irrigated (million ha)	% of total irrigated area
Ground water	3.814	79.67
Surface water		
LLP	0.838	17.51
Traditional	0.025	0.53
Gravity flow	0.110	2.29
Total		100.00

Source: BADC (2005); BBS (2005).

fatal for open-water fisheries. The ratio of monsoon to dry season water availability is 6:1 for the Ganga basin (pre-Farakka). Water demand in the dry season exceeds availability (Table 4.5) by 5,533 million m³. Thus, water is withdrawn from all kinds of water bodies — rivers, *beels* (marshes), *haors* (a haor is a bowl-shaped depression which is flooded every year during the monsoon in the north-east of Bangladesh), *baors* (synonymous to beel, familiar in the southwestern part of Bangladesh), canals and ponds. These water bodies generally dry out affecting the life-cycle of fishes. In an extreme drought year, the situation becomes more critical. Excessive withdrawal of groundwater causes reverse flow of water from the water bodies to groundwater aquifers resulting in further decline of surface water availability.

Cross-border Withdrawal of Water

Bangladesh has 54 common rivers with India while it shares only three rivers with Myanmar. Water-sharing problem of the common rivers between India and Bangladesh started six decades ago. Dispute over the Ganges water is the oldest. This dispute erupted in the early 1950s when India announced that it will build a barrage on the Ganges at Farakka. The purpose was to divert 1,134 m³/sec water to resuscitate the Calcutta Port which was dying due to excessive natural siltation process. The barrage was formally commissioned in April 1975. In the next two dry seasons until signing of a five-year agreement between the two countries in 1978, water was unilaterally diverted at full capacity of the feeder canal. Bangladesh and India finally signed a long-term (30-year) treaty in December 1996 to share the Ganges water.

Table 4.5: Gross Water Demand and Availability in the Critical Month of March (in Million m³)

Region	Total demand	Total availability*
North-west (NW)	4,980	3,722
North-east (NE)	5,405	3,845
South-east (SE)	2,347	1,706
South-central (SC)	3,162	3,253
South-west	3,004	836
Total	18,898	13,365

Source: MPO (1986).

Note: *Total availability includes full groundwater development. Water available in the main Ganges and Brahmaputra has not been included.

Due to this diversion, flow and water level of the Ganga and its distributaries in Bangladesh are substantially reduced in the dry season. In the dry season, the monthly mean discharge was reduced by minimum 14 per cent for November and maximum 55 per cent for March (Figure 4.2). Reduction of discharge in the Gorai River (the main distributary of the Ganga) is greater than that of the Ganga. The worst month is April when discharge decreases by 77 per cent, followed by 72 per cent and 57 per cent for March and May, respectively. The average change of discharge from December to February (Figure 4.3) was 67 per cent (Mirza 2004). The other disputed rivers are Teesta, Feni, Matamuhuri, Dharala, Dudhkumar, etc.

Reduction in the flow of river Ganga during the dry season and drying up of its principal tributary the Gorai have affected fish migration, reproduction, catch and consumption (Mirza 2004). Growth of aquatic organisms has been impeded due to decreased flow and stagnation of water in the form of a pool. Increased water temperatures have resulted in a shortage of oxygen and unfavourable conditions for the riverine fishery. The changes in the hydro-ecological condition in the lower reaches of the Ganges and its distributaries have affected *Hilsa* (*Tenualosa ilisha*) and other species along with 12 species

Figure 4.2. Mean Monthly Discharge of River Ganga at Hardinge Bridge

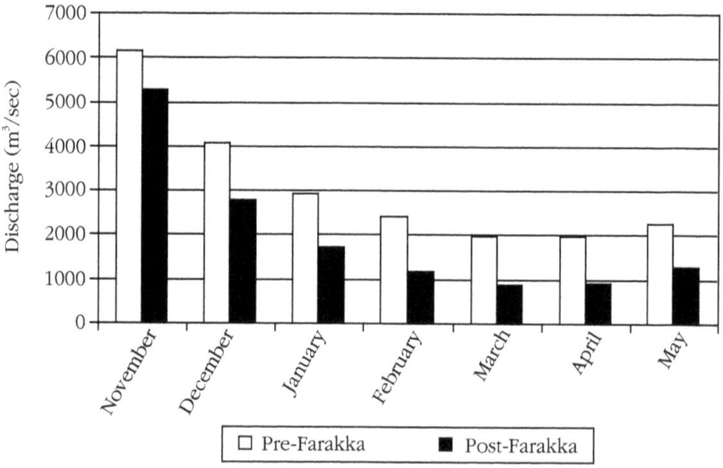

Source: Mirza (2004).
Note: Data from 1965–1998 was used for the analysis. Post-Farakka refers to the period after 1975.

Figure 4.3: Dry Season (November–May) Mean Monthly Discharge of the Gorai

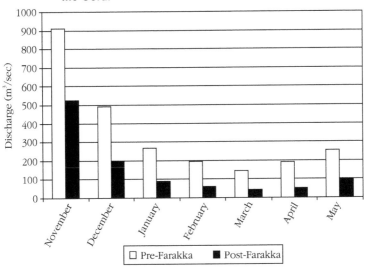

Source: Mirza (2004).
Note: Data from 1964–1995 was used for the analysis.

of prawns (Swain 1996). Adel (2001) reported the extinction of a number of open-water fish species in the Musa Khan sub-basin of the Ganges.

Institutional Weaknesses

The management of open-water fisheries in Bangladesh is plagued by the lack of institutional integration and weaknesses. The legal framework for fishery management in Bangladesh is based on the following two regimes (Kabir 2007)

1. The Doctrine of Public Trust which is traceable to Roman law concepts of common property. Under Roman law, the air, the rivers, the sea and the seashore were incapable of private ownership and they were dedicated to the public for free and unimpeded use.
2. Under the English Common Law, the sovereign could own these natural resources but it does not have the right to grant these to owners if any interference occur with public interest in navigation or fishing.

The Permanent Settlement Regulations (PSR) enacted in 1793 made *zamindars* (Landlords) owners of their land subject to payment of a fixed amount of revenue to the British India government. The landlords were empowered to collect rent from their subordinate tenants, who were again entitled to create subordinate interests. Under the PSR, landlords owned and managed flowing rivers, their tributaries and distributaries, and flood plains containing *beels* (water bodies). Under the State Acquisition and Tenancy (SAT) Act 1950, the government took over the right of receiving rent of the landlords in order to protect tenants from violation of their rights and privileges, and *jalmohals* (public fisheries) became an estate that cannot be retained under private ownership. In addition to this, various other acts and regulations were enacted in order to facilitate better management of fishery as well as providing protection to fishes from gradual depletion. However, these acts and regulations (Table 4.6) are enforced by various agencies. Most of the public fisheries are owned by the Ministry of Land although the conservation of water

Table 4.6: **Laws, Rules, Policies and Conventions for Fish and Fishery Management in Bangladesh**

Laws

Constitution of the People's Republic of Bangladesh
Bangladesh Fisheries Development Corporation Act, 1973
Bangladesh Water and Power Development Board Order, 1972
Bengal Tenancy Act, 1885
Environment Court Act, 2000
The Acquisition of Waste Land Act, 1950
The Agricultural Pest Ordinance, 1971
The Agriculture and Sanitary Improvement Act, 1920
The Canals Act, 1864
The Chittagong Port Authority Ordinance, 1976
The Coast Guard Act, 1994
The Culturable Waste Land (utilisation) Ordinance, 1959
The Dhaka City Corporation Ordinance, 1983
The Embankment and Drainage Act, 1952
The Environment Conservation Act, 1995
The Factories Act, 1965
The Fish and Fish Products (inspection and quality control ordinance), 1983
The Fisheries Research Institute Ordinance, 1984
The Forest Act, 1927
The Government Fisheries (protection) Ordinance, 1959
The Inland Shipping Ordinance, 1976
The Inland Water Transport Authority Ordinance, 1958

(Table 4.6 Continued)

(Table 4.6 Continued)

The Irrigation Act, 1876
The Land Reform Board Act, 1939
The Reform Ordinance, 1984
The Local Government (Union Parishads) Ordinance, 1983
The Marine Fisheries Ordinance, 1983
The Mongla Port Authority Ordinance, 1976
The Non-Agricultural Tenancy Act, 1947
The Open Space Protection Act, 2000
The Paurashava Ordinance, 1977
The Penal Code, 1860
The Private Fisheries Protection Act, 1889
The Protection and Conservation of Fish Act, 1950
The Shrimp Cultivation Taxation Act, 1992
The State Acquisition and Tenancy Act, 1950
The Tanks Improvement Act, 1939
The Territorial Water and Maritime Zones Act, 1974
The Water Supply and Sewerage Authority Act, 1996

Rules
Environmental Conservation Rules, 1977
Marine Fisheries Rules, 1983
Permanent Settlement Regulation, 1793
Territorial Water and Maritime Zones Rules, 1977
The Fish and Fish Products (Inspection and Quality Control) Rules, 1977
The Protection and Conservation of Fish Rules, 1985
The Shrimp Cultivation Taxation Rules, 1993

Policies
Environment Policy, 1992
Export Policy, 1997–2002
Fifth Five Year Plan, 1997–2002
Industrial Policy, 1991
Land Management Manual, 1990
Land Use Policy, 2001
National Environment Management Action Plan, 1995
National Fisheries Policy, 1998
New Agricultural Extension Policy, 1996
National Water Policy, 1999
Public Water Body Management Policy, 2005

Conventions
Convention for the Prevention of Pollution of the Sea by Oil, 1954
Convention on the Law of the Sea, 1982
The Convention on Biological Diversity, 1992
Wetlands of International Importance especially as Waterfowl Habitat, 1971

Source: Kabir (2007).

bodies and fishes are entrusted with the Ministry of Water Resources, Ministry of Environment and Forest and the Ministry of Fisheries and Livestock (MOFL). The fishery-related laws which have so far been formulated and enacted do not regulate the principles and practices of leasing or physical management of fisheries (Kabir 2007).

'Public goods' are generally indiscriminately used from the perception of 'tragedy of the commons'. The transfer of water bodies from private to public ownership under the 1950 SAT facilitated indiscriminate fishing in the openwater. While leasing of waterbodies is the responsibility of the Ministry of Land, it however does not have any mechanism or manpower to protect fishery resources from mismanagement. For example, the use of *current jal* (monofilament gill net) is illegal but these are extensively used in the open water bodies to catch fishes of all sizes including the fish fry. In the breeding season (April to August), it is illegal to catch fish fry or parent fish species Shol (*Channa striata*), Gazar (*Channa marulius*) and Taki (*Channa punctata*) to facilitate their augmentation and production. The sale of carps, Hilsha (*Tenualosa ilisha*), Pungus (*Pangasius pangasius*), Silon (*Silonia silondia*), Boal (*Wallago attu*) and Ayr (*Sperata aor*) below the size of nine inches (about 23 cm) has also been prohibited by law. However, they are often caught and sold in markets in front of the law enforcers. None of the concerned ministries has its own law-enforcing authority (magistracy power). There is also very little or no coordination among the ministries and departments who are the key stakeholders although the 'National Fisheries Policy' stated to include all institutions relevant to fisheries resources under this policy (see Box 4.1). For example, the MOFL is responsible for the management of fisheries resources but it has

Box 4.1: National Fisheries Policy of Bangladesh, 1998

In 1998, Bangladesh adopted its National Fisheries Policy (NFP). Including the 'Introduction', it has 12 major sections. Objectives of the NFP are:

1. Enhancement of the fisheries production.
2. Poverty alleviation through creating self-employment and improvement of socioeconomic conditions of the fishers.
3. Fulfil the demand for animal protein.
4. Achieve economic growth through earning foreign currency by exporting fish and fisheries products.
5. Maintain ecological balance, conserve biodiversity, ensure public health and provide recreational facilities.

Source: GoB (1998).

very little coordination with the Department of Fisheries (DOF) and Bangladesh Fisheries Research Institute (BFRI).

Degradation of Wetlands

Bangladesh is virtually a land of floodplains and it has enormous area of wetlands including rivers and streams, freshwater lakes and marshes, haors, baors, beels, jheels, water storage reservoirs, fish ponds, flooded cultivated fields and estuarine systems with extensive mangrove swamps. Every year, 21 per cent of Bangladesh and in an extreme flood year, up to 70 per cent of the country gets flooded (Mirza 2002). The haors, baors, beels and jheels are of fluvial origin and are commonly identified as freshwater wetlands.

Wetlands in Bangladesh are of very high economic values (Table 4.7). The first economic valuation of a Bangladesh wetland was conducted in Hail Haor, Molvi Bazaar, in the north-east of Bangladesh, and it was found that this wetland has an annual output of approximately Bangladeshi Taka BDT 30,000 (about US$ 450) per ha (World Bank 2007). Wetlands are depleting because of both natural and anthropogenic causes which has significant economic and financial implications. Sedimentation is the major natural cause of

Table 4.7: Economic Benefits of Wetlands

Use benefits		*Non-use benefits*	
Direct use benefits	*Indirect use benefits*	*Option benefits*	*Existence benefits*

• Commercial and subsistence harvest:
  • fish
  • trees
  • wild food plants
  • crops
  • fuel
  • fodder
• Recreation:
  • boating
  • birding and wildlife viewing
  • walking
  • fishing
	• nutrient retention
	• water filtration
	• flood control
	• shoreline protection
	• groundwater recharge
	• external ecosystem support
	• micro-climate stabilisation
	• erosion control
	• associated expenditures, e.g., travel, guides, gear, etc.
		• potential future uses (as per direct and indirect uses)
		• future value of information, e.g., pharmaceuticals, education
			• biodiversity
			• culture
			• heritage
			• bequest value

Source: Thompson and Colavito (2007) modified from Barbier et al. (1997)

loss of capacity of wetlands. Anthropogenic encroachments occur for a host of socioeconomic and institutional reasons. Due to growth of population and increased economic activity, wetlands are encroached for transforming them into agricultural lands, industrial areas and urban centers. Loss of wetland habitats occurs through abstraction and drainage resulting in depletion of aquatic fauna and flora and reduction in availability of water to the rural population (MoEF 2001). Lack of inter-institutional coordination also facilitates loss of wetlands. For example, experts opined that 'absence of coordination among the Ministry of Environment and Forests and the Ministry of Water Resources was the major cause of degradation of the country's wetlands (*New Age* 2006). Fish and aquatic resources in Bangladesh are also threatened due to indiscriminate catching and loss of breeding grounds (IUCN 2004).

The Threat of Alien Invasive Species (AIS) of Fish

Fourteen alien invasive species (AIS) of fish have been introduced in Bangladesh for aquaculture since 1953 without examining their potential impacts on the local fish species and the ecosystem at its entirety (Barua et al. 2000). AIS species usually escape to open water during an extreme flood event when closed water bodies are engulfed by flood waters. Khan et al. (2000) studied the impacts of AIS fish species and concluded that an extensive area of wetlands and over 200 highly diverse native fish species of Bangladesh are at risk from introduced fish. For example, the African catfish (*Clarias gariepinus*) introduced in 1989 from Thailand by the Department of Fisheries, Government of Bangladesh, was blamed for the depletion of 56 fish species in Bangladesh (Barua et al. 2000). Similar impacts have also been reported from neighbouring India where the Department of Forests, Ecology and Environment in Bangalore in 2005 issued directions to destroy the entire stock of the African catfish because they were found to be carnivorous and a voracious predator (Chhaparia, 2005; Lal et al. 2003). The governments of India and Bangladesh have recently banned its culture. This is to be noted that three important biological attributes — faster growth rate, resistance to diseases and possibility of high stocking density — made African catfish attractive to the aquaculturists (ibid.). Red-bellied Piranha (*Pygocentrus nattereri*) is another carnivorous species of South American origin that is being secretly bred in Bangladesh. They are very destructive to local varieties of fish and aquatic life. They move in a group and will charge a group of fish with great ferociousness, snapping at anything they can get a hold of and will chase individual prey until they catch it. They are locally sold under pseudo-names of Korch or Thai Pomfret or Thai Chanda and are attractive to lower and lower middle class buyers because of their cheaper prices compared to actual pomfret fish.

In addition to impacting population of local fish species by eating and destroying their habitat, the AIS fish species could cause outbreak of diseases of exotic nature. A deadly disease, 'Epizootic Ulcerative Syndrome' (EUS) caused by the exotic fish Silver Barb (*P. goniontus*) which was introduced in Bangladesh from Thailand affected numerous floodplain species (Ali 1997). In Bangladesh, EUS was first observed in Faridganj in Chandpur district, south-east of capital Dhaka in February 1988 (Ahmed and Hoque 1999). From a field survey conducted in a site located in the south-west region of Bangladesh, Sanaullah et al. (2001) sampled a total of 930 fish comprising 23 species to investigate outbreak of EUS. They found a diversity of the disease in the sample, *Channa striatus*, *Channa marulius* and *Channa punctatus* were severely affected (30 per cent), followed by *Puntius spp.* (24 per cent), *Mystus vittatus* (22.22 per cent), *Glossogobius giuris* (18 per cent), *Xenentodon cancila* (17.14 per cent), *Mastacembalus spp.* (16.30 per cent), *Ompok pabda* (15.38 per cent), and *Anabas testudineus* (14.28 per cent) of 930 sampled fish. The effect of the disease in fish population caused significant socioeconomic impact (Lilley et al. 2002). During 1988–1989, economic loss due to EUS was estimated to be US$ 3.4 million (Khan and Lilley 2002).

Impacts on Fisheries Resources, Livelihood Support, Protein Consumption and Sustainability

For a number of anthropogenic reasons (discussed earlier in the chapter), production of the inland capture fishery in Bangladesh has declined (World Bank 2007). The loss of production has implications on protein consumption and long-term sustainability of the resource base. Based on the availability of data, the World Conservation Union (IUCN) in its Red Book (published in 2001) for Bangladesh has raised an alarm that almost 30 per cent of all inland fish species are in some danger of extinction (IUCN 2001). The Red Book also reports that of the 266 fish species available in Bangladesh, 54 inland and estuarine fish species are endangered (World Bank 2007). From the data available for 1992–2002, it is estimated that inland Hilsha catch declined to half from the previous catch figures. The main reason behind the decline in catch is reduced flows in the major rivers. It is also estimated that catch of capture fishery has declined by 30 per cent (World Bank 2007). In economic terms, annual losses (Table 4.8) from decline in capture fishery could be as much as US$ 42 million which is very significant when livelihoods of millions of fishermen are taken into consideration.

In the annual flood cycle, 21 per cent of Bangladesh or three million ha goes under water. In an extreme case, it could be as high as 10 million ha. Fish catch from the annually flooded and other open water bodies is also a source of livelihoods for the poor. As mentioned earlier in the chapter, it is grossly estimated that about 9 per cent of the population and over one million full-time commercial fishermen and another 11 million part-time fishermen depend on fisheries for their livelihood. In addition, over 70 per cent of households in rural Bangladesh are involved in subsistence fishing for their daily protein supply (World Bank 1997, 2007). Grass-roots level studies reveal further insights of fishery and livelihoods issue. Surveys conducted by the Department of Fisheries demonstrate that poor are the main beneficiaries of public wetlands for their livelihoods. Field-level data collected by the Management of Aquatic Ecosystems through Community Husbandry (MACH) project from Hail Haor in Sylhet shows that 60 per cent of the benefits of this wetland go to the poor people of the communities. Another study from wetlands in Noakhali — a southern district — shows that 80–93 per cent of the community utilised the wetlands. The poorest members of the community received 15 per cent of their livelihood support from the harvested wetland products. A similar study conducted in Sherpur, situated north of capital Dhaka, demonstrates that over 80 per cent of the wetland users belong to the poorest group in the community (World Bank 2007).

The present per capita annual fish consumption in Bangladesh stands at about 14 kg/year (38 g/day) against a recommended minimum requirement of 18 kg/year (49 g/day) which is very much unevenly distributed among the various income groups. However, in the flood control projects, per capita consumption fishes is much lower than the national average because of drastic reduction in open-water fisheries as a result of loss of flood plains. Mirza and Ericksen (1996) estimated that per capita fish consumption dropped to 19 g/day in 1992 from 36 g/day in 1977 before the commissioning of the

Table 4.8: Estimated Physical and Financial Loss due to Declining Production of Selected Fish Species

Species	Maximum annual production from 1999–2002 (mt)	Average annual production from 1999–2002 (mt)	Production in 2002 (mt)	Physical production loss (mt)		Annual production loss ($)
Major Carps	9,639	2,780	1,443	8,196	1,337	1,586,271
Other Carps	3,594	1,345	382	3,212	968	984,203
Inland *Hilsha*	112,408	79,616	62,944	49,464	16,672	39,460,441
Indian Salmon	2,428	1,129	930	1,498	199	675,330
All Species						42,806,245

Source: World Bank (2007).

Box 4.2: Case Study — Chandpur Irrigation Project

The Chandpur Irrigation Project (CIP) in Bangladesh provides a case study for investigating the impacts of water control projects on fisheries, fish catch and consumption patterns. The Chandpur project was chosen as the site for studying for four reasons (Mirza 1996). First, before the project was implemented, the area was abundant with various types of freshwater and migratory estuarine fish. Second, a sufficient time has elapsed since the commissioning of the project in 1978 to investigate the impacts. Third, sufficient data on fish species and their production as well as other environmental parameters are available. Fourth, the area was readily accessible for research.

The CIP is one of the major water resources development project in Bangladesh (Figure 4.4). It covers an oval-shaped embanked area of about 57,000 ha in the Chandpur and Luxmipur districts. The rivers Meghna and Dakatia flow on the east and north side of the project area, respectively. The project was designed to provide flood control, drainage and irrigation benefit to the empoldered area. Before implementation of the project, the area was subject to frequent flooding in the monsoon due to high water levels in the adjacent rivers. On completion, intrusion of flood waters is prevented by embankments. Drainage of internally generated run-off is done by reversible pumps and regulators.

Figure 4.4: Open-water Fishery Production within the Chandpur Project (1981–1993)

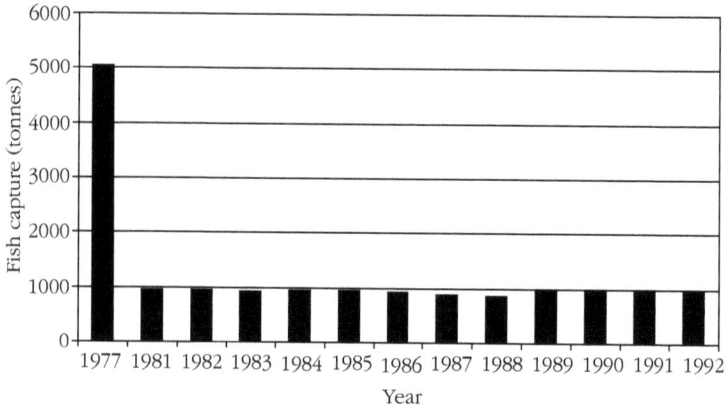

Note: The 1977 value refers to pre-project situation.

(*Box 4.2 Continued*)

(Box 4.2 Continued)

The empolderment of the CIP resulted in a series of changes in the ecosystem of the Dakatia river. The project blocked the connection of this river, an offtake of the Meghna river. The riverine habitats for resident species were destroyed and the passages of fry, juveniles and mature adults of migratory were blocked. As the extent of depth of flooded lands decreased, the production of floodplain fish was substantially reduced. Impacts on prawns included disappearance of two riverine varieties, *Macrobrachium villosimanus* and *Microbrachium mirabilis* and changes of sex ratios of some prawns. In 1978, the male–female sex ratio of *M. villosimanus* and *Microbrachium malcolmsonii* were 1.8:1 and 2.6, respectively. In 1993, the ratio changed to 2.2:1 for *M. villosimanus* and 3:1 for *Microbrachium malcolmsonii*. Some species of prawn benefited from the changed environment.

The CIP has seriously impacted population and production of five major carps and 18 species of tidal/estuarine fish. Fries of major carp species (*Labeo rohita, Catla catla, Cirrhonus mrigala, Labeo gonius* and *Labeo calbasu*) were previously not reported to have produced in the south Dakatia river. After commissioning of the project, juveniles were found to be absent inside the project area as the fry or fingerlings failed to immigrate into the Dakatia river system. Fishers reported a drastic reduction of carp fingerlings in the Dakatia river system after commissioning of the project and Mirza (1991) reported that they were extremely rare. The migratory routes of the tidal/estuarine fish species were blocked by the water control structures. Their population has greatly decreased in the project area.

The impacts of water control structures on fisheries were not considered in the pre-feasibility and feasibility studies of the project (Mirza 1991; Mirza and Ericksen 1996). MPO (1987) reported a pre-project (one year before commission) fish catch of 5,436 tonnes of which open-water fishery constituted 5,055 tonnes or 93 per cent of the total catch. In the post-project period (1981–1992), open-water fish catch on average annually was 943 tonnes, i.e., almost 81 per cent reduction (Figure 4.4). As the fish availability in the project area reduced, it also impacted number of fishermen. In 1978, an estimated 25,843 fishermen were present in the CIP but it declined to 8,400 in 1992 (Mirza and Ericksen 1996). Fish consumption in the project area also reduced to 11 gm/person/day from 36 gm/person/day (Mirza and Ericksen 1996).

Source: Prepared by the authors.

Chandpur Irrigation Project (See Box 4.2 for the case study). By 2010, loss of open-water fishery from the flood controlled area could be as high as 151,300 tons per year which would result in a substantial drop in daily protein consumption.

Due to lack of data, it is difficult to estimate the loss of open-water fisheries from various human interventions. Loss of production of open-water fish is being compensated by culture fishery. In some cases, aquaculture has contributed to reduction of poverty. In order to satisfy the growing demand for fish due to growth of population and income, culture fisheries are required but not at the cost of open-water fishery. This policy is wrong because increasing culture fish cannot meet the protein demand of the poor as they are expensive. Loss of open-water fisheries means loss of employment and livelihoods. The future aim should be to provide benefits to the fishermen and to ensure the long-term sustainability of the resources. To sustain the floodplain fisheries, both restoration of habitat and enhancement in fish population are important.

Human interventions have also impacted coastal fisheries as well as the environment of Bangladesh. Coastal aquaculture has been developed significantly in the decade of 1990s, particularly the shrimps (*Penaeus monodon* and *P. indicus.*) culture in medium to high saline water and prawn (*Machrobrachium sp.*) culture in less saline areas. Coastal shrimp farming is now spread over an area of 0.14 million ha (Mazid 2002). The environmental impacts of coastal aquaculture that have been listed are: increased soil salinity, reduced agricultural production, decreased livestock production and destruction of mangrove forests. Coastal aquaculture has also caused adverse impacts on biodiversity through the destruction of trees, grasses and crabs in the area of production (Asche and Khatun 2006). Decline in agricultural production and destruction of grazing lands due to high salinity content in the water are recorded. Shrimp cultivation is an important factor that causes deforestation and destruction of homestead vegetation. In Bangladesh, for example, at least 8,750 hectares of mangrove forests have been lost due to salinity and human intervention in the shrimp farming region that led to serious ecological imbalance (ibid.). Environmental problem such as destruction of mangroves can be reduced by moving shrimp aquaculture to non-forested low agricultural productive areas. The government can classify the lands suitable for shrimp aquaculture, crop agriculture and grazing lands.

Bangladesh is extracting its marine fisheries resources by only capturing the fishes from the sea. The country's coastline is approximately 714 km with an area of 166,000 km^2 Extened Economic Zone (EEZ). The EEZ is enriched with 1,093 aquatic marine organisms including finfish (44.35 per cent), shellfish (32.23 per cent), seaweeds (15.15 per cent) and other organisms including shrimps (8.32 per cent). It is estimated that the marine capture fish is declining at a rate of 5 per cent per annum. If this declining rate continues, the total fish production in the marine source will be exhausted in the near future. There is a huge potentiality of offshore mariculture which does not exist at present (Kabir n.d.).

Planning of Water Control Projects and Restoration of Wetland Habitats for Fisheries: Role of Engineers

Engineers are mainly involved in designing large-scale flood control, irrigation and drainage projects in South Asia. In the wake of increase in population and reduction in availability of fish nutrient, especially for the rural poor, it is imperative that already damaged wetlands and fish habitats should be restored. At the planning stage of water control projects and in the restoration phases, engineers can play a significant role along with experts of other disciplines who are involved with fisheries and socioeconomic development of flood plains.

Restoration of habitats can be defined as the activities undertaken to bring back the degraded or semi-degraded wetland habitats into their functional state. The restored habitats should be brought at such a stage so that the said habitats after restoration can support a species or group of species or a community or communities to perform its or their relevant required natural functions at different life stages namely, migration, reproduction, feeding, growth and refuge. For example, if a canal is restored, fish and other aquatic biota can perform their migration. Similarly, if a degraded beel is restored into a perennial stage, then fish and other aquatic biota can have year round habitat for feeding, growth and refuge (in the dry season).

Wetland restoration can be defined as the re-establishment of a disturbed or altered wetland to one with improved function. This may involve re-establishing original vegetation, hydrology, or other parameters to re-create original or closer-to-original wetland functions (Fields 1993). Degraded wetlands present restoration opportunities for improvements in water quality, habitat, water storage and other functions, and these opportunities can be particularly useful for watershed-scale environmental planning. The goal of restoration is typically to re-establish wetland ecosystems to levels that existed before the change brought about by human influence.

Actions at Planning Stage and Ways of Restoration

In addition to disconnecting the flood plains from the rivers/water courses, internal water bodies such as beels are reclaimed and perennial water bodies are completely drained for agricultural development. These should be controlled during the project planning and implementation phases. Sultana and Thompson (1997) suggested that to create wetland conservation areas, submersible bunds that would allow some flood waters to enter in a flood control project could be built around all or part of the beels, by digging earth from the beels during construction. Sufficient water in the dry season has to be maintained in the beels for survival of fish until the time when they migrate to spawn, and they would provide a all-season habitat for non-migratory species. Low-level bunds may also be required for the canals that connect the beels and rivers. However, their requirement is dependent on the land levels nearby, the system of operation and the crops at risk in the early monsoon (ibid.).

Restoration could be achieved in many ways. First, degraded wetlands/water bodies in the project area can be re-excavated for creation of habitat for fisheries. Major expenses are required for physical intervention including earthwork in a restoration programme. A very short time period is available for earthwork in wetlands because of withdrawal and onset of monsoon. To timely undertake the earth work, in some cases, de-watering may be required. In 2001, the Center for Natural Resources Studies (CNRS) found that restoration of degraded wetlands through re-excavation was economically viable. Second, in Bangladesh, there would be an annual loss of 68 kg to 202 kg fish per hectare (mean 119 kg fish) of flooded land lost (Scullion 1996). Fish yields inside a typical flood control compartment can be 50 per cent lower than outside, with up to 25 species of fish absent or less abundant (Halls et al. 2008). However, through controlled flooding, loss of fish habitat because of any new flood control projects could be avoided. In principle, instead of total prevention of flooding of an area, certain flood levels that are not harmful to agriculture, fishery and infrastructure, need to be maintained. Halls et al. (2008) examined both active and passive migration of fish through a sluice gate on the Lohajang river in Tangail, North-Central (NC) Bangladesh that controls inflowing water from the Jamuna river into the Compartmentalisation Pilot Project (CPP). They suggested eight measures to maximise the immigration of fish: (1) maximise inward flow rate; (2) maximise the frequency of gate openings; (3) maximise the turbulence of water outside the gate; (4) ensure that ebb flow velocities do not exceed the maximum sustainable swimming capacities of fish; (5) attempt to create ebb flows that attract the most fish towards the sluice gate; (6) control fishing activities along channels connecting the gate to the main rivers; (7) increase dry season water availability; and (8) reduce pressure on dry season water resources. Third, a managed capture or open-water fishery will not only require redesigned engineering structures and revised operating rules, but will also depend for its success on management by a fishing cooperative strong enough to regulate fishing activity (Sultana and Thompson, 1997). Engineers have a role to play in organising such cooperatives as well as train fishermen on better water management.

Conclusion

As a flat deltaic country, Bangladesh has significant potentials for open-water fishery. The loss of open-water fishery, especially for flood control, is a compromise between hazards, and agriculture and protein supply to the poorer section of the community. Encroachment to wetlands and other human interventions have put the future of open-water fishery in a challenging situation. Closed-water or culture fisheries can mitigate the problem of decline in supply. However, it cannot guarantee fish as a diet to the poor people because of their low-buying capacity. The management of open-water fisheries should start focusing on bringing human interventions under control and formulate effective policies for the gradual restoration of ecosystems and fish habitat.

References and Select Bibliography

Adel, M. M. 2001. 'Effect on Water Resources from Upstream Water Diversion in the Ganges Basin', *Journal of Environmental Quality*, 30 (2): 356–68.

Ali, Y., 1997. *Fish, Water and People: Reflection in Inland Openwater Fisheries Resources of Bangladesh*. Dhaka: Bangladesh Centre for Advanced Studies (BCAS).

Asche, F. and F. Khatun. 2006. *Aquaculture: Issues and Opportunities for Sustainable Production and Trade*. Geneva: International Centre for Trade and Sustainable Development (ICTSD).

Ahmed, G. U. and A. Hoque. 1999. 'Mycotic Involvement in Epizootic Ulcerative Syndrome of Freshwater Fishes of Bangladesh: A Histopathological Study.' *Asian Fisheries Science*, 12 (4): 309–99.

Bangladesh Agricultural Development Corporation (BADC). 2005. *Minor Irrigation Survey Report 2004-2005*. Dhaka: BADC.

Bangladesh Bureau of Statistics (BBS). 2005. *Information on Irrigation Practices*. Dhaka: BBS.

———. 2011. Table 1, Gross Domestic Product of Bangladesh at Current Prices, 2004–05 to 2008–09(p). http://www.bbs.gov.bd/webtestapplication/userfiles/image/subjectmatterdataindex/GScompend_09.pdf (accessed 3 May 2011).

Barbier, E. B., M. C. Acreman and D. Knowler. 1997. *Economic Valuation of Wetlands: A Guide for Policy Makers and Planners*. Gland: Ramsar Convention Bureau.

Barua, S. P., M. M. Khan and M. Ameen. 2000. 'The Status of Alien Invasive Species in Bangladesh and their Impacts in the Ecosystem' in P. Balakrishna (ed.), *Report of workshop on Alien Invasive Species*, Global Biodiversity Forum-South AND Southeast Asia Session, Colombo. IUCN Regional Biodiversity, Asia, Colombo, Sri, pp. 1–7. Dhaka: IUCN Bangladesh.

Centre for Environmental and Geographic Information Services (CEGIS). 2011. 'BWDP Completed Projects, Bangladesh', CEGIS Map Library, Dhaka.

Chadwick, M. T. and A. E. V. Clemett. 2003. *Managing Pollution from Small- and Medium-Scale Industries in Bangladesh: Inception Report*. York: Stockholm Environment Institute.

Chhaparia, P. 2005. 'African Catfish a Health Risk', *Times of India*, 2 July, Bangalore. http://timesofindia.indiatimes.com/articleshow/1158653.cms (accessed 13 June 2008).

Craig, J. F., A. S. Halls, J. J. F. Barr and C. W. Bean. 2004. 'The Bangladesh Floodplain Fisheries', *Fisheries Research*, 66 (2–3): 271–86.

Government of the People's Republic of Bangladesh (GoB). 1990. *Nutritional Survey of Bangladesh 1989–90*. Dhaka: Bangladesh Bureau of Statistics, Statistics Division, Ministry of Planning.

———. 1998. *National Fisheries Policy*, Ministry of Fisheries and Livestock. www.mofl.gov.bd/pdf/National_Fisheries_Policy.pdf (accessed 3 May 2011).

Halls, A.S., A. I. Payne, S. S. Alam and S. K. Barman. 2008. 'Impacts of Flood Control Schemes on Inland Fisheries in Bangladesh: Guidelines for Mitigation', *Hydrobiologia* 609 (1): 45–58.

Heijnen, H. 2001. 'Country Status Paper — Bangladesh'. Paper presented at the Seventh Global Information Network on Chemicals (GINC) Tokyo Meeting for Information Exchange and Collaboration in Asia on Chemical Management and Pesticide Poisoning, 18–20 April, Tokyo.

Islam, M. R., S. F. Begum, Y. Yamaguchi and K. Ogawa. 1999. 'The Ganges and Brahmaputra Rivers in Bangladesh: Basin Denudation and Sedimentation', *Hydrological Processes*, 13 (17): 2907–23.

Japan International Cooperation Agency (JICA). 1999. *Country Profile on Environment: Bangladesh*. Tokyo: JICA.

Kabir, H. (n.d.). *Marine Fisheries in Bangladesh: An Overview*. Dhaka: PMTC (Bangladesh) Limited.

Kabir, I. 2007. 'The Legal Background to Community Based Fisheries Management', Booklet No. 4, The WorldFish Center, Dhaka.

Khan, M. H. and J. H. Lilley. 2002. 'Risk Factors and Socio-economic Impacts Associated with Epizootic Ulcerative Syndrome (EUS) in Bangladesh', in J. R. Arthur, M. J. Phillips, R. P. Subasinghe, M. B. Reantaso and I. H. MacRae (eds), *Primary Aquatic Health Care in Rural Small-Scale Aquaculture Development*, pp. 27–39. Rome: FAO.

Khan, S. 2004. 'African Catfish Turn Predators in Bangladesh', *One World South Asia*. http://128.227.186.212/fish/InNews/africancatfish2004.html (accessed 13 June 2008).

Lal, K. K., R. K. Singh, V. Mohindra, B. Singh and A. G. Ponniah. 2003. 'Genetic Make-up of Exotic Catfish *Clarias gariepinus* in India', *Asian Fisheries Science*, 16 (3/4): 229–34.

Lilley, J. H., R. B. Callinan, and M. H. Khan. 2002. 'Social, Economic and Biodiversity Impacts of Epizootic Ulcerative Syndrome (EUS)', in J. R. Arthur, M. J. Phillips, R. P. Subasinghe, M. B. Reantaso and I. H. MacRae (eds), *Primary Aquatic Health Care in Rural Small-Scale Aquaculture Development*, pp.127–39. Rome: FAO.

Master Plan Organisation (MPO), 1986. *National Water Plan — Summary Report*. Dhaka: Ministry of Irrigation, Water Development and Flood Control, Government of Bangladesh.

———. 1987. *Fisheries and Flood Control, Drainage and Irrigation Development*, Technical Report 17. Dhaka: Ministry of Irrigation, Water Development and Flood Control.

Mazid, M. A. 2002. *Development of Fisheries in Bangladesh: Plans and Strategies for Income Generation and Poverty Alleviation*. Dhaka: Nasima Mazid.

Ministry of Environment and Forests (MOEF). 2001. *Bangladesh: State of the Environment 2001*. Dhaka: MoEF.

Minkin, S., S. Halder, M. Rahman, D. Schuy and M. Rahman. 1993. *Flood Control and the Nutritional Consequences of Biodiversity of Fisheries, Bangladesh Flood Action Plan (FAP 16)*. Dhaka: ISPAN.

Mirza, M. M. Q., 1991. 'Environmental Impacts of Water Development Projects: A Case Study of Chandpur Irrigation Project'. Unpublished thesis, Department of Water Resources Engineering, Bangladesh University of Engineering and Technology, Dhaka.

———. 2002. 'Global Warming and Changes in the Probability of Occurrence of Floods in Bangladesh and Implications', *Global Environmental Change*, 12 (2): 127–38.

———. 2004. 'Hydrological Changes in Bangladesh.' in M. M. Q. Mirza (ed.), *The Ganges Water Diversion: Environmental Effects and Implications*, pp.13–38. Dordrecht: Kluwer Academic Publishers.

Mirza, M. M. Q. and S. Paul. 1992. *Pakritik Durjog O Bangladesher Poribesh*. Dhaka: Centre for Environmental Studies and Research (CESR).

Mirza, M. M. Q. and N. J. Ericksen. 1996. 'Impact of Water Control Projects on Fisheries Resources in Bangladesh', *Environmental Management*, 20 (4): 523–39.

New Age. 2006. 'World Wetland Day Observed', 3 February. http://www.newagebd.com/2006/feb/03/nat.html (accessed May 3 2011).

Parveen, S. and I. M. Faisal. 2003. 'Open-water Fisheries in Bangladesh: A Critical Review'. Paper presented at the Second International Symposium on the Management of Large Rivers for Fisheries: Sustaining Livelihoods and Biodiversity in the New Millennium, February 11–14, Phnom Penh, Cambodia.

Rahman, M. M., M. M. Rahman, and M. M. Rahman. 2005. 'Training Module on Integrated Floodplain Management: Options and Approaches', Final Technical Report Annex F, Centre for Natural Resources Studies (CNRS), Dhaka.

Sanaullah, M., B. Hjeltnes and A. T. A. Ahmed. 2001. 'The Relationship of Some Environmental Factors and the Epizootic Ulcerative Syndrome Outbreaks in Beel Mahmoodpur, Faridpur, Bangladesh', *Asian Fisheries Science*, 14 (3): 301–15.

Scullion, J. 1996. 'Flood Control, Rice and Fish in Bangladesh: Some Lessons for the Lower Mekong Basin', *Mekong Fisheries Network Newsletter*, 2 (2), November. http://www.mekonginfo.org/mrc_en/doclib.nsf/0/52 F0532D5CC7609647256DDC00135BCF/$FILE/FULLTEXT.pdf (accessed 3 May 2011).

Steinbronn, S., C. Geiss, A. Fangmeier, N. N. Tuan, U. Focken and K. Becker. 2005. 'The Use of Pesticides in Paddy Rice and Possible Impacts on Fish Farming in Yen Chau/Son La Province, Northern Vietnam'. Paper presented at a conference on The Global Food and Product Chain — Dynamics, Innovations, Conflicts, Strategies, Deutscher Tropentag, 11–13 October 2005, Hohenheim.

Sultana, P. and P. M. Thompson. 1997. 'Impact of Flood Control and Drainage on Fisheries in Bangladesh and Design of Mitigating Measures', *Regulated Rivers*, 13 (1): 43–55.

Swain, A. 1996. *Environmental Trap: The Ganges River Diversion, Bangladesh Migration and Conflicts in India*. Report No. 41. Upsala: Department of Peace and Conflict Research, Upsala University.

Television Trust for Environment. 2004. 'Down the Drain — Bangladesh'. http://www.tve.org/ho/series5/07_Green%20Currents_reports/programme_7_mm/Down%20the%20Drain%20-%20Bangladesh%20pdf.pdf (accessed 3 May 2011).

Thompson, P. and L. Colavito. 2007. 'Economic Value of Bangladesh Wetlands', MACH Technical Paper 6. Dhaka: Winrock International, Bangladesh Centre for Advanced Studies, Center for Natural Resource Studies and CARITAS Bangladesh.

USAID. 2008. 'Bangladesh: Current Conditions: Environment'. http://www.usaid.gov/bd/programs/environ.html (accessed 3 May 2011).

United Nations Environment Programme (UNEP). 2002. *Bangladesh State of Environment 2001*. Nairobi: UNEP.

Valbo-Jørgensen, J. and P. Thompson. 2007. *Culture-based Fisheries in Bangladesh: A Socio-economic Perspective*. Rome: Food and Agriculture Organisation.

Water Resources Planning Organisation (WARPO). 2000. *Annex G: Environment*. National Water Management Plan Project. Dhaka: Ministry of Water Resource, Government of Bangladesh.

———. 2001. *Presentation on National Water Management Plan*. Dhaka: WARPO.

World Bank. 1997. *A Review of Key Environmental Issues in Bangladesh*. Dhaka: World Bank.

———. 2007. *Bangladesh Country Environmental Analysis*. Dhaka: World Bank.

World Conservation Union (IUCN). 2001. *Red Book of Threatened Fishes of Bangladesh*. Gland: IUCN.

———. 2004. *Introduction to Community Based Haor and Floodplain Management*. Dhaka: IUCN, Bangladesh.

5

Management of Water Quality and Biodiversity of the River Ganga

R. K. Sinha and K. Prasad

The Ganga river basin is one of the most densely populated river basins in the world. The Ganga river has 29 Class I cities, 23 Class II cities (Class I cities: population greater than 100,000 people; Class II cities: population between 50,000 to 99,999 people), and 48 towns, and thousands of villages along its course. Over 500 million people were estimated to be living in the entire Ganga river basin in 2000, and this number is expected to grow to 1 billion by 2030 (Markandya and Murty 2000). According to an estimate by the Central Pollution Control Board (CPCB), 2,600 million liter per day (MLD) of sewage from Class I and II cities and 365 MLD of industrial effluents are discharged into the Ganga (personal communication Dr R. Dalwani). Rapid developmental activities coupled with demographic explosion have accelerated irrational exploitation of river resources together with their floodplains at an ever-increasing pace.

The Ganga basin is also home to a wide variety of relict species, including the Ganges river dolphin (*Platanista gangetica gangetica*), the Ganges river shark (*Glyphis gangeticus*), Ganges soft-shell turtle (*Aspideretes gangeticus*), gharials (*Gavialis gangeticus*) and several species of endemic freshwater crabs. The Ganga harbours 265 fish species (Talwar and Jhingran 1991). The impacts of increased population growth, industrial development, deforestation and construction of dams and barrages as well as embankments have had serious adverse impacts on fisheries and other biodiversity. A steady decline has been seen in populations of prized carp and *Tenualosa ilisha* (hilsha) fishes, as well as minnows (Ray 1998). The construction of Farakka Barrage (commissioned in 1975) had a significant impact on fisheries as far up as Allahabad. Catches are reported to have declined from an average of 19.2 ton hilsa/year to 0.9 ton hilsa/year (ibid.). The yield of the major carps has been reduced from 26.62 kg/ha/yr in 1958–1961 to a dismal low of 2.55 kg/ha/yr in 1995 (M. Sinha 2000).

To meet the challenges of an increasing demand on water, declining resources, fisheries and threatened biodiversity, there is an urgent need for an ecosystem approach of integrated water resources management. It requires taking into account the ecosystem goods and services, incorporate the functioning of the entire catchments into planning and management and focus on managing both water and land resources within the river basins. Managing upper-watershed forests in the Himalaya wisely, adopting alternative non-structural measures to control floods and droughts and regulating discharge of pollutants into the river waters are characteristic of this management approach. Efforts would have to be made for sustainable utilisation of the river water and its bio-resources. Also since the river basin covers China, Nepal, India and Bangladesh the best way to protect and manage the river would be by close international co-operation between all the countries within the river basin — bringing together all interests upstream and downstream. This chapter provides an overview of these issues and suggests approaches to water management in the Ganga basin.

The Ganga River Basin

Physiography and Hydrography

The Ganga river basin is one of the most dynamic, productive and diverse river basins in the world and is a part of the composite Ganga-Brahmaputra-Meghna basin, which lies in China, Nepal, India and Bangladesh. The Ganga basin is bounded on the north by the Himalaya, on the west by the Aravalli as well as the ridge separating it from the Indus basin, on the south by the Vindhyas and Chotanagpur Plateau and on the east by the Brahmaputra ridge. The basin drains an area of 1,086,000 km^2 out of which 861,000 km^2 lies in India. In India, the basin catchments lies in the states of Uttar Pradesh (294,364 km^2), Madhya Pradesh (198,962 km^2), Bihar (143,961 km^2), Rajasthan (112,490 km^2), West Bengal (71,485 km^2), Haryana (34,341 km^2), Himachal Pradesh (4,317 km^2) and Delhi (1,484 km^2). The basin comprises three main physiographic regions of the Indian subcontinent namely:

1. The young Himalayan fold mountains in the north with dense forests, as well as the sparsely forested Shiwalik Hills;

2. the central highland and peninsular shield in the south consisting of mountains, hills and plateaus intersected by valleys and river plains, and in between;
3. the Gangetic alluvial plain.

In the south, the central highland and peninsular plateaus comprise distinctive physiographic subdivisions from the west to east: Aravalli Range, running in a north-east direction divides the Ganga basin drainage on the east and Indus basin drainage on the west. In the east, Bundelkhand Plateau and Malwa Plateau are located. The 1000-km-long Vindhyan Range runs west to east and rises to height of 550–600 m. The major tributaries originate from the Vindhyan mountains in the south. In geographic order from west to east the Chambal river, the Sind river, the Betwa river, and the Ken river join with the Yamuna river. The Tons, and the Sone discharge directly into the Ganga. Further east of the Vindhyan lies the extensive upland tract of the Chotanagpur Plateau which attains height of 700 m (Singh 1996).

The middle part of the basin occupies some 400,000 km² of vast, alluvial plain at an elevation of 300 m above sea level (Singh 1996). The important soil types found in the basin are sand, loam, clay and their combinations such as sandy loam, silt clay, etc. The Ganga alluvial valley acts as a pathway for sediment-water discharge from the basin to the Ganga delta and the Bay of Bengal.

The Ganga River System

The river Ganga, originates as Bhagirathi from the Gaumukh ice-cave (30°55 N, 79°07 E) of the Gangotri glacier system at an altitude of 4,100 m and discharges into the Bay of Bengal after traversing for over 2,700 km. The river Bhagirathi flows for more than 200 km to meet the Alaknanda, originating from the Alkapuri glacier near Badrinath shrine, and gets the name Ganga at Dev Prayag (520 m asl). Both Bhagirathi and Alaknanda receive a number of tributaries. After flowing for about 50 km, the Ganga descends to the plains at Rishikesh and after another 1,900 km it arrives at Farakka near India-Bangladesh border, where a barrage was commissioned in 1975. About 40 km downstream from the Farakka Barrage, the Ganga divides into two — the main channel flows along the India-Bangladesh border for about 90 km before entering into Bangladesh where it is called the Padma. It discharges into the Bay of Bengal forming an extensive Sundarban delta. The other channel, again called the Bhagirathi, flows for about 320 km in West Bengal, after which it is known as Hooghly which is its estuarine reach and flows 190 km down to the Sagar Island to merge with the Bay of Bengal. Until the end of fifteenth century, the river Ganga flowed through the Bhagirathi river in Bengal, and only later the main flow got diverted to the east and left the river as a spill channel (Basu 1967).

The Ganga, along with its tributaries the Ramganga, Yamuna, Ghaghara, Sone, Gandak, Kosi and Damodar, has a combined length of about 12,000 km. The river breaks into several interlaced channels in the plains forming meanders, oxbow lakes and swamps in its northern tributaries.

The river Ganga can be divided into following physical sections (Payne et al. 2004):

1. Upland reaches — 250 km from the sources at 4,100 m asl, to Rishikesh at 360 m asl, includes union of Bhagirathi and Alaknanda rivers at Dev Prayag at 520 m asl, gradient 1:67.
2. Equivalent range in northern tributaries to point where they enter the Gangetic plain.
3. Upper plains — 850 km from Rishikesh to Allahabad at 58 m asl with main intersection with plain at Haridwar at 310 m asl with mean gradient 1:4100.
4. Middle plains — 1,060 km from Allahabad to Farakka at 19 m asl through the lowlands of Uttar Pradesh, Bihar and West Bengal, with fringing floodplains, includes the large floodable area of Bihar wetlands where the river Kosi joins the main river, with mean gradient 1:13000.
5. Lower plains — 560 km from Farakka to Sagar Island, comprising the delta in India and Bangladesh (including the Sundarbans and an extensive floodplain) with mean gradient 1:24000.

Climate and Hydrology

The climate of the Ganga basin is humid subtropical type. Usually 70–80 per cent of the total annual rainfall occurs during July–September. The main sources of water in the basin are rainfall and melting of snow during the summer. More than 70 per cent of the flow at Haridwar is due to rainfall and the river has significant amount of base flow downstream of Haridwar. The whole basin receives 1 million m^3 of water km^{-2} annually. Out of this, 50 per cent is available to surface flow, 30 per cent is lost through evapo-transpiration and the rest is lost through groundwater seepage (Singh 1996). The annual mean discharge rate in the Ganga is 18.7×10^3 m^3 sec^{-1} (Welcomme 1985). Among the tributaries of the Ganga, the Yamuna river contributes about 61 per cent of the total flow at Allahabad, and just 16 per cent comes from Haridwar. Four mighty rivers join the Ganga in Bihar: Ghaghara, Gandak, Kosi and Sone. The contribution of these four rivers is 246,740 MCM, which is 1.62 times the flow at Allahabad. The average annual flow at Patna is about 364,000 MCM, which is nearly 17 times the flow at Dev Prayag (Jain 2008). Extreme variations in flow exist within the catchment area, to the extent that the mean maximum flow is 52.3 times greater than the mean minimum flow. The total annual flow of the Ganga is 468.7×10^9 m^3 which is 25.2 per cent of India's total water resources and is the fifth highest in the world.

Sediments

The river with a drainage basin of 1.1 million km^2 carries an annual load of 1.46 million ton of sediment (Natarajan 1989), second only to the Yellow river of China (Lisitzen 1972). A suspended sediment budget for the Ganga-Brahmaputra catchment shows that of the 794×10^6 t/yr transported in the rivers of the Ganga catchment, 80 ± 10 per cent comes from the High Himalaya. About 8 per cent of the river sediment is deposited on floodplains and delta plains in Bangladesh. The floodplains in the Ganga catchment appear to be aggrading, thereby exacerbating the annual over-bank flood. Aggradation may be the result of enhanced sediment delivery to the rivers as a result of land use, change in rainfall change, or neotectonics in the Himalaya. The relationship between land use, erosion and sedimentation is not clear. High altitude grazing, forest management, limited cultivation and road building in the High Himalaya probably jointly result in dominant source of sediment (Wasson 2003). The Ganga-Brahmaputra rivers contribute almost all of the sediment making up the Bengal delta and Submarine fan, a vast structure that extends from Bangladesh to south of the Equator, is up to 16.5 km thick and contains at least 1.13×10^{16} ton of sediment (ibid.).

The river Ganges receives sediments from the Himalayas as well as the Peninsular region. The total measured flow of suspended sediment in the tributaries to the Ganga river is 488×10^6 t/yr, while the quantity of sediment moving in the Ganga at Farakka is 729×10^6 t/yr of which 328×10^6 t/yr is transported down the Hooghly river (Wasson 2003). In the active channel, sedimentation forms mid-channel bars, side sand bars deposits of different size and floodplains. The river receives high sediment load from Ghaghara, Gandak and Sone rivers in and around Patna. The stretch between Patna and Farakka displays sediment accumulation rather than sediment transportation. This part exposes extensive development of sand bars, the meander loops and high unstable active channels. The exposed monadnocks (rock islands and rocks jutting out into the river) are the characteristic features of the river in this segment.

The water-related issues of the basin are both due to high and low flow. In India, the states of Uttar Pradesh, Bihar and West Bengal are affected by recurrence of annual floods. As Bangladesh lies at the confluence of Brahmaputra and Ganga, it experiences terrible floods almost every year. Many of the flood problems in the Ganga are caused by its northern tributaries such as Gandak, Kosi and Mahananda. However, ecological and economic roles of flood must be understood. Floods import woody debris into the channel, where it creates new, high-quality habitat. Floodplain wetlands provide important nursery grounds for fish and export organic matter and organisms back into the main channel (Junk et al. 1989; Sparks 1995; Welcomme 1992). A range of flows is necessary to scour and revitalise gravel beds, to import wood and organic matter from the floodplain and to provide access to productive riparian wetlands.

Water Development and Flood Control Projects

A network of canals totaling about 9,500 km distributes Ganga water to irrigate about 7 million ha (Dasgupta 1984) of its basin (see Figure 5.1). Multipurpose dams and barrages have been constructed on the main stem of the Ganga and all the tributaries to divert water (Smith et al. 2000). The major barrages on the main river are Bhimgoda barrage at Haridwar, Middle Ganga barrage at Bijnor, Lower Ganga barrage at Narora and the Farakka barrage near India-Bangladesh border. Barrages have also been built on the four major river systems of Nepal — Mahakali, Karnali, Narayani and Sapta Kosi (respectively known as Sharda, Ghaghra, Gandak and Kosi in India) near the India-Nepal border. Similarly, the flow of all the southern tributaries has been affected by dams and/or barrages for abstracting water.

Map 5.1: Ganga River System Showing Water Development Works across the Rivers

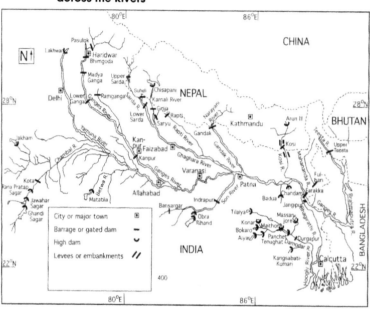

Source: Smith et al. (2000).

Construction of embankments as a flood control measure in eastern Uttar Pradesh, Bihar and West Bengal has not only disrupted the linkages of the river with its flood plain wetlands, but has also increased the flood-prone area, defeating the purpose for which these embankments were constructed. About 3,500 km embankment was constructed in Bihar alone after 1947 resulting in increase in flood-prone area from 2.5 million ha to almost 7 million ha and recurrence of annual devastating floods. Similarly, 1,811 km and 1,184 km long embankments were constructed in Uttar Pradesh and West Bengal respectively. The Kosi flood due to breach of the eastern embankment at Kusaha in Nepal (13 km inside from the border) on 18 August 2008 was probably the worst of its kind as 3.5 million people became homeless, one million cattle perished and 125,000 ha agricultural land damaged in 979 villages in five districts. The Kosi created a 150-km-long and 36 km wide water mass.

Water Quality

The Ganga has been a source of holy water, supposed to be the purest water, since time immemorial. People believe that the river water never spoils even if stored for decades and has germ-killing quality. But their faith has been belied in recent decades and the water quality degraded rapidly until remedial measures were initiated (Table 5.1) in the 1980s (Pahwa and Mehrotra 1966; Khanna 1993; Sinha et al. 1998). The data show that the water quality of river Ganga up to Haridwar is still good. There was no perceptible change in the values of water quality parameters between Kanpur and Patna from those in 1960. The water quality of river Ganga is monitored every month under the National River Action Plan of the Government of India. After the implementation of the Ganga Action Plan, there has been marked improvement in Dissolved Oxygen (DO) values (between 1990 and 1995–1996) at Kanpur, Allahabad, Varanasi and Patna to Rajmahal (Sinha and Khan 2001). Inflow of freshwater in the Hooghly estuary after the commissioning of the Farakka barrage in 1975 resulted in increased DO content at Uluberia (Sinha et al. 1998).

The pH values (Table 5.1) in the entire stretch of Ganga did not vary much in the last 3–4 decades indicating its high buffering capacity. Phosphate-P concentrations in river Ganga are higher than in other large rivers of the world (Allen 1995). Between 1960 and 1995–1996, the PO_4-P concentrations increased at Kanpur, Allahabad and Varanasi whereas the nitrate concentrations decreased (Sinha and Khan 2001). However, the 2006 data show that the water quality again deteriorated relative to that in 1996 (Tare et al. 2003a). An increase in the BOD and nitrogen in the reach between Kannauj and Kanpur is attributed to unexpected rise in the urban population and poor operation and maintenance of the treatment facilities (Tare et al. 2003). Further downstream in Bihar, the water quality did definitely improve except that the faecal coliform count still exceeds the desirable limit everywhere.

Toxic Chemicals

It is estimated that annually, 2,573 ton of pesticides (Mohan 1989) and 1.15 million ton of chemical fertilisers (Dasgupta 1984) end up in the river system. Intensive farming practices in over 46 million hectares of land utilise varieties of agrochemical some of which do not degrade at all (e.g., DDT, lindane, or polychlorinated biphenyls); these are instead transported to the water (Schwarzenbach et al. 2006), and also tend to bioaccumulate.

During 1986–1992, the DDT concentration in Ganga water (0–5.8082 ppb) exceeded the EPA limit of 0.001 ppb for freshwater organisms (ITRC 1992). Heavy metals, organochlorine pesticides and

Table 5.1: Physico-chemical Characteristics of Ganga River Water during Two Time Intervals

Centre	Water temperature (°C)	pH	D.O. mg/l	Free CO_2 mg/l	Phosphate mg/l	Nitrate mg/l	Silicate mg/l
Haridwar							
1984–1985	11.25–19.75	7.60–8.00	7.60–12.50	0.75–4.65	NA	NA	NA
1995–1996	12.50–26.00	7.90–8.30	8.30–9.60	Nil–3.00	0.60–4.40	.067–0.21	.090–0.19
Kanpur							
1960	16.50–30.50	7.70–8.30	5.00–10.50	0.60–4.50	.067–0.21	.090–0.190	8.20–20.30
1985–1990	NA	6.10–7.90	3.73–8.60	NA	0.01–2.10	.08–1.90	NA
1995–1996	16.00–30.00	7.10–8.30	5.00–9.00	Nil	Tr.–2.50	Tr.–0.24	2.4–14.2
Allahabad							
1960	17.50–31.50	7.95–8.25	6.00–10.80	1.10–3.70	0.09–0.20	0.11–0.22	6.70–17.0
1985–1990	NA	7.48–8.40	7.33–8.00	NA	0.11–0.32	0.10–0.33	NA
1995–1996	17.00–32.00	7.00–8.40	5.00–11.90	Nil	Tr.–0.80	0.06–0.24	1.60–14.2
Varanasi							
1960	18.50–31.50	7.60–8.40	5.00–8.90	Nil–6.50	0.08–0.12	0.08–0.14	4.00–12.6
1985–1990	NA	7.13–8.50	2.04–9.00	NA	0.12–0.73	0.16–12.49	NA
1995–1996	20.00–31.00	7.40–8.30	4.50–10.20	Nil–2.00	Tr.–1.00	Tr.–0.28	0.6–11.2
Buxar–Ballia							
1960	18.00–31.00	7.80–8.30	5.50–10.30	2.90–5.00	0.05–0.12	0.08–0.18	9.10–18.40
1985–1990	NA	NA	NA	NA	NA	NA	NA
1995–1996	19.00–33.00	7.30–8.30	3.40–10.00	Nil	Tr.–0.40	Tr.–0.28	0.80–9.20
Patna							
1960	18.50–31.00	7.70–8.20	5.40–8.60	2.25–10.0	0.07–0.11	0.09–0.18	7.20–14.00
1985–1990	NA	7.80–8.00	4.70–7.90	NA	NA	NA	NA
1995–1996	19.50–31.00	7.20–8.80	5.00–10.80	Nil–1.00	Tr.–0.01	Tr.–0.86	0.80–7.10

Source: Khanna (1993); Pahwa and Mehrotra (1996); Sinha et al. (1998).
Note: NA = Data not available.

polychlorinated biphenyls were recorded in the Gangetic dolphins and fishes (Kannan et al. 1993). On the other hand, the DDT concentration in the blubber of dolphins ranged from 4.7 to 13 μg/g wet wt, the tissues had 360 to 620 ng PCBs per g wet wt and 2000 ng butyltin per g wet wt, that were 5–10 times higher than in their diet (Kannan et al. 1997; see also Senthilkumar et al. 1999). The concentrations in fish and benthic invertebrates were 3–10 times greater than sediment. Kumari et al. (2001a and 2000b) found high levels of DDT and HCHs in the water and fishes at Patna whereas even mercury was recorded to bioaccumulate in benthic fauna and fishes (up to 2.64 ppm) at Varanasi (Sinha et al. 2007). Further downstream, high concentration of a variety of heavy metals including Cr, Cd and Hg has been reported (Tables 5.2 and 5.3) from the Hooghly estuary (Saha et al. 2004).

Biodiversity of the Ganga River System

Although the flora and fauna of river Ganga have been investigated for about a century, mainly by fishery scientists (Datta et al. 1954; Bose 1956; Shetty et al. 1961; Lakshminarayana 1965; Pahwa and Mehrotra 1966; Pahwa 1979; Ghosh et al. 1982; Dasgupta 1984; Bilgrami 1988; Jhingran 1988; Natarajan 1989; Khan et al. 1996; and Sinha 1998), and then under the Ganga Action Plan along the entire river length during 1985–1988 (Krishnamurti et al. 1991), a comprehensive holistic understanding of the river biota and their functions in providing the ecosystem services is not yet complete. Considerable work has been done in recent years by the researchers at Patna University, focusing particularly on dolphins as the keystone species of Ganga river systems in Nepal and India (Smith et al. 1994, Sinha 1996, 1997, 1999; Sinha et al. 2000; Sinha 1999). The Central Inland Fisheries Research Institute (CIFRI) explored the fisheries of river Ganga at 43 centers from Tehri in the Himalaya to Kakdwip near Sagar Island during 1995–1996 (Sinha et al. 1998). The available information on the biodiversity of Ganga is summarised in the section that follows.

Fungal Flora of the Ganga

Fungi are decomposer organisms which enter the water bodies along with decaying litter, soil and other organic matter. Varying number of taxa has been reported from river Ganga: 10 species from Rishikesh to Garhmukteshwar, 16 species between Kalakankar and Phaphamau

Table 5.2:   Average Heavy Metals Content (ppm) in Sediment of Hooghly Estuary at Various Stations in 2002–2003

Stations	Distance	Pb	Fe	Cd	Cr	Zn	Ni	Co	Hg	As
Kalyani	0	8.17 ± 3.87	28250 ± 8500	BDL	15.50 ± 3.88	45.95 ± 15.48	17.15 ± 6.92	7.92 ± 2.94	0.19 ± 0.03	2.89 ± 0.71
Dunlop	6	9.25 ± 5.02	26125 ± 8262	0.25 ± 0.0	11.26 ± 3.91	48.70 ± 18.49	17.96 ± 7.81	8.11 ± 2.96	0.21 ± 0.06	3.55 ± 1.37
Barrackpore	22	9.38 ± 3.42	30458 ± 8430	BDL	14.92 ± 8.34	55.85 ± 16.85	19.23 ± 4.90	8.78 ± 3.05	0.18 ± 0.03	4.03 ± 1.09
Sreerampore	2	11.46 ± 1.92	31583 ± 11567	BDL	13.67 ± 4.91	61.14 ± 9.44	19.93 ± 3.91	12.78 ± 5.94	0.21 ± 0.05	3.50 ± 0.72
Kamarhati	9	17.42 ± 6.55	35867 ± 13816	1.08 ± 1.0	7.33 ± 6.63	80.26 ± 25.89	19.67 ± 7.29	9.58 ± 8.31	0.26 ± 0.07	3.62 ± 0.68
Cossipore	6.5	20.0 ± 7.26	26958 ± 10762	0.50 ± 0.25	14.36 ± 4.55	89.46 ± 30.45	19.74 ± 6.06	8.27 ± 8.21	0.25 ± 0.04	4.72 ± 1.37
Ferrighat	5	15.83 ± 2.07	41802 ± 27723	3.67 ± 2.09	15.19 ± 3.74	90.95 ± 48.0	20.86 ± 1.87	8.81 ± 1.55	0.21 ± 0.05	5.07 ± 1.37
B.Garden	7	13.75 ± 5.85	29500 ± 4861	0.46 ± 0.17	16.78 ± 4.58	64.02 ± 23.63	16.95 ± 6.06	5.59 ± 4.06	0.28 ± 0.08	4.61 ± 1.11
Garden Reach	1	35.79 ± 31.25	28917 ± 12547	0.33 ± 0.12	18.23 ± 4.82	71.8 ± 40.80	17.05 ± 2.54	7.93 ± 1.73	0.18 ± 0.05	4.0 ± 0.62
Birlapur	30	6.60 ± 1.81	26150 ± 6536	0.25 ± 0.0	17.0 ± 11.39	44.3 ± 6.23	15.35 ± 1.72	8.10 ± 1.45	0.28 ± 0.06	3.42 ± 1.15

Source: Saha et al. (2004).

Table 5.3: Average Heavy Metals Content (ppm) in Hooghly Water at Various Stations in 2002–2003

Stations	Pb	Fe	Cd	Cr	Zn	Ni	Co	Hg	As
Kalyani	0.004 ± 0.004	1.27 ± 0.8	BDL	0.02 ± 0.02	0.02 ± 0.01	0.002 ± 0.001	0.002 ± 0.0009	BDL	0.001 ± 0.0002
Dunlop	0.002 ± 0.001	1.52 ± 1.16	BDL	0.01 ± 0.007	0.02 ± 0.01	0.002 ± 0.0009	0.002 ± 0.001	BDL	0.002 ± 0.0002
Barrackpore	0.007 ± 0.009	0.96 ± 0.85	0.0005	0.008 ± 0.006	0.02 ± 0.02	0.002 ± 0.001	0.002 ± 0.001	BDL	0.002 ± 0.00005
Sreerampore	0.005 ± 0.006	1.23 ± 0.58	BDL	0.02 ± 0.02	0.02 ± 0.02	0.002 ± 0.0009	0.002 ± 0.001	BDL	0.002 ± 0.0001
Kamarhati	0.003 ± 0.0006	0.92 ± 0.56	BDL	0.01 ± 0.009	0.02 ± 0.02	0.002 ± 0.002	0.002 ± 0.001	BDL	0.001 ± 0.0003
Cossipore	0.006 ± 0.006	1.47 ± 1.13	BDL	0.02 ± 0.008	0.02 ± 0.01	0.004 ± 0.002	0.002 ± 0.001	BDL	0.002 ± 0.0002
Ferrighat	0.003 ± 0.0001	0.48 ± 0.22	0.0005	0.010 ± 0.007	0.02 ± 0.02	0.002 ± 0.0008	0.002 ± 0.002	BDL	0.002 ± 0.0002
B.Garden	0.005 ± 0.003	0.73 ± 0.49	BDL	0.01 ± 0.01	0.02 ± 0.02	0.002 ± 0.0006	0.002 ± 0.001	BDL	0.002 ± 0.0004
GardenReach	0.004 ± 0.002	0.69 ± 0.50	BDL	0.01 ± 0.008	0.02 ± 0.02	0.002 ± 0.0009	0.002 ± 0.001	BDL	0.002 ± 0.0005
Birlapur	0.004 ± 0.001	1.03 ± 0.54	BDL	0.01 ± 0.02	0.01 ± 0.004	0.006 ± 0.01	0.002 ± 0.001	BDL	0.002 ± 0.0002

Source: Saha et al. (2004).

(Krishnamurti et al. 1991), 51 species in water and 54 species in sediment from Buxar to Barh in Bihar (Sharan and Sinha 1988). However, many fungi are pathogenic, e.g., *Aspergillus flabus, A. fumigatus, A. niger, Histoplasma capsulatum, Curvularia lunata, Aspergillus nidulans, Scopulariopsis brevicaulis, Fusarium sp., Penicileum sp.* and *Candida albicans* cause a variety of skin infections, dermatitis and keratitis, etc.

Algal Flora of the Ganga

Numerous species of planktonic algae have been reported in various studies on different stretches of river ganga. The mountain reaches (Gangotri to Rishikesh) are dominated by diatoms. Downstream up to Garhmukteshwar, phytoplankton density increases with increase in other groups of algae. About 100 species and a very high density (8,30,000 u/l) has been recorded between Mirzapur and Ballia. A total of 182 species of phytoplankton were recorded from the Munger to Farakka reach. Below Farakka, between Berhampur and Katwa, the phytoplankton density was quite low (8200 u/l) as also the number (42 species) of species (Krishnamurti et al., 1991). Pahwa and Mehrotra (1966) also reported decreasing density of phytoplankton (comprising mainly of diatoms, green and blue green algae) from Kanpur to Rajmahal. Sinha and Khan (2001) reported a decline in plankton density and also the abundance of pollution indicator taxa in the middle and lower freshwater reaches of Ganga between 1960 and 1995. In recent years, 523 species of algae has been recorded from the lower stretch of the Ganga in West Bengal (personal communication from B. C. Jha of CIFRI, Barrackpore).

Macrophytes of the Ganga

Macrophytic vegetation includes both aquatic and semi-aquatic plants in shallow water, on river banks and the floodplains. These plants are important in soil binding, checking bank erosion, providing shelter and breeding sites for fish, birds, reptiles, etc. They are also economically important as they are used for fodder, thatch, fuel, etc. Krishnamurti et al. (1991) report the occurrence of 31 species of trees, 17 climbers, 36 shrubs, 139 herbs, 8 grasses, 5 sedges and 32 aquatic macrophytes from different parts of the basin.

Invertebrate Fauna

Studies during 2000–2007 recorded 59 rotifers, 25 cladocerans and 13 copepods, 1 porifera, 1 platyhelminthes, 7 oligochaetes, 4 polychaetes,

2 leeches, 6 crustaceans, 41 insects, 29 gastropods and 18 pelecypods from the 500 km stretch of Ganga in Bihar. In another survey of Ganga from Haridwar to Farakka (c.1900 km), Sinha (1999) documented 87 species of zooplankton (54 rotifers, 21 cladocerans, and 12 copepods), 66 species of benthic macroinvertebrates (3 polychaetes, 2 oligochaetes, 1 leech, 3 crustacea, 39 insects, 11 gastropods and 9 bivalves), 83 species of fishes and 2 mammals. In the upper reaches (up to Allahabad), rotifers, cladocerans and copepods contributed 65, 25 and 10 per cent respectively, whereas in the middle reach (Allahabad to Patna) it was 61, 22 and 17 per cent respectively. In the lower reaches (Patna to Farakka), there were 72 per cent rotifers, 15 per cent cladocerans and 13 per cent copepods. Among the benthic macroinvertebrates, insects were the dominant group — 85 per cent and 46 per cent in lower and upper reaches respectively. But in the middle reach gastropods were the dominant group (44 per cent), followed by insects (32 per cent) and pelecypods (20 per cent). The dominance of mollusks can be related to higher calcium concentration in water whereas insects are favoured in the lower reaches by the sluggish flow and clayey sediments (less than 79 per cent sand between Haridwar and Patna; Sinha and Khan 2001). Another recent survey of Ganga from Haridwar to Kanpur has reported 40 species of zooplanktons, 4 crustaceans, 15 molluscs, 51 insects, 83 fishes, 12 turtles, 2 crocodiles, 48 aquatic birds and two mammals (Rao 2001).

An improvement in water quality after the implementation of the Ganga Action Plan was reflected by the density of macrozoobenthos which had declined by 1995–1996 in the middle and lower Ganga but increased manifold in the Hooghly estuarine stretch. Oligochaete and chironomid decline was reported by Sinha (1999) and Sinha and Khan (2001).

Recent surveys for benthic fauna (except insects) have brought to light several new species (27 oligochaeta, 10 hirudinea and 2 polychaeta; Nesemann et al. 2003 and 2004) from Patna. Further, Nesemann et al. (2007) have recorded 36 bivalves, 40 gastropds, 6 polychaetes, 37 oligochaetes, 14 leeches and 8 crustaceans from Ganga in Bihar and Nepal.

Vertebrate Fauna of the Ganga

The first ever scientific document on Ganga recorded 268 fish species from the river system (Hamilton 1822). Roxburgh (1801) reported

Platanista gangetica for the first time from Hooghli near Calcutta. Anderson (1879) gave a detailed account of distribution and biology of the Gangetic dolphin *Platanista gangetica.* Zoological Survey of India (ZSI) has listed 375 species of fish, 11 species of amphibia, 27 of reptiles, 177 of birds and 11 of mammals from the river Ganga (ZSI 1991). Remarkably, amphibians are the least known group among vertebrates of the Ganga.

Fish and Fishery

The Ganga fisheries comprising mainly of cyprinids (176 spp.) and silurids (Sinha and Khan 2001) include the Indian Major Carps (*Labeo rohita, Catla catla, Cirrhinus mrigala,* and *Labeo calbasu*), large catfishes (*Aorichthys aor, A. seenghala, Wallago attu* and *Bagarius bagarius*), featherbacks (*Notopterus notopterus* and *N. chittala*) and murrels (*Channa marulius* and *C. punctatus*) which are of commercial interest. In the rhithron region (origin to Haridwar), the fishery comprise of *Schizothorax spp., Glyptosternum spp., Tor tor, Tor putitora* and *Labeo dero.*

The riverine fisheries in the middle stretch have declined sharply, both qualitatively and quantitatively. The major carp yield at Allahabad has decreased from 44.5 per cent (1958–1966) to 8.3 per cent in 1996–1997 and *Tenualosa ilisha* (hilsa) from 9.7 to 4.2 per cent, whereas the proportion of large catfishes has increased slightly from 22.7 to 24.1 per cent. The yield of miscellaneous fishes (*Setipinna phasa, Chela spp., Mastacembelus spp., Puntius spp., Eutropiichthys vacha, Clupisoma garua, Notopterus spp., Rita rita, Mystus vittatus, Ailia coila, Nandus nandus,* etc.) also increased from 23.1 to 63.4 per cent during this period. Similar declining trends have been observed at other fishing centers also (Sinha 2000; Hassan 1999; Payne et al. 2004).

The average annual yield of prawn and fish from the estuary increased from 9481.5 ton in pre-Farakka barrage period (1966–1997 to 1974–1975) to 33,341.4 ton during post-barrage period (1984–1985 to 1994–1995) and further up to 42,703.2 ton during 1995–1997 and to 61,032 ton during 1997–2000. This is due to increased influx of freshwater in the Hooghly after the barrage was commissioned in 1975. Hilsa fishery increased in Hooghly from 1,457.1 ton in 1975 to 9,576.9 ton in 1997–2000. Freshwater species like *Eutropiichthys vacha, Clupisoma garua, Rita rita, Aorichthys seenghala, A. aor, Catla catla* and *Labeo bata* have made their appearance in the upper zone of the estuary. These species were not reported prior to 1975 from these locations.

Reptiles

Endemic species like *Gavialis gangeticus,* a fish-eating crocodile, and *Aspideretes gangeticus,* a large size soft-shell turtle are endemic to the Ganga basin. Large-scale exploitation in the past few decades has caused near extinction of these species. The gavialis is now rarely seen. Occasionally water snakes and monitor lizards are also sighted. Rao (2001) reported 12 species of turtles from the Ganges. Within the Ganga basin, National Chambal Sanctuary and Katarniaghat Gharial Sanctuary in Girwa river (about 25 km stretch of the river Ghaghra, between India-Nepal border and Girijapuri barrage in Bahraich district of Uttar Pradesh) have been set up for conservation of the gavialis. The river Karnali in Nepal is a part of Royal Bardia National Park and the river Girwa has the Katarniaghat Gharial Sanctuary as a protected area.

Birds

The floodplain wetlands along the river support a rich diversity of avifauna. The ZSI recorded 177 species of which 97 occur in the Vikramshila Gangetic Dolphin Sanctuary (VGDS) between Sultanganj and Kahalgaon in Bihar, whereas 162 species occur between Buxar and Maniharighat, Katihar. Indian Skimmer, *Rynchops albicollis,* is a vulnerable species which is virtually extinct in Southeast Asia and is declining in South Asia (Crosby 2004). However, a flock of less than 25 Indian Skimmer was sighted in the river Sarda, a tributary of the river Ghaghara (Sinha and Sharma 2003a) and in the VGDS. The river Sarda, with a large number of sand bars, vast floodplains and dense filamentous algal flora during the low water season (February–March) attracts a very rich and diverse migratory and resident avifauna. Besides poaching, degradation of wetlands, pesticides and loss of riparian vegetation are major threats to the avifauna.

Mammals

One of the most rare, endemic and endangered mammals of the Ganga basin is the Gangetic dolphin, *Platanista gangetica gangetica* (Sinha 1999, 2000; Sinha et al. 2000; and Sinha and Sharma 2003b). It is listed in Schedule-I of Indian Wildlife (Protection) Act, 1972. It is found in the Ganga-Brahmaputra-Meghna river systems in India, Nepal and

Bangladesh and in Sangu-Karnaphuli river system of Bangladesh. An estimated population of 2,000 individuals has been reported in the Ganga-Brahmaputra river systems and 500 individuals in Bangladesh. The species is almost extinct in rivers of Nepal (Smith et al. 1994).

The Gangetic dolphin has disappeared between Haridwar and Bijnor barrage whereas a small population of 35 dolphins survives between Bijnor and Narora barrage (Sinha 1999). Because of low water levels, there are only few dolphins between Narora and Allahabad. The population increases considerably downstream and is regularly monitored (Sinha 1997). A total of 225 individuals have been recorded from Sundarban area of Bangladesh (Smith et al. 2005). No dolphins could be sighted in the river Sarda upstream of Sarda Barrage (Sinha and Sharma 2003). There are small populations of dolphins also in the lower reaches of river Yamuna, Kosi and Gandak (Sinha and Sharma 2003b). Other mammals of the Ganga are confined to the deltaic region in the Sundarban (ZSI 1991). However, Indian smooth coated otters (*Lutra perspicilata*) have been sighted in parts of Bihar and near Middle Ganga Barrage at Bijnor.

Human Impacts in the Ganga Basin

Major human impacts on river Ganga, as in other tropical streams, are excessive water abstraction for irrigation and human consumption, and related alterations in longitudinal, lateral and vertical riverine connectivity; over-exploitation (mainly over-fishing) of other resources; deforestation in the entire basin, particularly catchments; and high levels of pollution from different sources (Dudgeon et al. 2005). The Ganga basin is also facing the problems arising from receding glaciers. Some of these impacts are examined in the section that follows.

Receding Glaciers: Climate Change and Flow in the Ganges

According to Cruz et al. (2007) by year 2070–2099, surface temperature in South Asia is likely to rise by 1.56 °C and 5.44 °C. Some fear that the Ganga may become a seasonal river in the near future as a consequence of climate change as the Gangotri glacier may disappear by the year 2035. In the late 1960s, the glacier retreated about 30 m/yr (Thayyen 2008). Assuming the recession rate of 40 m/yr, the

30-km-long glacier will take about 700 years to melt away. Thus, the glaciers are unlikely to disappear in the near future (Jain 2008). However, long-term studies on glacier mass balance and glacier dynamics are needed to understand the impact of climate change on Himalayan glaciers. Global warming will also affect other variables like precipitation intensity and quantity, which will have greater influence on the river hydrology.

Habitat Alteration

The major causes of habitat alteration are construction of dams and barrages, embankments, drainage channels, sedimentation, etc. Dams and barrages on the river Ganga have been described earlier in the chapter. The Upper Ganga Canal diverts more than 60 per cent of the annual flow and almost 100 per cent of dry seasons flow at Haridwar. The Lower Ganga Canal at Narora downstream diverts water again. The dams and barrages act as physical barrier for migratory aquatic species (Reeves and Leatherwood 1994) and block the downstream transport of sediments which replenish the delta and are an important source of nutrients for aquatic biota (Chen 2002). The Farakka Barrage traps the silt and the 'silt free' water diverted to the Bhagirathi-Hooghli river system has destroyed the hydrogeomorphological complexities of the river resulting in the decline of biodiversity. A small tributary, river Ajay, brings large amount of sediments to Bhagirathi at Katwa after which the river supports relatively rich faunal diversity and more dolphins.

Recently, the Uttarakhand government has approved construction of more than 35 hydel projects on the Alaknanda and Bhagirathi rivers. The cessation of free flow of the river will have an irreversible effect on the rive biota and its functions. The cost-benefit analyses of hydel projects have not accounted for the goods and services provided by the rivers, nor their effects on the livelihood of the people downstream.

Embankments and drainage channels also disrupt the movement of animals and materials along a river network and from the river to the floodplain and hence, have profound negative implications for ecosystem integrity (Pringle 1997; Pringle et al. 2000; Bunn and Arthington 2002; Agostinho et al. 2004). Most lowland river biota depends on lateral connectivity with the floodplain, which are inundated periodically.

Sedimentation

Changing land use and loss of vegetation cover in the catchment and floodplains have accelerated erosion and increase in the rivers sediment load. In conjunction with river regulation measures, this is resulting in rise of river bed and decrease in discharge capacity. Sedimentation has direct bearing on the physical and biological characteristics of the river basin. It chokes spawning sites and destroys benthic food sources for fishes. Floodplain wetlands function to improve water quality by acting as sinks for nutrients such as nitrogen and phosphorus. Changes in floodplain vegetation and chemistry affect water quality as well productivity.

Pollution

The river Ganga receives an estimated 12,000 mld domestic sewage, with partial or no treatment (CPCB) from 29 Class I and 23 Class II cities. Delhi alone discharges 3,300 mld sewage into river Yamuna (personal communication from Dr R. Dalwani). The Ganga basin is a hub of industrial activity. Practically all kinds of industries discharge their effluents (after some treatment) into the river. Agriculture contributes nutrients and pesticides (Sinha and Khan 2001). Also, hundreds of human corpses and thousands of animal carcasses are disposed into the river every day. The waste discharged into the river often exceeds the river flow during the pre-monsoon period (Ray 1998). Despite many reports on bioaccumulation of toxic substances in organisms, the full impact on the river ecosystem is not yet well understood.

Introduction of Exotic Species

Introduced species occasionally replace native species in natural habitats through competition or because the altered environments provide the introduced species an ecological advantage. Thai Mangur (*Clarias gariepinus*), Chinese Grass Carp (*Ctenopharyngodon idella*) and common carp (*Cyprinus carpio*) are some of the introduced species in the Ganga river system. Details of impact on the native fish fauna are not well understood. Exotic invertebrates such as *Physa (Haitia) mexicana*, a north American snail, have also been reported from the Ganga and Yamuna (Sinha et al. 2003). Among plants, water hyacinth has spread throughout the basin within a century of its introduction, and has greatly affected both lotic and lentic water bodies.

Over-exploitation of Biological Resources

Fish are the most exploited fauna of the Ganga resulting in steep decline in fish catch; for example, it declined by 75 per cent at Buxar during 1958 to 1984 (Natrajan 1989). Turtles, gavialis, crocodiles, birds, otter and freshwater dolphins are exploited heavily throughout the Ganga basin. Among invertebrates crabs, bivalves and freshwater prawns are highly exploited for food. Mollusc shells were used on a large scale for making lime as a substitute for cement. Thick shells of the bivalves are commercially exploited from the Gandak river for making buttons for garments.

Management Challenges and Mitigation Measures

The Ganga river basin is one of the most important river basins that sustain a significant proportion of humankind. It has been inhabited by humans for millennia but the human activities are increasing at an unprecedented rate affecting the flow regimes, water quality and biodiversity. Flow regime regulates habitat variability, essential for rich and abundant biodiversity. Five critical components of the flow — the magnitude, frequency, duration, timing and rate of change — regulate ecological processes in river ecosystems (Richter et al. 1996; Walker et al. 1995).

Most of the 500 million inhabitants of the Ganga basin are rural poor with livelihoods directly dependent on the availability of water for the production of food. Certain water resource interventions have helped in raising the living standards, whereas others have not realised their poverty-reduction objectives. Basin level upstream–downstream linkages, where land and water-related decisions in one part of the basin impact other human and environmental uses elsewhere, are difficult to address in water resources management, particularly in a transboundary system. So the best way to protect and manage water quality and the biodiversity is by close international co-operation between all countries within the natural geographical and hydrological unit of the river basin — bringing together all interests upstream and downstream. However, there is no ecosystem-based approach to river management among the riparian countries except some political agreements for sharing of river water between India and Bangladesh and some treaties with Nepal. We need to achieve improved trans-boundary management based on ecosystem approach between the riparian countries. The ecosystem approach is a strategy for the integrated management of land, water and living resources that promotes conservation and sustainable use in the equitable way. The application of the ecosystem approach in water management has led to the development of objectives for safeguarding the functional integrity of aquatic ecosystems. The functional integrity of aquatic ecosystems is characterised by a number of physical, chemical, hydro-logical and biological factors and their interactions.

The trans-boundary issues in the Ganga basin include use of river water for alleviating poverty and sharing of river water in lean season for which it is essential to increase river flow in such periods by technical interventions, especially in upper reaches in India and Nepal. Such interventions are expected to reduce the annual flood fury. It is important to develop mutual confidence. Trans-boundary protected areas may be declared for conservation and protection of endangered species in particular and biodiversity in general, e.g., in the Karnali-Girwa river system. Besides, pollution issues may also be addressed in a better way through international cooperation. Some years ago, one such programme called South Asia Water Analysis Network (SAWAN) was initiated among India, Nepal, Bangladesh and Pakistan. But unfortunately, it could not continue due to lack of interest of the concerned governments. Such schemes may be of great help in addressing issues of water quality management.

The poor management of land and water resources in the Ganga river basin, like many river basins of the world, has led to major floods, water shortages, pollution and loss of biodiversity adversely affecting development and increasing poverty. There is a need for an integrated approach to river basin management which incorporates ecosystem functions and values. Guidelines and decisions under the Convention on Biological Diversity (CBD) and Ramsar Convention on Wetlands have set out a strategic approach to ensure the proper management and sustainability of ecosystems and associated biodi-versity within river basins.

Water has conventionally been considered a free commodity and government policies have provided little incentive to encourage ef-ficient use of the resource. Moreover, water being a state subject, the price for its use in different sectors is fixed by the state governments and varies from state to state. Typically, water rates for agriculture and domestic consumption do not cover even working expenses of providing the services, let alone capital costs. Planning and implementation of water development projects is currently cumbersome with a number of organisations involved, both at state and centre levels, in the duplication and ambiguity of functions. While there is an extensive legislative framework to address water pollution, there are no regulations on surface water abstraction. Projections for water requirements and water pollution loads must be worked out well in advance for undertaking new projects in the river basin.

Conclusion

The Ganga Action Plan launched in 1985 aimed at abatement of pollution in the Ganga and some of its tributaries. The mandate did not include ecological restoration or maintenance of ecological integrity. The Government of India's policy stipulated withdrawal of 37 per cent of the annual flow but unfortunately even during the lean flow season, 90 and 95 per cent of the flow is withdrawn at Haridwar and Narora, respectively (*The Hindu*, 11 May 2008). The Ganga has several stakeholder agencies in the government and outside but there is no one to manage the whole basin. The new initiative of the Prime Minister of India on 4 November 2008 to declare the river Ganga as a National River and look at the entire basin as a single ecological entity will, hopefully, serve the purpose.

It is however a matter of concern that the surplus water from the Ganga basin is proposed to be transferred to water-starved rivers of peninsular India. This is part of a larger proposal on interlinking of rivers throughout India. The solution to water shortage in deficient areas lies in better management of existing water resources rather than importing water for irrigation through interlinking of rivers. The irrigation practices need to be made more efficient by using drip irrigation in place of flood irrigation. It is hoped that the proposed Ganga River Basin Authority will consider the issue holistically.

For the conservation of river Ganga, the topmost priority should be to allow the river to flow with enough water that is essential for maintaining its ecological integrity. Floodplain must be recognised as an integral part of the river. The river basin approach must be followed for river management; and trans-boundary issues must be addressed in coordination with the basin-sharing countries. Both point and non-point sources of pollution must be considered for abatement of pollution.

These goals require serious coordination among the concerned ministries such as those devoted to water resources, agriculture, environment and forests, urban development, industries, etc., and the state governments of the river basin. Watershed management, more vegetation in the catchment area, restoration of wetlands and practicing water efficient agriculture will ensure more water in the river basin and improve livelihood of the people. The hydrological and biological data required for the management of water quality and biodiversity of the Ganga basin are still poor and need to be strengthened. Participation by the people in the entire process of management must be ensured. Also, there is a need for improved communication between scientists/technocrats and policy makers so that the expertise is translated into action, and that the requirements of policy makers and society are properly understood (Oki and Kanae 2006). The exponential growth of human population in the river basin is the biggest challenge which can be addressed by considering human being as part of the ecosystem and by realising the new ecological reality in the basin.

References and Select Bibliography

Agostinho, A. A., S. M. Thomaz and L. C. Gomes. 2004. 'Threats for Biodiversity in the Floodplain of the Upper Prana River: Effects of Hydrological Regulation by Dams', *Ecohydrology and Hydrobiology*, 4 (3): 267–80.

Ahmad, Q. K., A. K. Biswas, R. Rangasari and M. M. Sainju. eds. 2001. *Ganges-Brahmaputra-Meghna Region: A Framework for sustainable Development*. Dhaka: The University Press.

Allen, J. D. 1995. *Stream Ecology: Structure and Function of Running Waters*. New York: Chapman and Hall.

Anderson, J. 1879. *Anatomical and Zoological Researches: Comprising an Account of Zoological Results of the Two Expeditions to Western Yunnan in 1868 and 1875; and a Monograph of Two Cetacean Genera, Platanista and Orcaella*. London: B. Quaritach.

Annandale, N. 1922. 'The Marine Element in the Fauna of Ganges, Bird', *Bijdragen tot de Dierkunde*, 22: 143–54.

Basu, S. R. 1967. 'On Some Aspects of Fluvial Dynamics of River Bhagirathi, with Special Reference to its Physical and Hydraulic Characteristics', *Indian Journal of Power River Valley Development*, 17 (11): 32–42.

Bilgrami, K. S. 1988. 'Biological Monitors of Rivers: Problems and Prospects', *Proceedings of Indian Academy of Sciences*, B 54 (2, 3): 171–74.

Bose, B. B.1956. 'Observation the Hydrology of the Hooghly Estuary', *Indian Journal of Fisheries*, 3 (1):101–18.

Bunn, S. E. and A. H. Arthington. 2002. 'Basic Principles and Ecological Consequences of Altered Flow Regimes for Aquatic Biodiversity', *Environmental Management*, 30 (4): 492–507.

Chen, C. A. 2002. 'The Impact of Dams on Fisheries: Case of the Three Gorges Dam', in W. Steffen, J. Jager, D. J. Carson and C. Bradshaw (eds), *Challenges of a Changing Earth*, pp.97–99. Berlin: Springer.

Crosby, M. 2004. 'Threatened Birds and IBAs in Asia: Setting Priorities for Conservation Action', *Mistnet*, 5 (3 and 4): 23–25.

Dasgupta, S. P. 1984. *Basin Sub-basin Inventory of Water Pollution: The Ganga Basin*, Part I. Delhi: Central Pollution Control Board.

Dudgeon, D., A. H. Arthington, M. O. Gessner, Z. Kawabata, D. J. Knowler, C. Leveque, R. J. Naiman, A. Prieur-Richard, D. Soto, M. L. J. Stiassny and C. A. Sullivan. 2005. 'Freshwater Biodiversity: Importance, Threats, Status and Conservation Challenges', *Biological Review*, 81: 163–82.

Datta, N., J. C. Malhotra and B. B. Bose. 1954. 'Hydrology and seasonal fluctuations of the plankton in the Hooghly Estuary', in IPFC (ed.), Proceedings of the *Symposium on Marine and Freshwater Plankton in the Indo-Pacific*, pp. 35–47. Bangkok: Indo-Pac. Fish Council (IPFC).

Editorial. 2008. 'Making the Most of the Worlds Dwindling Water Supply', *Nature*, 452 (7185): 253.

Ghosh, B. B., M. K. Mukhopadhyaya and M. M. Baghchi. 1982. 'Some Observations on Bioaccumulation, Toxicity, Histopathology of Some Fishes in Relation to Heavy Metal Pollution in the Hooghly Estuary between Nabadwip and Kakdwip', First National Environment Congress. Delhi: Indian Environment Congress Association and Department of Environment, Government of India.

Gopal, B. 2000. 'River Conservation in the Indian Subcontinent', in P. J. Boon, B. R. Davies and G. E. Pelts (eds), *Global Perspectives on River Conservation: Science, Policy and Practice*, pp. 233–61. New York: Wiley and Sons Ltd.

Hamilton, B. 1822. *An Account of Fishes Found in the River Ganges and its Branches*. Edinburgh: A. Constable and Co.

Hassan, S. S. 1999. 'The Current Status of the Fish Stock of Commercial Importance in River Ganga in and around Patna'. Unpublished Ph.D. Thesis, Patna University.

Hassan, S. S., R. K. Sinha, S. N. Ahsan and N. Hassan. 1998. 'Impact of Fishing Operations and Hydrological Factors on Recent Fish Catch in Ganga Near Patna, India', *Journal of Inland Fisheries Society of India*, 30 (1): 1–12.

Hassan, S. S., R. K. Sinha, N. Hassan and I. Ahsan. 1998. 'The Current Seasonal Variation in Catch Diversity and Composition of Fish Communities vis-à-vis Various Factors in the Ganges at Patna (India) and Strategies for Sustainable Development', *Journal of Freshwater Biology* 10 (3–4): 141–57.

Hinrichsen, D. and H. Tacio. 1997. *The Coming Freshwater Crisis is Already Here from Finding the Source: The Linkages Between Population and Water*. Washington, D.C.: Woodrow Wilson Centre.

Industrial Toxicology Research Centre (ITRC). 1992. *Sixth Annual Progress Report: Measurements on Ganga River Quality — Heavy Metals and Pesticides*. Lucknow: ITRC.

Jain, S. K. 2008. 'Impact of Retreat of Gangotri Glacier on the Flow of Ganga River, *Current Science*, 95 (8): 1012–14.

Jhingran, A. G. 1988. *Impact of Environmental Perturbations on the Fisheries Ecology of River Ganga — A Synopsis*. Barrackpore: CIFRI.

Junk, W. J. P. B. Bayley and R. E. Sparks. 1989. 'The Flood Pulse Concept in River-floodplain Systems', in D. P. Dodge (ed.), *Proceedings of the International Large River Symposium, Canadian Special Publication of Fisheries and Aquatic Sciences*, 106: 110–27.

Kannan, K., R. K. Sinha, S. Tanabe, H. Ichihasi and R. Tatsukawa. 1993. 'Heavy Metals and Organochlorine Residues in Ganges River Dolphin', *India Marine Pollution Bulletin*, 26 (3): 159–62.

Kannan, K., K. Senthilkumar and R. K. Sinha. 1997. 'Sources and Accumulation of Butyltin Compounds in Ganges River Dolphin', *Platanista gangetica Applied Organometallic Chemistry*, 11 (3): 223–30.

Khan, M. A., R. S. Panwar, A. Mathur and R. Jetly. 1996. 'Investigation of Bomonitoring and Eco-restoration Measures in Selected Stretches of the River Ganga and Yamuna'. Final Technical Report (1993–1995) submitted to the Ministry of Environment and Forests, Government of India, Delhi.

Khanna, D. R. 1993. *Ecology and Pollution of the Ganga River*. Delhi: Ashish Publishing House.

Krishnamurti, C. R., K. S. Bilgrami, T. M. Das, and R. P. Mathur. 1991. *The Ganga: A Scientific Study*. Delhi: Northern Book Centre.

Krishnaswami, S. and S. K. Singh. 2005. 'Chemical Weathering in the River Basins of the Himalaya, India', *Current Science*, 89 (5): 841–49.

Kumari, A., R. K. Sinha and K. Gopal. 2001a. 'Concentration of Organochlorine Pesticide Residues in Ganga Water in Bihar, India', *Environment and Ecology*, 19 (2): 351–56.

Kumari, A., R. K. Sinha and K. Gopal. 2001b. 'Organochlorine Contamination in the Fish of the River Ganges, India', *Journal of Aquatic Ecosystem Health and Management*, 4 (4): 505–10.

Lakshminarayana, J. S. S.1965. 'Studies on the Phytoplankton of the River Ganges, Varanasi, India. II. The Seasonal Growth and Succession of the Plankton Algae in the River Ganges', *Hydrobiologia* 25 (1/2): 138–65.

Lisitzen, A. P. 1972. 'Sedimentation in the World Oceans', *Society of Economic Paleontology Mineralogy Special Publication*, 17: 218.

Markandya, A. and M. N. Murty. 2000. *Cleaning-up the Ganges: A Cost-Benefit Analysis of the Ganga Action Plan*. Delhi: Oxford University Press.

Mohan, R. S. L. 1989. 'Conservation and Management of the Ganges River Dolphin, Platanista gangetica', in W. F. Perrin, R. L. Brownell Jr., Z. Kaiya and L. Jiankang (eds), *Biology and Conservation of River Dolphins*, pp. 64–69. Gland: IUCN.

Natarajan, A. V. 1989. 'Environmental Impact of Ganga Basin Development on Gene-pool and Fisheries of the Ganga River System', in D. P. Dodge (ed.), *Proceedings of the International Large River Symposium, Canadian Special Publication on Fisheries Aquatic Sciences*, 106: 545–60.

Nesemann, H., G. Sharma and R. K. Sinha. 2003. 'The Bivalvia Species of the Ganga River and Adjacent Stagnant Water Bodies in Patna (Bihar, India) with Special Reference in Unionacea', *Acta Conchyliorum*, 7: 1–43.

———. 2004. 'Aquatic Annelida (Polychaeta, Oligochaeta, Hirudinea) of the Ganga River and Adjacent Water Bodies in Patna (India: Bihar), with Description of a New Leech Species (Family Salifidae)', *Annals of Natural History Museum Wien*, 105 B: 139 –87.

Nesemann, H., S. Sharma, G. Sharma, S. N. Khanal, B. Pradhan, D. N. Shah and R. D. Tachamo. 2007. *Aquatic Invertebrates of the Gangariver System [Mollusca, Annelida, Crustacea (in part)]*, vol. 1. Kathmandu: H. Nesemann.

National River Conservation Directorate (NRCD). 2006. 'Water Quality Bulletin on River Water Quality Monitoring and Performance Monitoring of Sewage Treatment Plants', Ministry of Environment and Forests, Government of India, Delhi.

Oki, T. and S. Kanae. 2006. 'Global Hydrological Cycles and World Water Resources', *Science*, 313 (5790): 1068–72.

Pahwa, D. V. 1979. 'Studies on the Distribution of the Benthic Macrofauna in the Stretch of River Ganga', *Indian Journal of Animal Science*, 49 (3): 212–19.

Pahwa, D. V. and S. N. Mehrotra. 1966. 'Observations of Fluctuations in the Abundance of Plankton in Relation to Certain Hydrobiological Conditions of River Ganges', *Proceedings of National Academy of Sciences*, B36 (3): 157–89.

Payne, A. I., R. K. Sinha, H. R. Singh and S. Haq. 2004. 'A Review of the Ganges Basin: Its Fish and Fisheries', in: R. L. Welcomme and T. Petr (eds), *Proceedings of the 2nd Large River Symposium* (LARS), pp. 229-251. Bangkok: FAO and MRC, RAP Publication.

Pringle, C. M. 1997. 'Exploring How Disturbance is Transmitted Upstream: Going Against the Flow', *Journal of the North American Benthological Society*, 16 (2): 425–38.

Pringle, C. M., F. N. Scatena, P. Paaby-Hansen and M. Nunez-Ferrera. 2000. 'River Conservation in Latin America and the Caribbean', in P. J. Boon, B. R. Davies and G. E. Pelts (eds), *Global Perspectives on River*

Conservation: Science, Policy and Practice, pp. 41–77. New York: Wiley and Sons Ltd.

Raina, V., A. Chowdhury and S. Chowdhury. 1997. *The Dispossessed: Victims of Development in Asia*. Hong Kong: Arena Press.

Rao, P. 2007. 'Himalayas Retreat of the Glaciers', *The Hindu Survey of the Environment 2007*, special issue, pp. 19–23, Chennai.

Rao, R. J. 2001. 'Biological Resources of the Ganga River, India', *Hydrobiologia*, 458 (1–3): 159–68.

Ray, P. 1998. *Ecological Imbalance of the Ganga River System: Its Impact on Aquaculture*. Delhi: Daya Publishing House.

Reeves, R. R. and S. Leatherwood. 1994. 'Dams and River Dolphins: Can They Co-exist?', *Ambio*, 23 (3): 172–75.

Richter, B. D., J. V. Baumgartner, J. Powell and D. P. Braun. 1996. 'A Method for Assessing Hydrologic Alteration within Ecosystems', *Conservation Biology*, 10 (4): 1163–74.

Roxburgh, W. 1801. 'An Account of a New Species of Delphinus, an Inhabitant of the Ganges', *Asiatick. Res.* 7: 170–74.

Saha, T., P. B. Ghosh and M. Banerjee. 2004. 'Investigation on the Metal Concentration in the Water and River Bed Sediments of River Hooghly'. Final Technical Report submitted to the State Pollution Control Board, West Bengal. Kolkata: Institute of Wetland Management and Ecological Design.

Schwarzenbach, R. P., B. I. Escher, K. Fenner, T. B. Hofstetter, C. A. Johnson, U. V. Gunten and B. Wehrli. 2006. 'The Challenge of Micropollutants in Aquatic Systems', *Science*, 313 (5790): 1072–77.

Senthilkumar, K., K. Kannan, R. K. Sinha, S. Tanabe and J. P. Giesy. 1999. 'Bioaccumulation Profiles of Polychlorinated Biphenyl Congeners and Organochlorine Pesticides in Ganges River Dolphins', *Environmental Toxicology and Chemistry*, 18 (7): 1511–20.

Sharan, R. K. and R. K. Sinha. 1988. *Ganga Basin Research Project: Final Technical Report*. Patna: Patna University.

Shetty, H. P. C., S. B. Saha and B. B. Ghosh. 1961. 'Observations on the Distribution and Fluctuations of Plankton in the Hooghly-Matlah Estuarine System with Notes on their Relation to Commercial Fishlandings', *Indian Journal of Fisheries*, 8 (2): 326–63.

Shiva, V. 2002. *Water Wars: Privatisation, Pollution and Profit*. Cambridge: South End Press.

Singh, P., U. K. Haritashya, K. S. Ramasastri and N. Kumar. 2005. 'Prevailing Weather Conditions during Dummer Seasons Around Gangotri Glacier', *Current Science*, 88 (5): 753–60.

Singh, M. 1996. *The Ganga River: Fluvial Geomorphology, Sedimentation Processes and Geochemical Studies*. Heidelberg: Heidelberger Beitr. Umwelt-Geochemie.

———. 1998. 'Fisheries in the Coastal Areas of West Bengal and the Required Conservation'. Paper presented at National Seminar on Management of

Coastal Ecosystem in India, 8–10 August, CARI, Port Blair, Andaman Island.

Singh, M. 2000. 'Inland Aquatic Resources of India, Issues and Threats — A Fisheries Perspective', in M. Sinha, B. C. Jha and M. A. Khan (eds), *Summer School on Environment Impact Assessment of Inland Waters for Sustainable Fisheries Management and Conservation of Biodiversity*, pp. 1–9. Barrackpore: CIFRI.

Sinha, M., D. K. De and B. C. Jha. 1998. *The Ganga: Environment and Fishery*. Barrackpore: CIFRI.

Sinha, M. and M. A. Khan. 2001. 'Impact of Environmental Aberrations on Fisheries of the Ganga (Ganges) River', *Journal of Aquatic Ecosystem Health and Management*, 4 (4): 493–504.

Sinha, R. K. 1995. 'Commercial Exploitation of Freshwater Turtle Resource in the Middle Ganges River System in India', Proceedings of International Congress of Chelonian Conservation – Gonfaron, 6–10 July, France.

———. 1996. 'Bioconservation of the Gangetic Dolphin, *Platanista gangetica*'. Final Technical Report submitted to the National River Conservation Directorate, Ministry of Environment and Forests, Government of India, Delhi.

———. 1997. 'Status and Conservation of Ganges River Dolphin in the Bhagirathi-Hooghly River Systems in India', *International Journal of Ecology and Environmental Science*, 23 (4): 343–55.

———. 1999. 'The Ganges River Dolphin — A Tool for Baseline Assessment of Biological Diversity in River Ganges, India'. Final Technical Report for The Biodiversity Support Programme (BSP) — A Consortium of Wildlife Fund, The Nature Conservancy and The World Resources Institute, Washington, D.C.

———. 2000. 'Status of the Ganges River Dolphin (Platanista gangetica), in the Vicinity of Farakka Barrage, India', in R. R. Reeves, B. D. Smith and T. Kasuya (eds), *Biology and Conservation of Freshwater Cetaceans in Asia*, pp. 42–48. Gland: IUCN.

Sinha R. K., B. D. Smith, G. Sharma, K. Prasad, B. C. Choudhary, K. Sapkota, R. K. Sharma and S. K. Behera. 2000. 'Status and Distribution of the Ganges Susu Platanista gangetica, in the Ganges River System of India and Nepal', in R. R. Reeves, B. D. Smith and T. Kasuya (eds), *Biology and Conservation of Freshwater Cetaceans in Asia*, pp. 54–61. Gland: IUCN.

Sinha, R. K. and G. Sharma. 2001. 'New Records of a Gastropod, Stenothyra ornate Annandale and Prashad, 1921 from the River Ganga in Bihar', *Journal of Bombay Natural History Society*, 98: 485–89.

Sinha, R. K., K. Prasad, G. Sharma and R. Dalwani. 2001. 'Ecological Restoration of the River Ganga', *International Journal of Ecology and Environmental Science*, 27: 127–35.

Sinha, R. K. and G. Sharma. 2003a. 'Faunal Diversity of the River Sarda, Uttar Pradesh, India', *Journal of Ecophysiology and Occupational Health*, 3 (1–2): 103–16.

Sinha, R. K. and G. Sharma. 2003b. 'Current Status of Ganges Dolphin, Platanista Gangetica in River Son and Kosi in Bihar', *Journal of Bombay Natural History Society*, 100 (1): 27–37.

Sinha, R. K., H. Nesemann and G. Sharma. 2003. 'New Records of *Physa* (Gastropoda: Physidae) from Indian Sub-continent', *Club Conchylia Informationen* 34 (4/6): 3–11.

Sinha, R. K., S. K. Sinha, D. K. Kedia, A. Kumari, N. Rani, G. Sharma and K. Prasad. 2007. 'A Holistic Study on Mercury Pollution in the Ganga River System at Varanasi India', *Current Science*, 92 (9): 1223–28.

Smith, B. D., R. K. Sinha, U. Regmi and K. Sapkota. 1994. 'Status of Ganges River Dolphins (Platanista Gangetica) in the Karnali, Mahakali, Narayani and Sapta Kosi Rivers of Nepal and India in 1993', *Marine Mammal Science* 10 (3): 368–75.

Smith, B. D., R. K. Sinha, Z. Kaiya, A. A. Chaudhry, L. Renjen, W. Ding, B. Ahmed, A. K. M. Aminul Haque, R. S. L. Mohan and K. Sapkota. 2000. 'Register of Water Development Projects Affecting River Cetaceans in Asia', in R. R. Reeves, B. D. Smith and T. Kasuya (eds), *Biology and Conservation of Freshwater Cetaceans in Asia*, pp. 22–39. Gland: IUCN.

Sparks, R. E. 1995. 'Need for Ecosystem Management of Large Rivers and their Floodplains', *BioScience*, 45 (3): 168–82.

Talwar, P. K. and A. G. Jhingran. 1991. *Inland Fisheries of India and Adjacent Countries*, 2 vols. Delhi: Oxford and IBH Publishing Co.

Tare, V., P. Bose and S. K. Gupta. 2003a. 'Suggestions for a Modified Approach Towards Implementation and Assessment of Ganga Action Plan and other Similar River Action Plans in India', *Water Quality Research Journal of Canada*, 38 (4): 607–26.

Tare, V., A. V. S. Yadav and P. Bose. 2003b. 'Analysis of Photosynthetic Activity in the Most Polluted Stretch of River Ganga', *Water Research* 37 (1): 67–77.

Thayyen, R. J. 2008. 'Lower Recession Rate of Gangotri Glacier during 1971–2004', *Current Science*, 95 (1): 9–10.

Walker, K. F., F. Sheldon and J. T. Puckridge. 1995. 'A Perspective on Dryland River Ecosystems', *Regulated Rivers: Research & Management*, 11 (1): 85–104.

Wasson, R. J. 2003. 'A Sediment Budget for the Ganga-Brahmaputra Catchment', *Current Science*, 84 (8): 1041–47.

Welcomme , R. L. 1985. *River Fisheries*. Food and Agriculture Organization (FAO) Fisheries Technical Paper No. 262. Rome: FAO.

———. 1992. 'River Conservation: Future Prospects', in P. J. Boon, P. Calow and G. E. Petts, (eds), *River Conservation and Management*, pp. 454–562. New York: John Wiley and Sons.

Zoological Survey of India (ZSI). 1991. *Faunal resources of Ganga. Part I.* Calcutta: Zoological Survey of India.

6

Impact of Climate Change on Water Resources in South Asia with Special Reference to Bangladesh

Md. Golam Rabbani, A. Atiq Rahman
and Shymal Chandra Bhadra

South Asia is home to the largest population in the world. Millions of people live along with Ganges, Brahmaputra and Meghna (GBM) and other river basins. It has been reported that over 500 millions people depend on GBM for irrigation, fisheries, forestry, livestock, domestic water use, etc. In reality, livelihoods of human population, development, economic sustainability, etc., of all South Asian countries are mostly dependent on water resources. Availability of safe water for drinking and sanitation is critically linked with the health of South Asian people. According to a World Bank report, about 20 per cent of South Asian population lacks access to water services (World Bank 2008). This might become higher in future with increasing growth rate of population and impacts of climate change. The South Asian population is expected to rise to 1,800 million in 2015 and 2,000 million in 2025 (GWP 2006). Another report states that over 27 per cent of South Asian citizen are without adequate food; this number may be higher in India and Bangladesh (Rahman et al. 2007). Climate change and climate variability may aggravate the situation in the future. Changes in climate may directly and indirectly cause decreased water availability and deterioration of water quality, decrease in reliability of hydropower and biomass production, have adverse impacts on fisheries biology and aquatic ecosystem, cause increase incidences of waterborne diseases (diarrhea, cholera, dysentery, etc.), reduction of agriculture crops, increase morbidity and mortality, increase demand and consumption of water due to increase of temperature, increase pressure on groundwater, etc. Ultimately, the most vulnerable population would be the poor of this region, particularly the women and children.

This chapter is composed of six sections with a number of sub-sections. Section 1 presents a very brief overview of the impacts of climate change on water and relevant sectors in South Asia. Section 2 is an overview of the basic concepts and science of climate change. Section 3 includes the observed changes in the climate system and the parameters in South Asia including Bangladesh. Section 4 highlights the impacts of climate change on water resources and the connected ecosystems in South Asia with special reference to Bangladesh. Section 5 gives an overview on the global and regional responses to address climate change. Section 6 provides concluding remarks.

A Brief on the Science of Climate Change

Climate Change is basically the changes of climate over a time period that usually ranges from decades to centuries. The term climate change refers to both natural and human-induced changes. On the other hand, the term 'climate variability' refers to short-term changes of climate parameters. The two official definitions of climate change are as follows:

- Climate change refers to any change in the climate over time, whether due to natural variability or as a result of human activity (Intergovernmental Panel on Climate Change or (IPCC).
- A change of climate which is attributed directly or indirectly to human activity that alters the composition of the global atmosphere and which is in addition to natural climate variability observed over comparable time periods (United Nations Framework Convention on Climate Change or UNFCCC).

There are many natural and human origin factors which determine the climate of the earth. A number of gases in the atmosphere absorb most of the infrared heat energy transmitted by the earth towards space. This phenomenon is called as the 'greenhouse effect' and the gases include water vapor (H_2O), carbon dioxide (CO_2), nitrous oxide (N_2O), methane (CH_4) and ozone (O_3) (Jain et al., 2004). Moreover, a number of human-made gases including Sulphur hexafluoride (SF_6), Hydro fluorocarbons (HFCs) and Per fluorocarbons (PFCs) are also responsible for changing the climate system. In fact, there are two gases including water vapor and carbon dioxide that contribute most significantly to global warming. The other gases including methane, nitrous oxide, ozone and sulfur hexafluoride contribute even in small quantities to the greenhouse effect.

The IPCC predicts that the globally averaged net effect of human activities since 1750 has been one of warming. It is estimated that the average global temperature will rise by 1.8°C to 4.0°C by the year 2100 (UNFCCC 2008). Hence it is the contention of scientists that climate change is a reality and quite inevitable because of past and currents emissions of Greenhouse Gases (GHGs). The climate does not respond immediately to external changes. However, after 150 years of industrialisation, global warming has gained momentum, and it will continue to affect the earth's natural systems including the water ecosystem for hundreds of years even if GHG emissions are reduced and temperature levels stop rising.

Since scientists predict that there will be continued warming and increases in sea levels with significant impacts on natural and human systems, the globe may suffer various impacts such as coastal flooding; heat waves; storms and droughts; less frost, snow and polar ice; more people at risk of food and water shortage; and reduced habitat for many plant and animal species and more people exposed to infectious diseases such as malaria, diarrhea, dengue, etc.

Some of the key messages of the Fourth Assessment Report (FAR) which provides assessment and evidences of changes in the climate system around the world may include:[4]

- Eleven of the last 12 years (1995–2006) rank among the 12 warmest years in the instrumental record of global surface temperature (since 1850).
- The total temperature increase from 1850–1899 to 2001–2005 is 0.76°C (0.57°C to 0.95°C).
- The average atmospheric water vapour content has increased since 1980s over land and ocean as well as in the upper troposphere. The increase is broadly consistent with the extra water vapour that warmer air can hold.
- Observations since 1961 show that the average temperature of the global ocean has increased to depths of at least 3,000 m and that the ocean has been absorbing more than 80 per cent of the heat added to the climate system. Such warming causes seawater to expand, contributing to Sea Level Rise (SLR).

- Mountain glaciers and snow cover have declined on an average in both hemispheres. Widespread decreases in glaciers and ice caps have contributed to SLR (ice caps do not include contributions from the Greenland and Antarctic ice sheets).
- Global average sea level rose at an average rate of 1.8 (1.3 to 2.3) mm per year over 1961 to 2003. The rate was faster over 1993 to 2003: about 3.1 (2.4 to 3.8) mm per year.
- Widespread changes in extreme temperatures have been observed over the last 50 years. Cold days, cold nights and frost have become less frequent, while hot days, hot nights and heat waves have become more frequent.

Observed Climate Change

Climate change related impacts including flood, drought, SLR, salinity, temperature and rainfall variations, etc., have become major concerns for most countries of the world due to its long-term implications and adverse effects on development activities. Although both the developed and developing countries are being affected, the developing and underdeveloped countries are more vulnerable to climate change and climate variability as this has a direct impact on economic, social and development sectors. It has further put additional pressure on the limited natural resources like water, land and biodiversity. It has been predicted that 'climate change impacts will be differently distributed among different regions, generations, age classes, income groups, occupations and genders' (IPCC 2001). The IPCC also notes: 'the impacts of climate change will fall disproportionately upon developing countries and the poor persons within all countries, and thereby exacerbate inequities in health status and access to adequate food, clean water and other resources'. It became an unkind or terrible reality for the world community to face extreme climatic events, e.g., recent prolonged and devastating floods in Bangladesh, India, China and in the UK; Cyclone Sidr in Bangladesh (2007), Cyclone Nargis in Myanmar (2008), Typhoon in Philippines (2008); severe drought in Asia and Africa, extreme heat waves in central Europe (e.g., 48°C temperature in Hungary on 24 July 2007, which killed over 500 people there), etc. However, Climate change may have adverse impacts on natural systems and sub-systems in different degrees in different countries of South Asia. Bangladesh would be the most vulnerable to the adversities of climate change in the region.

The Fourth Assessment Report (AR4) of IPCC and scientists state that different countries/sub-regions of Asia will show changes in climate parameters including variation in temperature and precipitation. The variability in rainfall trend has been observed during the past few decades all across Asia. Decreasing trends in annual mean rainfall are being observed in many parts of Asia including coastal belts and arid plains of Pakistan and parts of north-east India. But in Bangladesh, the annual mean rainfall exhibits increasing trends. However, a number of ill-effects — changes in extreme events and severe climate anomalies — have also been observed in South Asian countries. Tables 6.1. and 6.2 show some of the observed past and present climate trends and observed ill-effects in the recent past.

On the other hand, Bangladesh is exposed to adverse impacts of climate change due to its geophysical location, hydrological influence by erratic monsoon rainfall and changes in regional water flow patterns, etc. Most parts of the country face too much water in the monsoon and many parts of it face scarcity of water in the dry season. This situation may be aggravated with the warmer climate causing severe droughts and increasing floods. It is predicted that the temperature in Bangladesh may rise by 0.7°C in summer and 1.3°C during winter by 2030 whereas it would be 1.1°C in summer and 1.4°C in winter by 2050 (World Bank 2000). It was also predicted that the sea level is likely to rise to 30 and 50 cm by 2030 and 2050 respectively. It has been revealed that by the year 2030, an additional 14.3 per cent of the country would become extremely vulnerable to floods, while the existing flood-prone areas will face higher levels of flooding. Analysis of past floods suggest that about 26 per cent of the country is subject to annual flooding and an additional 42 per cent is at risk of floods with varied intensity (IPCC 2002).

For Bangladesh, to date, climate models have generally been consistent in simulating warming throughout the country in all seasons, moderate increases in monsoon rainfall and moderate decreases in dry season rainfall (Table 6.3). Most of the climate models estimate that precipitation will increase during the summer monsoon because they estimate that air over land will get warmer than air over oceans in the summer. This will deepen the low pressure system over land that happens anyway in the summer and will enhance the monsoon. It is notable that the estimated increase in summer precipitation appears to be significant. This does not mean that increased monsoon is certain, but this increases the chance that it is likely to happen.

Table 6.1: Summary of Key Observed Past and Present Climate Trends and Variability

Region	Country	Change in temperature	Change in precipitation
South Asia	India	0.68°C increase per century, increasing trends in annual mean temperature	Increased extreme rains in north-west in recent decades, lower number of rainy days along east coast
	Nepal	0.09°C rise in per year in Himalaya, and 0.04°C in Terai region, more in winter	No distinct long-term trends in precipitation records for 1948 to 1994
	Pakistan	0.6°C to 1°C rise in mean temperature in coastal areas since early 1900s	10% to 14% decrease in coastal belt and hyper arid plains, increase in summer and winter precipitation over the last 40 years in Northern Pakistan
	Bangladesh	Increasing trend of about 1°C in May and 0.5°C in November during the 14 year period from 1985 to 1998	Decadal rain anomalies above long term averages since 1960s
	Sri Lanka	0.016°C increase per year between 1961 and 1990 over entire country	Increasing trend in February and decreasing trend in June

Source: Cruz et al. (2007).

Table 6.2: Summary of Observed Ill Effects, Changes in Extreme Events and Severe Climate Extremes in South Asia

Sl.	Extreme events	Key trend
1	Drought	50% of droughts associated with El Nino
		Consecutive drought in 1999 and 2000 in Pakistan and north-west India led to sharp decline in water table
		Drought of Orissa in 2000 and 2002 affected 1 million people
		Drought in recent years affects millions of people in northern Bangladesh
2	Flood	Serious and recurrent floods in Bangladesh, Nepal and north-east states of India during 2002, 2003, 2004 and 2007
		A record of 944 mm of rainfall in Mumbai during 26–27 July 2005 caused deaths of about 1,000 people
		17 May 2003 floods in southern Sri Lanka was triggered by 730 mm of rain
3	Cyclone	Frequency of monsoon depressions and cyclone formations in the Bay of Bengal
4	Heat waves	Frequency of hot days and multiple-day heat wave has increased in the past century in India; increase in deaths due to heat stress in recent years in India Heat waves of June 2005 killed 30 people in Bangladesh

Source: Cruz et al. (2007); Chiang (2008).

Table 6.3: GCM Projections of Changes in Temperature and Precipitation for Bangladesh

Year	Temperature change (0°C) mean (Standard deviation)			Rainfall change (%) mean (Standard deviation)		
	Annual	*DJF*	*JJA*	*Annual*	*DJF*	*JJA*
Baseline average 2030	1.0 (0.11)	1.0 (0.18)	0.8 (0.16)	3.8 (2.30)	−1.2 (12.56)	4.7 (3.17)
2050	1.4 (0.16)	1.6 (0.26)	1.1 (0.23)	5.6 (3.33)	−1.7 (18.15)	6.8 (4.58)
2100	2.4 (0.28)	2.7 (0.46)	1.9 (0.40)	9.7 (5.8)	−3.0 (31.6)	11.8 (7.97)

Source: Agrawala et al. (2003).

Note: DJF represents the months of December, January and February, usually the winter months; JJA represents the months of June, July and August, the monsoon months.

Bangladesh will be particularly affected by climate change through many ways. Some of the main ones are given below (Wassmann et al. 2004; GoB, 2005; Stern 2007; Cruz et al. 2007):

1. Serious and recurrent floods due to melting of glacier and increased intensity of rainfall.
2. Increased intensity and frequency of droughts will affect agriculture the most.
3. Projected rise in sea level would affect millions of people in the coastal areas.
4. Frequency of monsoon depressions and cyclones formation in the Bay of Bengal has decreased but the rate of intensity is increasing causing severe damages to life and property.
5. Degradation of wetlands, biodiversity and ecosystems.
6. Climatic changes in Bangladesh may contribute to land degradation, shortfalls in food production, rural poverty and urban unrest.
7. Diarrhoeal diseases and outbreaks of other infectious diseases (e.g., cholera, hepatitis, malaria, dengue fever) to be influenced by climate-related factors such as severe floods, droughts, temperature variation, etc.
8. An increase of 10 to 20 per cent in intensities of tropical cyclone will lead to a rise in sea-surface temperature of 2 to 4°C.
9. Salinity intrusion is already affecting both land and water areas along the coastal zone of the country. About 1.2 million hectares out of 2.85 million hectares have been affected by varying degrees of soil salinity.

Climate Change Impacts on Water Ecosystem in South Asia

Geographically, South Asia is dominated by the largest of the South Asian countries, India, along with the Himalayan mountains and the river networks and huge river basins of the GBM as mentioned above (Rahman et al. 2007). This river system dominates the water flow of India, Bangladesh, Nepal and Bhutan. The Indus river basin is shared between Pakistan and India which dominates northern part of the subcontinent.

Water ecosystem in South Asia has many components such as coastal and marine resources, rivers, lakes and huge numbers of ponds and smaller water bodies which are accessed by the communities across South Asia. There are cultural practices that have developed around these water bodies. Groundwater also is a major source of drinking water and it is also used by different sectors including agriculture, industry, etc. On the other hand, about 15,000 (about 11 per cent of the total in the world) Himalayan glaciers support the GBM basin with its stored 12,000 km³ freshwater (Cruz et al. 2007). Coincidently, about 11 per cent of the total human population of this region (including Bangladesh, Nepal, Bhutan, India and Pakistan) depends on this 11 per cent of the total glaciers in the world for their livelihood support. Regarding Bangladesh, it is basically criss-crossed by many rivers and its peoples lives are dominated by such rivers as the country is situated in the lower riparian of the GBM basins. These rivers drain more than 90 per cent of the total run-off from the GBM basins (BCAS 2003).

The rapid growth of population in South-Asia is puttting additional pressure on the natural resources including water. Extreme levels of pollution, water scarcity, natural calamities, infrastructure development coupled with weak governance, management issues and also international water-sharing conflicts have made water resources sector the poorest in terms of development management and conservation in this region. In addition, climate change has put further pressure on water resources or the water ecosystem in the region and also in the country. The following sections state the ways in which climate change may affect the water ecosystem of South Asia including Bangladesh.

Anomalies of Climate Induced Water Related Hazards

Flood

Increased rate of sedimentation, melting of glaciers and snow from the Himalayan permafrost and increasing volume of water in the sea due to rise in temperature will force more water to flow through the GBM river systems and their river networks. This may create additional flooding extending over the flood plain areas across South Asia including Bangladesh. Rainfall may be another factor for causing and intensifying flooding in this region. This not only adds to the existing flood and water-logging problems but also causes flash floods in the hilly areas of different countries, especially in Bangladesh. However, a brief description of the flood-prone areas in different countries with examples of flood events in the past and also predictions of what is going to happen in future are given below.

In India, out of 40 million hectares of its flood-prone area, floods affect an area of about 7.5 million hectare per year on an average (MoWR 2002). Of the total estimated flood-prone area in India, 68 per cent lies in the GBM states, mostly in Assam, West Bengal, Bihar and Uttar Pradesh. The Ganges, located in northern India, receives water from its northern tributaries originating in the Himalayas, and causes high levels of flood, especially in Uttar Pradesh and Bihar. Likewise, the Brahmaputra and the Barak (headwaters of the Meghna) causes flood and drainage congestion in northeastern India with overflows of rainwater.

In Nepal, sudden cloudburst causes floods in hill valleys. It has been observed that major floods are induced by glaciers in higher mountains, i.e., Glacier Lake Outburst Floods (GLOF) (Ahmad et al. 2001). In addition, rockslides and landslides aggravate flooding problems by reducing river flow capacity. The Nepalese *Terai* (valley) region is prone to flash floods, which also produce spillover effects in northern India.

Floods and landslides are two major threats for Bhutan. GLOF have increasingly become a threat for this country due to global warming. This might have serious impact on life, property and development of future infrastructure.

Pakistan and Sri Lanka are already facing flood problems and these may increase in future due to climate change. Flash floods have recently caused damages amounting to millions in the wet zone in Sri Lanka (Government of Sri Lanka 2008). The flood caused by excessive rainfall (730 mm) in southern province of Sri Lanka on 17 May 2003 affected lives and livelihoods. The flood of 1992 in Pakistan was one of the major extreme events (Government of Pakistan 2008).

Maldives has identified coastal floods as one of the main environmental problems in the country (Government of Maldives 2008). This country has already been predicted to be adversely affected by inundation in future.

Flood is an annual event in Bangladesh. The general flood plain topography (see Annexure 1) is the main reason for inundation of large areas of the country. In fact, the country is a low-lying delta formed at the confluence of three large rivers: the Ganges, the Brahmaputra and the Meghna. Both experience and experiments show that about 20 per cent of the country gets inundated by overflowing rivers during monsoon, which is termed as a normal flood year. In case of about 35 per cent inundation, it is a moderate flood year, and for more than 60 per cent inundation, it is a major flood year (Salehin et al. 2007). On the other hand, it is predicted that by the year 2030, an additional 14.3 per cent of the country would become extremely vulnerable to floods, while the existing flood-prone areas will face higher levels of flooding. Analysis of past floods suggests that about 26 per cent of the country will be subject to annual flooding and an additional 42 per cent will be at risk of floods with varied intensity (IPCC 2002). In addition, a 10 per cent increase in monsoon rainfall in Bangladesh could increase overflow depth by 18 to 22 per cent, resulting in an extremely wet year (Qureshi and Hobbie 1994). Another study indicates that the monsoon rainfall could increase by 11 and 28 per cent by 2030 and 2050 respectively, causing surface run-off increases by 20 and 45 per cent, respectively (Ahmed and Alam 1998). To sum it up, the normal flood event is predicted to be aggravated by climate change and affect the country.

Drought

Drought is one of the major climate-related hazard for many countries of the world. India, Pakistan, Bangladesh are the most drought affected countries in South Asia. It has been predicted that drought would throw up water-resource challenges in affected areas, and will lead to shifts in locations of population and economic activities, and additional investments in water supply (Parry et al. 2007).

In India, climate change may exacerbate water shortages, especially during the dry season. The country is already struggling with water scarcity. It has been predicted that the gross per capita water availability in India will decline from about 1,820 m3/year in 2001 to as low as about 1,140 m3/year in 2050 (Gupta and Deshpande 2004). On the other hand, droughts in 1999 and 2000 caused sharp decline of water tables in north-west India. It also caused mass starvation, damaged crops and affected 11 million people in Orissa (Cruz et al. 2007).

Drought is also affecting Balochistan, Sindh, Punjab and some other areas in Pakistan. The human lives, livelihood opportunities, human health problems and eco-biological diversity are being severely affected. It has been predicted that climate change would aggravate the water shortage, water supply and water quality problems in the country (Cruz et al. 2007). It appears that the consecutive droughts in 1999 and 2000 in Pakistan led to a sharp decline in water tables. It resulted in the drying up of the wetlands and caused degradation of the ecosystem. A study indicates that about 2 million people, 9.31 million livestock and 2 million acres of land were affected by the 2001 drought (Brohi 2002).

Drought is one of the major climate-related hazards for Bangladesh. In the last 60 years, the country experienced about 24 drought events. Of them, 11 were very severe (WARPO 2005). Bangladesh faces long spells of dry weather, and moderate to severe droughts affect a huge area in the northern part of the country including Rajshahi, Natore, Chapai Nawabganj, Rangpur, Dinajpur, Bogra, Kushtia, Jessore and Dhaka. The government made a major effort and invested in agricultural development over the last two decades in the Barind area (in Rajshahi). This effort shows an increasing trend and has been considered as a successful measure in boosting agricultural productivity. However, most of these efforts may be challenged by the prediction made regarding increase drought in northwest Bangladesh. In Bangladesh, about 18 per cent of Rabi (mid-October to mid-March) and 9 per cent kharif (mid-July to mid-October) crops are highly vulnerable to the drought, and it may increase in future due to climate change. Drought actually affects rice, wheat, pulses and potatoes, especially where irrigation possibilities are limited. However, existing and future drought-prone areas of Bangladesh are shown in Annexure 2.

SLR and Intrusion of Salinity

The SLR and/or salinity intrusion is a problem being faced by many of the countries in South Asia. It is projected that the possible SLR may severely affect this region including Bangladesh, India, Sri Lanka and Pakistan. It has been reported that a vast and diverse coral reef of South Asia were lost in 1998 due to coral bleaching induced by the 1997–1998 El Niño event (Wilkinson 2000; Arceo et al. 2001; Wilkinson 2002). A recent report shows that Bangladesh would face the largest impacts due to SLR (World Bank 2007). An overview of the impacts of SLR and salinity intrusion in South Asia countries is given in the following section.

The possible SLR may affect Bangladesh by inundating coastal areas. It has been predicted that by 2030 and 2050, the sea level will rise by atleast 30 and 50 cm, respectively (World Bank 2000).

The recent report shows that if the sea level rises by 25 cm, then 40 per cent of Sundarban will be submerged, and in case of SLR by above 60 cm, the whole of Sundarban will disappear (Hare 2003). In fact, the SLR is likely to inundate the coastal wetlands, lowlands, accentuate coastal erosion, increase frequent and severe floods, create drainage and irrigation problems and finally dislocate millions of people from their homes and occupation (Rahman et al. 2008). This may catalyze the increasing rate of rural–urban migration within the country. An estimation based on a coarse digital terrain model and global population distribution data shows that more than 1 million people will be directly affected by SLR in 2050 in each of the GBM delta in Bangladesh (Ericson et al. 2005; Cruz et al. 2007). On the other hand, increase in salinity became one of the major problems for the coastal zones of Bangladesh. This may be happening due to low flow of freshwater from the Ganges and ingress of salt water from Bay of Bengal. So the compound effect of SLR and salinity may disrupt agriculture (e.g., reduction of rice), lead to destruction of mangroves including the Sunderbans and coastal ecosystem and create additional health problems in the local communities. The recent reports state that the coastal community may suffer more from water-borne diseases and other physical problems (e.g., menstruation problems of the women from drinking of saline water) due to SLR and salinity intrusion (BCAS 2007). However, the poor and marginal groups would be critically affected by the possible SLR and salinity intrusion in coastal zone of Bangladesh (see the map in Annexure 3 for predicted SLR and salinity intrusion in Bangladesh).

Like Bangladesh, projected SLR would cause losses of coastal ecosystems and dislocate many people in India too, especially from Kolkata and Mumbai. It has been reported that the potential impacts of one metre SLR include inundation of 5,763 km^2 in India (TERI 1996). However, according to the recent report of the World Bank, India would be the least affected country in terms of impact on area, wetlands, damage to agriculture, etc., due to predicted SLR (World Bank 2007).

In Pakistan, SLR may affect different coastal areas including Karachi, Sindh and Baluchistan. The recent World Bank report states that Pakistan would be the least impacted country due to SLR in terms of urban damage, affected population, GDP and total land area. On the other hand, agricultural impacts due to SLR in Pakistan may be higher than in India (ibid.).

In Sri Lanka, it has been reported that about 60 per cent of the total population lives in the coastal region, more than 80 per cent of the industries are located along the coastal zone and huge utility services and infrastructures have been developed and established here. Now the climate change predictions say that SLR and salinity intrusion would severely affect these areas of Sri Lanka (Government of Sri Lanka 2008). These may affect availability of freshwater, cause loss of livelihoods and may lead to inundation resulting in increase suffering of poor people, increase in rate of rural–urban migration, loss of both agricultural and industrial production, etc.

Maldives, which is only 2.4 meter above the sea level, considers the climate change issues as grave. Predictions of global warming causing melting of ice caps and glaciers and SLR, threaten the existence of entire Maldives (Hoffman 2007).

Increase in Cyclones and Storm Surges

One of the predictions of the IPCC AR4 is the intensification of the extreme weather events such as cyclones and associated storm surges. There are evidences of decreasing frequency of monsoon depression and formation of cyclone but increase of intensity in the Bay of Bengal since 1970 (Lal 2003; Cruz et al. 2007). The acceleration in intensities of wind velocity is expected to inflict greater losses to vulnerable communities and ecosystems in South Asia including Bangladesh. In addition, increase in SLR will bring the waterline further inwards. Consequently, the affect of storm surge will penetrate deeper into the landmass. As population increases, both numbers of affected people and investment in infrastructure will incur greater losses. These are going to affect agricultural production, health, and loss of livelihoods and increase in poverty of this region. The vulnerability of community and ecosystem in Bangladesh is again higher than the other countries.

The most recent Cyclone Sidr of 2007 killed 2,997 people and affected over 6 million in Bangladesh. Nearly 0.3 million homes were destroyed and about 0.9 million damaged. Also, crops on about 0.35 million hectares of land had been destroyed. The super cyclone of 1999 with the storm surge of 26 feet (8 meters) struck the coast of Orissa, travelling up to 20 km inland. The Orissa cyclone killed 10,000 people in India and destroyed crops on about 17,110 km^2 of land and an additional 90 million trees were either uprooted or snapped. It destroyed about 275,000 homes.

The cyclone and storm surges also hit Pakistan and Sri Lanka at different times but the impacts were lower than those in Bangladesh and India. However, the climate change could increase the intensity of cyclone events and threatens this whole region, not only Bangladesh or India.

Quality of Water Resources

Pollution of both surface water and groundwater is a matter of concern all over the world. Industrial waste, municipal solid waste, sewage, etc., discharged into freshwater and coastal areas cause changes in the water ecosystem. It may serve as a reservoir for pathogens that may result in causing human health problems, increase in suspended solids, nutrient inputs and higher levels of biochemical oxygen demand (BOD). In addition, huge amounts of agricultural waste, including chemical fertilisers and pesticides, reach the water stream throughout the year, especially during the rainy season. The combined effects of increase in temperature of surface water and chemical and biological waste may be favourable for eutrophication but quite unfavourable for all aquatic organisms and the ecosystem. In addition, salinity intrusion would be another factor for the deterioration in water quality in different countries of the region. In fact, it is already affecting Bangladesh, India, Sri Lanka and Maldives to different extents. On the other hand, climate-related events like flood, cyclone and storm surge bring water from pollution point to non-pollution point within the country or even outside. These extreme events act like a vehicle for transferring pollution risk. Many ponds or water bodies become unusable due to flood and cyclone events in Bangladesh. For example, over 6,000 ponds have been contaminated with saline water due to Cyclone Sidr. This is basically a common feature of the areas which get inundated by flood water in any country of South Asia including Bangladesh and India (West Bengal). All these ultimately affect agriculture, domestic, industrial and other relevant sectors, e.g., Water Supply and Sanitation (WATSAN) in the region. However, the quality of water resources may severely deteriorate due to climate change and climate variability factors in this region including Bangladesh.

Excessive Exploitation/Utilisation of Water for Development

Due to rapid urbanisation, industrialisation, population growth, increased consumption, inefficient use, irrigation uses, domestic use, etc., the demand for water has tremendously increased in each of the South Asian countries. To meet this demand, groundwater is being overexploited in many countries including Bangladesh. In fact, irrigation, industry, domestic sectors, etc., are mostly dependant on groundwater sources in Bangladesh and probably in other South Asia countries as well. This may be the cause of declining groundwater level in many places of the country, especially in the urban areas, e.g., Dhaka. On the other hand, utilisation of surface water is limited in almost all the countries in the region. This is probably due to lack of service and infrastructure facilities, contamination of surface water, difficulties (e.g., too far) in accessing surface water, etc. However, it has been predicted that due to increase of temperature, demand for water would be higher, again specifically in urban areas.

Threat to Biodiversity

The changes in surface water temperature, SLR, salinity intrusion and relevant climate factors would result in putting pressure on the aquatic ecosystem of different countries of the South Asia region. As mentioned in the earlier section, the combined effects of increase of temperature and waste from agriculture and domestic sectors may facilitate growth of algal bloom and eutrophication and this might have an adverse impact on sensitive species. It has been reported that corals are very sensitive in sea surface temperature. In 1998, vast and diverse coral reefs of South Asia were lost due to El Nino. According to the IPCC, drought and decline of precipitation in South Asia including Bangladesh, India and Pakistan are causing drying up of wetlands and degradation of the aquatic ecosystem (Cruz et al. 2007). These are in fact affecting aquatic biodiversity of the region.

Reduction of Agricultural Production

One of the major challenges for the South Asia countries would have to face is reduction of agricultural products which may occur due to climate-related events including variation in temperatures, rainfall, drought, flood, cyclone and storm surge, salinity intrusion, etc. In fact, all the countries are more or less facing this extreme conditions of the climate.

The most recent report states that the major food growing regions of India including Haryana, Punjab, western Uttar Pradesh, Tamil Nadu, etc., are facing climate-related problems (Kelkar and Bhadwal 2008). It has been predicted that the production of wheat in central India

Box 6.1: Climate Change Impacts on Sundarban!

The World Heritage and the largest mangrove ecosystem, The Sundarban located in Bangladesh (62 per cent) and India (38 per cent) are threatened by climate change. It is the home of the Royal Bengal tiger (*Panthera tigris tigris*). It is the habitat of a great diverse plant flora and the forests, known to support 425 species of wildlife — 49 mammal species, 315 bird species, 53 reptiles and 8 amphibians. This unique ecosystem is predicted to be adversely impacted by sea level rise (SLR), salinity intrusion, cyclone and storm surge and other climate parameters, e.g., temperature rise. It has been reported that at least 15 per cent of the Sundarban may be submerged due to 10 cm SLR while the whole Sundarban would disappear in case of over 60 cm SLR. However, the freshwater systems and forests would be inundated slowly, harming the growth and reproduction of species that rely on freshwater. The productivity of the system would decline at one stage and later the closed canopy forests would be replaced by shrubs and bushes, leading to loss of species.

Source: Hare (2003).

Box 6.2: Climate Change and Food Production!

Over 50 per cent of the children in South Asia less than 5 years are malnourished. Recently, it has been reported that 79 per cent of the children of Chitori Khurda village of Madhya Pradesh are malnourished. It also states that 60 per cent of the children of Madhya Pradesh state, India suffer from malnutrition. According to the local experts, it is being happened due to two major factors — lack of rainfall in last four years that led reduction of food production in the area and rise of food prices. Similar pictures may exist in the northern part of Bangladesh and also in other countries of South Asia. It clearly indicates that water is quite essential for the poor people for their livelihood, agriculture and ultimately for their sustenance. But climate change may affect these vital resources in South Asia through many ways as mentioned above (details in chapter 4).

Source: Damian Grammaticas, BBC News, 2008.

may drop by 2 per cent in a pessimistic scenario of climate change (GoI 2004). Another study states that an increase of temperature of 0.5 to 1.5°C may cause decrease of yields of wheat and maize by 2 to 5 per cent in India (Aggarwal 2003).

In Bangladesh, there is a prediction that the yields of rice and wheat may drop by 8 and 32 per cent respectively by 2050 (Faisal and Parveen 2004). Studies suggest that wheat production would face the highest impacts of climate change in Bangladesh, followed by rice.

In fact, the *aus* rice, usually grown under rain-fed condition, seems to be the most vulnerable compared to the other two varieties, *aman* and *boro*, produced in Bangladesh (MoEF 2005). On the other hand, it shows that under a severe climate change scenario (4°C temperature rise), the potential shortfall in rice production could exceed 30 per cent from the trend while that for wheat and potato could be as high as 50 per cent and 70 per cent respectively (Karim 1996, quoted in MoEF 2005). In addition, drought causes damage to crops on about 3.52 m ha during the Rabi, pre-Kharif and Kharif seasons every year. Flood, cyclone and storm surge, salinity, etc., are already affecting agricultural products in almost entire Bangladesh. SLR is going to be an additional factor that would increase existing vulnerability to all the sectors, including agriculture, in a synergistic way.

In Pakistan, the higher temperature is likely to result in decline of crop yields. It has been predicted that wheat crops may drop by 6–9 per cent in sub-humid, semi-arid and arid areas in the country due to increase of 1°C (Kelkar and Bhadwal 2008). It also states that the rise of temperature by 0.3°C may affect a number of cash crops like cotton, mango and sugarcane.

In Sri Lanka, rice production is being predicted to drop by 6 per cent due to increase of 0.5°C (MENR 2000, quoted in ibid.). Dryness and deficiency of moisture may also cause reduction of many crops including tea, rubber, coconut and vegetables in the country. A recent study indicates that variation in both temperature and monsoon rainfall may significantly affect coconut production in the future in Sri Lanka (Kelkar and Bhadwal 2008).

In South Asia, production of non-irrigated crops including wheat and rice would significantly drop due to an increase of temperature of over 2.5°C (Lal 2007). It has been predicted that the net production of cereal in South Asian countries would decrease by 4 to 10 per cent before the end of century, even in the most conservative climate change scenario.

Threats to Human Health — Increase of Water-borne Diseases

Health risk due to climate change is predicted to increase all over the world including most of the countries of South Asia. South Asian countries may be affected mostly by vector and water-borne diseases. Studies suggest that a number of diseases and health problems including malaria, dengue, kalazar, cholera, malnutrition and diarrhoeal diseases are associated with climate-related factors such as temperature, rainfall, flood, drought, etc. These diseases were also found to be associated with non-climatic factors including poverty, lack of access to safe drinking water and poor sewerage system, etc. (Cruz et al. 2007). It may be noted that Global burden (mortality and morbidity) of climate-change attributable diarrhoea and malnutrition are already the largest in Southeast Asian countries including Bangladesh, Bhutan, India, Maldives, Myanmar and Nepal in 2000, and the relative risks for these conditions for 2030 is expected to be also the largest (McMichael et al. 2004, quoted in Cruz et al. 2007).

Global and Regional Response to Climate Change

The Inter-governmental Panel on Climate Change (IPCC)

The IPCC is a scientific body which provides information, facts and figures on climate change and relevant issues through reports. This body was established by the World Meteorological Organisation (WMO) and United Nations Environment Programme (UNEP) in 1988 to respond to the consequences of climate change and climatic variabilities in the world (IPCC 2008). The reports that this body publishes are based on scientific evidences on issues related to climate throughout the world. This report follows a rigorous review process done by experts, scientists and governments.

The IPCC provides its reports at regular intervals (in every 5/6 years since 1990) and these reports are widely being used by policy makers, experts, scientists, journalists, researchers, development partners, students, etc. Since its inception, four reports have been published by the IPCC in the last two decades. These are:

1. The IPCC First Assessment Report (FAR) published in 1990. It highlighted the fact that GHGs because of human activities (mainly burning of fossil fuels for energy, industry and transport) has been growing at a considerable rate for the last 150 years. It stated that if the world fails to reduce the GHGs emissions very rapidly, there is possibility of major global warming with potentially catastrophic impacts on the planet.

2. The Second Assessment Report (SAR) of the IPCC was produced in 1995. This has mentioned the potential impacts of climate

change on both human and natural ecosystems and specifically looked at some of the more vulnerable areas of the world such as the small island states, Africa and Asia (IPCC, 1996). It also highlighted the fact that there was increasing evidence that human-induced climate change may be occurring already and therefore, the need to act has became more imperative. One major result of the second assessment's outcomes was the negotiation and signing of the Kyoto Protocol at the third conference of parties (COP3) of UNFCCC held in Kyoto, Japan in December 1997.

3. The Third Assessment Report (TAR) of IPCC was released in November 2001 and once again highlighted the fact that there was considerable evidence that human-induced climate change was occurring already and that the impacts were likely to be felt most severely in the tropical regions of the world where most of the poorer developing countries are located (IPCC 2001). One of the major outcomes of Working Group II of the TAR was the emphasise on the need for adaptation to climate change as a response option, specifically for the developing countries.

4. The Fourth Assessment Report (AR4), the final part of AR4, was released in Valencia, Spain, on 17 November 2007. Earlier in 2007, the contributions of the three IPCC Working Groups to the AR4 were released. This report clearly states that climate change is occurring now and is mostly happening due to human activities. This report provides greater attention to impacts of climate change and possible adaptation strategy at the regional level, identifying the most vulnerable areas.

United Nations Framework Convention on Climate Change (UNFCCC)

Since August 1996, the secretariat has been located in Bonn, Germany. It moved from its previous location in Geneva, Switzerland, following an offer from Germany to host the secretariat, an offer which was accepted by Conference of the Parties (COP) 1.

UNFCCC is basically committed to (UNFCCC 2007):

1. Making a contribution to sustainable development through support for action to mitigate and to adapt to climate change at the global, regional and national level.

2. Providing high-quality support to the inter-governmental process in the context of the Convention and the Kyoto Protocol.
3. Creating and maintaining necessary conditions for an early, effective and efficient implementation of the Kyoto Protocol.
4. Providing and disseminating high-quality, understandable and reliable information and data on climate change and on efforts to address it.
5. Promoting and enhancing the active engagement of NGOs, business and industry, the scientific community and other relevant stakeholders in our work and processes, including through effective communication.
6. Creating and maintaining a caring working environment that is conducive to self-actualisation of staff, information sharing and teamwork and allows the delivery of the highest quality products.

Kyoto Protocol

The Kyoto Protocol was adopted at the third Conference of the Parties to the UNFCCC (COP 3) in Kyoto, Japan, on 11 December 1997 after two and a half years of intense negotiations. The Protocol shares the objective and institutions of the Convention. The major distinction between the two, however, is that while the Convention encouraged developed countries to stabilise GHG emissions, the Protocol made them commit to do so. The detailed rules for its implementation were adopted at COP 7 in Marrakesh in 2001, and are called the 'Marrakesh Accords'.

The Kyoto Protocol is considered to be the most far-reaching agreement on environment and sustainable development ever adopted as it will affect virtually all major sectors of the economy. Following ratification by Russia, the Kyoto Protocol entered into force on 16 February, 2005.

Bali Declaration (December 2007)

The conference, hosted by the Government of Indonesia, took place at the Bali International Convention Centre and brought together more than 10,000 participants, including representatives of over 180 countries together with observers from inter-governmental and non-governmental organisations and the media.

The Bali Declaration basically recognises a number of issues including rapid population growth, changes in demographic structure, constraints on social and economic development efforts, the environment and natural resources (UNPOPIN 2008). The governments had been requested to implement programmes that promote greater harmony among population, resources, environment and development.

The conference culminated in the adoption of the Bali roadmap with a number of core elements as decisions. They include the launch of the Adaptation Fund as well as decisions on technology transfer and on reducing emissions from deforestation.

Dhaka Declaration — SAARC (July 2008)

The Dhaka Declaration on climate change was adopted by South Asian Ministers on Thursday, 3 July 2008, under the auspices of the South Asian Association for Regional Cooperation (SAARC). The ministers of eight SAARC countries declared that they will work together to build regional capacity to reduce the impacts of climate change and consult each other before international negotiations. The declaration came out with the SAARC action plan on climate change, approved by the ministers of all the member countries. The action plan includes the following tasks to be done to address climate change in South Asia region (*The Daily Star*, (2008):

1. Capacity building for Clean Development Mechanism (CDM) projects.
2. Exchange of information on disaster preparedness and extreme events.
3. Exchange of meteorological data.
4. Monitoring climate change impacts including SLR, glacial melting and threats to biodiversity.
5. Mutual consultation in international negotiation process.
6. Media briefing as and when required.

Conclusion

It appears that climate change and climate variability has both direct and indirect impacts on water and relevant resources in all South Asian countries. Impact of climate change on water is going to affect crop agriculture and food security, water supply and sanitation, human health and other relevant sectors. Eventually, MDGs and the sustainable development process of all the countries of this region will face adversity of various climate phenomena including increase of temperature, rainfall variation, floods, drought, cyclone and storm surge, SLR and salinity intrusion, etc.

The present capacity of adapting to such changes of climate system has to be rapidly strengthened at local, sub-national, national and regional level. Though responses exist to address climate threat at global, regional and at the national level in each country, the farmers and rural communities still need to be prepared to adapt to this change.

References and Select Bibliography

Aggarwal, P. K. 2003. 'Impact of Climate Change on Indian Agriculture', *Journal of Plant Biology*, 30: 189–98.

Agrawala S., T. Ota, A. U. Ahmed, J. Smith and M. van Aalast. 2003. *Development and Climate Change in Bangladesh: Focus on Coastal Flooding and the Sundarbans*. Paris: Organisation for Economic Cooperation and Development (OECD).

Ahmad, Q. K., A. K. Biswas, R. Rangachari and M. M. Sainju. eds. 2001. *Ganges-Brahmaputra-Meghna Region: A Framework for Sustainable Development*. Dhaka: The University Press Limited.

Ahmed, A. U. and M. Alam. 1998. 'Development of Climate Change Scenarios with General Circulation Models', in S. Huq, Z. Karim, M. Asaduzzaman and F. Mahtab (eds), *Vulnerability and Adaptation to Climate Change for Bangladesh*, pp. 13–20. Dordrecht: Kluwer Academic Publishers.

Arceo H. O., M. C. Quibilan, P. M. Alino, G. Lim and W. Y. Licuanan. 2001. 'Coral Bleaching in Philippine Reefs: Coincident Evidences with Mesoscale Thermal Anomalies', *B. Mar. Sci.*, 69: 579–93.

Bangladesh Centre for Advanced Studies (BCAS). 2003. *Dialogue on Water and Climate*. Dhaka: Bangladesh Centre for Advanced Studies.

———. 2007. *Climate Change and Human Health of Bangladesh*. Dhaka: Bangladesh Centre for Advanced Studies.

Brohi, S. 2002. 'Balochistan Drought: Quick Official Action Awaited'. http://www.dawn.com/2002/09/09/ebr7.htm (accessed 20 July 2008).

Chiang, S. 2008. 'Heat Waves, the Other Natural Disaster: Perspectives on an Often Ignored Epidemic'. http://www.globalpulsejournal.com/2007_chiang_silvia_heat_waves.html (accessed 16 October 2008).

Cruz, R. V., H. Harasawa, M. Lal, S. Wu, Y. Anokhin, B. Punsalmaa, Y. Honda, M. Jafari, C. Li and N. Huu Ninh, 2007. 'Asia Climate Change 2007: Impacts,

Adaptation and Vulnerability', in M. L. Parry, O. F. Canziani, J. P. Palutikof, P. J. van der Linden and C. E. Hanson (eds), *Contribution of Working Group II to the Fourth Assessment Report of the Intergovernmental Panel on Climate Change*, pp. 469–506. Cambridge: Cambridge University Press.

Ericson, J. P., C. J. Vorosmarty, S. L. Dingman, L. G. Ward and M. Meybeck. 2005. 'Effective Sea-level Rise and Deltas: Causes of Change and Human Dimension Implications', *Global Planetary Change*, 50 (1–2): 63–82.

Faisal, I. M. and S. Parveen. 2004. 'Food Security in the Face of Climate Change: Population Growth and Resource Constraints: Implications for Bangladesh', *Environmental Management*, 34 (4): 487–98.

Global Water Partnership (GWP). 2006. 'Strategic Plan 2006–2010 (Draft version)', Global Water Partnership South Asia. Battaramula: GWP.

Government of Bangladesh (GoB). 2005. *National Adaptation Programmes of Action (NAPA) for Bangladesh*. Dhaka: Ministry of Environment and Forests, Government of Bangladesh.

Government of India (GoI). 2004. *India's initial National Communication to the United Nations Framework Convention on Climate Change*. New Delhi: Ministry of Environment and Forests, Government of India.

Government of Maldives. 2008. 'Maldives Country Presentation'. Paper presented at SAARC Experts Group Meeting on Climate Change, 1–2 July, Dhaka.

Government of Pakistan, 2008. 'Presentation on Climate Change Impacts due to Global Warming'. Paper presented at SAARC Experts Group Meeting on Climate Change, 1–2 July, Dhaka.

Government of Sri Lanka, 2008. 'Country Paper-Sri Lanka'. Paper presented by Government of Sri Lanka at SAARC Experts Group Meeting on Climate Change, 1–2 July, Dhaka.

Gupta, S. K. and R. D. Deshpande. 2004. 'Water for India in 2050: First-order Assessment of Available Options', *Current Science*, 86 (9): 1216–24.

Hare, W. 2003. *Assessment of Knowledge on Impacts of Climate Change — Contribution to the Specification of Article 2 of the UNFCCC*. Berlin: wBGU.

Hoffman, Justin. 2007. 'The Maldives and Rising Sea Levels'. http://www.american.edu/ted/ice/maldives.htm (accessed 2 October 2008).

Intergovernmental Panel on Climate Change (IPCC). 2001. *Climate Change 2001: Impacts, Adaptation and Vulnerabilities*. Intergovernmental Panel on Climate Change, Working Group II. Cambridge: Cambridge University Press.

———. 2002. *Climate Change 2001: Impacts, Adaptation, and Vulnerabilities*. Intergovernmental Panel on Climate Change, Working Group II, Cambridge University Press, 2002.

Intergovernmental Panel on Climate Change (IPCC). 2007. 'Summary for Policy Makers', in S. Solomon, D. Qin, M. Manning, Z. Chen, M. Marquis, K. B. Averyt, M. Tignor, H. L. Miller (eds), *Climate Change 2007: The Physical Science Basis*, pp. 1–18. Contribution of Working Group I to the Fourth Assessment Report of the Intergovernmental Panel on Climate Change. Cambridge: Cambridge University Press.

———. 2008. 'Assessment Reports'. http://www.ipcc.ch/publications_ and_data/publications_and_data_reports.shtml (accessed 20 October 2008).

Jain, S. L., S. D. Ghude and B. C. Arya. 2004. 'Green House Gases Measurement at Maitri, Antarctica'. Paper presented at the 35th COSPAR Scientific Assembly, 18–25 July, Paris.

Karim, Z. 1996. 'Agricultural Vulnerability and Poverty Alleviation in Bangladesh', in T. E. Downing (ed.), *Climate Change and World Food Security*, pp. 307–46. NATO ASI Series 137. Berlin: Springer-verlag.

Kelkar, Ulka and S. Bhadwal. 2008. *South Asian Regional Study on Climate Change Impacts and Adaptation: Implications for Human Development*. A Background Paper for Human Development Report 2007/2008. New York: UNDP.

Lal, M. 2007. *Implications of Climate Change on Agricultural Productivity and Food Security in South Asia: Key Vulnerable Regions and Climate Change — Identifying Thresholds for Impacts and Adaptation in Relation to Article 2 of the UNFCCC*. Dordrecht: Springer.

McMichael, A. J., D. Campbell-Lendrum, S. Kovats, S. Edwards, P. Wilkinson, T. Wilson, R. Nicholls et al. 2004. 'Global Climate Change', in M. Ezzati, A. Lopez, A. Rodgers and C. Murray (eds), *Comparative Quantification of Health Risks: Global and Regional Burden of Disease due to Selected Major Risk Factors*, pp. 1543–1649. Geneva: World Health Organization.

Ministry of Environment and Natural Resources (MENR). 2000. *Initial National Communication under the UNFCCC: Sri Lanka*. Colombo: Government of Sri Lanka.

Ministry of Environment and Forests (MoEF). 2005. *National Adaptation Programmes of Action to Climate Change*. Dhaka: Government of Bangladesh.

Ministry of Water Resources (MoWR). 2002. *National Water Policy*. Delhi: Government of India.

Parry, M. L., O. F. Canziani, J. P. Palutikof and Co-authors 2007. 'Technical Summary', in M. L. Parry, O. F. Canziani, J. P. Palutikof, P. J. van der Linden and C. E. Hanson (eds), *Climate Change 2007: Impacts, Adaptation and Vulnerability. Contribution of Working Group II to the Fourth Assessment Report of the Intergovernmental Panel on Climate Change*, pp. 23–78. Cambridge: Cambridge University Press.

Pun, S. B. 2004. 'Overview: Conflicts over the Ganges?', in B. Subba and K. Pradhan (eds), *Disputes over the Ganga*, pp. 3–20. Kathmandu: Panos Institute South Asia.

Qureshi, A. and D. Hobbie. 1994. *Climate Change in Asia*. Manila: Asian Development Bank.

Rahman, A. A., M. Alam, S. S. Alam, M. R. Uzzman, M. Rashid and M. G. Rabbani. 2008. *Risks, Vulnerability and Adaptation in Bangladesh*. A Background Paper Prepared for Human Development Report 2007/2008. New York: UNDP.

Rahman, A. A., A. S. Huda and M. G. Rabbani. 2007. *Situational Analysis of Capacity Building Needs for IWRM in South Asia*. Dhaka: Bangladesh Centre for Advanced Studies.

Rangachari, R. and B. G. Verghese. 2001. 'Making Water Work to Translate Poverty into Prosperity: The Ganga-Brahmaputra-Barak Region', in Q. K. Ahmad, A. K. Biswas, R. Rangachari and M. M. Sainju (eds), *Ganges-Brahmaputra-Meghna Region: A Framework for Sustainable Development*, pp. 81–142. Dhaka: The University Press Limited.

Salehin, M., A. Haque, M. R. Rahman, M. S. A. Khan and S. K. Bala. 2007. 'Hydrological Aspects of 2004 Floods in Bangladesh', *Journal of Hydrology and Meteorology*, 4 (1), 33–44.

Stern, N. 2007: *Stern Review on the Economics of Climate Change*. Cambridge: Cambridge University Press.

Tata Energy Research Institute (TERI). 1996. 'Report No 93/GW/52', submitted to the Ford Foundation, TERI, Delhi.

The Daily Star. 2008. 'SAARC Nations Vow to Fight Climate Fallout', 4 July 2008. http://www.thedailystar.net/newDesign/news-details. php?nid=44145 (accessed on 10 July 2008).

United Nations Educational, Scientific and Cultural Organisation (UNESCO). 2003. 'International Year of Freshwater', World Water Assessment Programme. http://www.un.org/events/water/brochure.htm (accessed 15 October 2008).

United Nations Framework Convention on Climate Change (UNFCCC). 2008. The Year of Climate Change. Intergovernmental Negotiation on Climate Change towards Copenhagen. http://unfccc.int/files/meetings/ sb30/press/application/pdf/090608_pres_yvo.pdf (accessed 20 October 2011).

United Nations Population Information Network (UNPOPIN). 2008. 'The Bali Declaration'. http://www.un.org/popin/icpd/newslett/92_12/ The+Bali+Declaration.html (accessed 10 October 2008).

Water Resources Planning Organisation (WARPO). 2005. 'National Adaptation Program of Action (NAPA): Water, Coastal Areas, Natural Disaster and Health Sector', Formulation of Bangladesh Programme of Action for

Adaptation to Climate Change Project. Dhaka: Water Resources Planning Organisation with Center for Environmental and Geographic Information Services (CEGIS).

Wassmann, R., N. X. Hien, C. T. Hoanh and T. P. Tuong. 2004. 'Sea Level Rise Affecting the Vietnamese Mekong Delta: Water Elevation in the Flood Season and Implications for Rice Production', *Climatic Change*, 66 (1–2): 89–107.

Wilkinson, C. ed. 2000. *Status of Coral Reefs of the World: 2000*. Townsville: Australian Institute of Marine Science.

———. ed. 2002: *Status of Coral Reefs of the World: 2002*. Townsville: Australian Institute of Marine Science.

World Bank. 2000. *Bangladesh: Climate Change and Sustainable Development*. Report no 21104-BD. Dhaka: World Bank.

———. 2007. 'The Impact of Sea Level Rise on Developing Countries: A Comparative Analysis', World Bank Policy Research Working Paper 4136, 2007. Washington D.C.: World Bank.

———. 2008. *Coping with Water Scarcity in South Asia*. http://web.worldbank.org/WBSITE/EXTERNAL/COUNTRIES/SOUTHASIAEXT/0, contentMDK:21259974~pagePK:146736~piPK:146830~theSitePK:22354 7,00.html (accessed 25 October 2008).

Annexure 1: General Floodplain Topography of Bangladesh

Map A6.1: Flood-prone Areas of Bangladesh

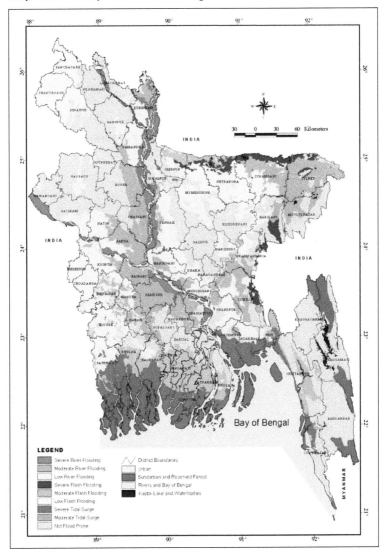

Source: Bangladesh Agricultural Research Council, Dhaka, Bangladesh, 2000.

Annexure 2: Drought-prone Areas of Bangladesh

Map A6.2a: Drought-prone Area, Rabi and Pre-kharif Season — Existing Situation

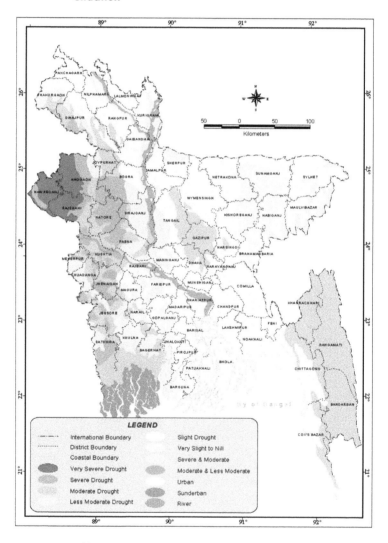

Source: Prepared by GIS Division, Bangladesh Centre for Advanced Studies (BCAS), Dhaka, Bangladesh, 2008.

Map A6.2b: Drought-prone Area, Rabi and Pre-kharif Season — Future Situation (2075)

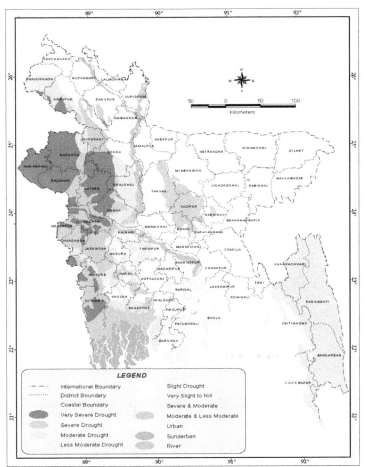

Source: Prepared by GIS Division, Bangladesh Centre for Advanced Studies (BCAS), Dhaka, Bangladesh, 2008.

Map A6.2c: Drought-prone Area, Kharif Season — Existing Situation

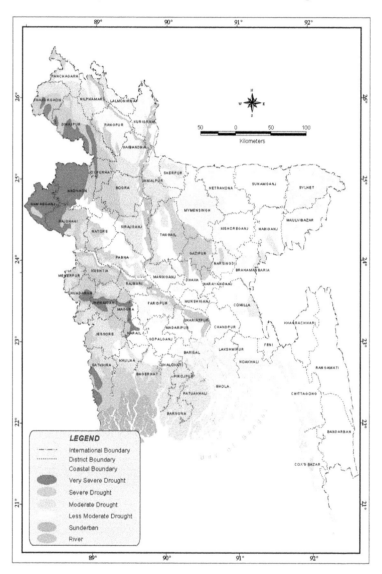

Source: Prepared by GIS Division, Bangladesh Centre for Advanced Studies (BCAS), Dhaka, Bangladesh (2008).

Map A6.2d: Drought-prone Area, Kharif Season — Future Situation (2075)

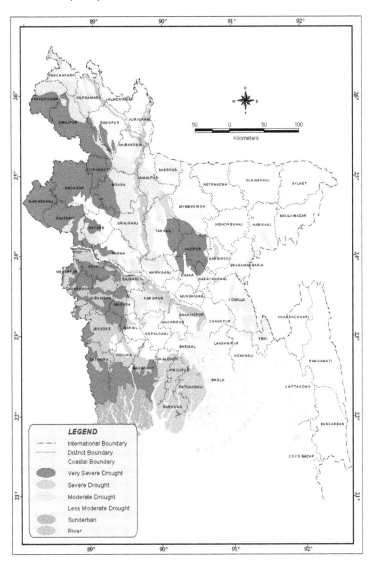

Source: Prepared by GIS Division, Bangladesh Centre for Advanced Studies (BCAS), Dhaka, Bangladesh, 2008.

Annexure 3: Sea Level Rise (1 Meter Prediction) and Salinity Intrusion in Bangladesh

Map A6.3: Predicted Salinity Intrusion in Bangladesh Due to 1 Meter Sea Level Rise

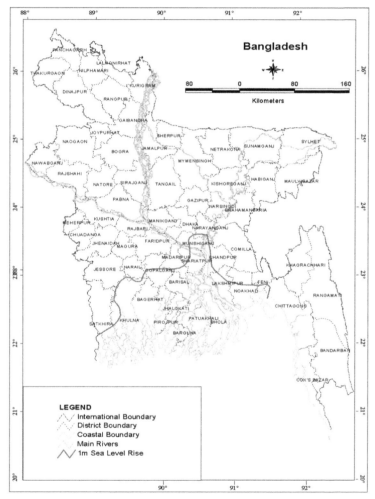

Source: Prepared by GIS Division, Bangladesh Centre for Advanced Studies (BCAS), Dhaka, Bangladesh, 2008.

7

Assessment of Environmental Impact of Development of Water Resources

Bishnu P. Das and Subhadarshi Mishra

Judicious conservation of land and water resources must receive prime attention and involvement of civil society across the globe. Water-related development primarily results in abstraction from rivers utilising dams and diversion structures such as weirs and barrages for providing irrigation to dry lands, generate hydropower, protect fertile lands and property from flood inundation and effecting drainage in critically waterlogged areas. Such abstraction in turn can impact ecosystem adversely. According to the Report of World Commission on Dams (WCD), there are 47,655 large dams in the world. All these dams and water-related development including ground water exploitation have been planned, designed and constructed after due appraisal of their impact on environment following the norms prevalent during their formulation. Concern is however felt as competitive demand on water increases, with population increase and related food need, and the nature gets progressively deprived of water for its sustainability. Ill-planned water abstraction, either from surface or ground, leads to an environmental and social imbalance.

During the UN Conference on Environment and Development at RiodeJenerio in 1992, Global Consensus and Political Commitment on Development and Environment was categorically expressed. Agenda 21 was adopted which addresses today's problems and preparedness for the next century with regard to water development, duly incorporating the environmental concerns. Increasing abstraction and flow modification disturb the overall water environment which, if not contained, will seriously impair the capacity of the resource to meet the need of the future generation. This is precisely why an assessment has to be made of the environmental impact of development to mitigate the consequent damages. The concept of assessing the environmental impact of development initiatives, which had originated in the United States in 1969, is rapidly being adopted all over the world, particularly in the south and South East Asia, since the 1980s.

The International Commission on Irrigation and Drainage (ICID) developed a Country Policy Support Programme (ICID 2004) in respect of water for a vision year 2025, relating to five countries — China, India, Pakistan, Egypt and Mexico. This study categorised the need of water to be broadly allocated for food (for agriculture), for people (for drinking and industrial need, sanitation) and for nature (for ecosystem). What is required is allocating the legitimate need of each sector without unduly stressing other sectors, particularly for the nature, which is possible by an environmental impact assessment study.

Impact of Development of Water Resources

Water Resources development projects primarily provide economic benefits of irrigation, hydropower generation and flood control with ancillary benefits of drinking water and industrial supply. Almost 300 million ha of the 1,500 million ha of arable land worldwide is irrigated through large and small dams. The environmental impact of the change in riverine flow has both positive and negative effects on the civil society, flora and fauna, terrestrial and aquatic biodiversity.

Positive Impacts

The economic benefits accruing from water projects which concurrently may also bring environmental benefits is described below.

Increase of Food Production

Expansion of irrigation with assured water availability increases crop production leading to increase in the economic standard of people and food security.

Flood and Drought Control

The twin specters of drought and flood have been largely controlled by dams. The flood storage reservoirs act as moderator by regulating the peak flood discharge. The reservoirs also supplement the requirement of water for crops in dry years to counter drought. In arid and semi-arid areas, erosion due to wind and rain gets controlled when irrigation is provided to the area accompanied by significant land leveling and shaping. This is a major environmental benefit.

Increase of Groundwater Table

The storage and regulated release of water for irrigation projects helps in recharging the groundwater table.

Dependable Drinking Water Source

Water resources projects provide a dependable source of drinking water. With rapid urbanisation in South Asia, the demand for drinking water has grown exponentially leading to requirement of more and more impounding through reservoirs. In the case of Delhi, the capital of India, the drinking water need is now substantially met from Tehri Dam on Bhagirathi (Ganga), situated 200 km away. This will be augmented by the proposed 148 m high Renuka dam on river Giri, a tributary of river Yamuna in Himachal Pradesh. This dam is primarily a captive storage to meet the drinking water need to the tune of 437 million gallons (mgd) daily to the national capital territory of Delhi during 8 months of lean season (September–June). However, the dam is environment friendly as after power generation at the dam, it will release 23 cumec perennially to the river which flows down for 250 km before utilisation, thereby contributing to environmental flow over a large stretch of the river.

Hydropower and Industrial Growth

The phenomenal industrial growth, crucial for a healthy economy in most parts of the globe, has been possible by appropriate harnessing of water, generation of non-polluting hydropower, which is a renewable energy.

Improvement in Health and Sanitation Facility

Quite often, people living in command areas of irrigation projects enjoy better health and sanitation. Thus, people in the command area of Rajasthan canal which irrigates over a million hectares enjoy better health and sanitation facility.

Micro-changes in Climatic Condition

The command area of large projects experience a micro-climatic change for the better, particularly in the arid region. The prevailing intense heat in Punjab, Haryana and Uttar Pradesh and the Indo-Gangetic plain of India has subsided after construction of related reservoirs, particularly the massive Bhakra dam on Sutlej river (CWC 2003).

Tourism and Recreation Facility

Reservoirs-linked tourism and recreation are natural benefits accruing from large irrigation projects. Both capture and culture fishing in reservoirs have resulted in favourable growth of fish species and population.

Soil Conservation and Afforestation

Catchment area of water resources projects often are treated to conserve soil, improve its water holding capacity, help wildlife and promote economic activities through water harvesting. Afforestation taken up to compensate for submerged forests restores barren land by improving the forest status, rejuvenating degraded forest along with development of a green belt in the reservoir periphery.

Improvement in Environmental Flow

The water resource projects could release the stored water in lean period which results in an incremental environmental flow.

Negative Impacts

Two major adverse impacts are related to: (a) rehabilitation and resettlement consequent to impounding and (b) altered flow regime affecting the aquatic biodiversity, in particular fish, both upstream and downstream of the dam/barrage. Whereas the benefits have been astounding, the negative impact has also surfaced in almost all categories of water resources development; the major demerits are listed below.

Rehabilitation and Resettlement

Rehabilitation and resettlement of a large number of people displaced due to submersion, particularly upstream of dams. Displacement is considered the most agonising adverse consequence of large or even medium dams.

Submergence

Submersion of places of worship; archaeological, historical and cultural monuments; scenic stretches of river rapids; mineral deposits of commercial value; and forest land, which directly affects the native flora and fauna, rare and endangered species.

Nature

Impact on protected areas, neighbouring wildlife sanctuaries, avifauna and water quality.

Morphology

Possible degradation of the river downstream due to silt abstraction at impoundments and aggradation of reach upstream in the parent river and tributaries causing raising of flood levels. Morphological changes also occur downstream due to the alteration of 2 year return period flood, which represents the bankful discharge.

Decrease in Flood Carrying Capacity

Flood-moderated rivers below dam deteriorate with irregular deposition and erosion. The river becomes incapable of carrying discharge and charge (sediment) efficiently over a large stretch downstream under altered flow regime.

Lowering of Ground Water Table

Reduction of inundation of flood plain and their frequency affects wetlands. Groundwater regime is also altered. Significant lowering of Ground Water Table (GWT) and increase in salinity also occurs downstream.

Aquatic Biodiversity

Impact related to migration of fish in particular is felt: (a) in the upstream due to modification from lotic to lentic environment (Sugunan 1995), (b) in the downstream due to partial drying up of the river and change from the pattern to which the native fish is habituated, and (c) absence of an attracting flow due to altered outflow from dam at the right time (cue) for fish to migrate.

A fluvial hydro system is a continuum from the source to the mouth because the nutrient inputs progressing along a river sustains the aquatic biodiversity. A river continuum can be characterised by three distinct phases: the headwater channels, the middle river and the low land flood plain river. In the head water, which comprises first to third order, streams solute and sediment supplies from the hill slope constitute the source of nutrients providing a trophic pathway. The environmentally heterogeneous middle river reach is a zone with high species richness. The lowland flood plain characterises the deposition or storage zone. If a river is dammed, a major impact is disruption of connectivity for the migrant fish from flood, sediment and nutrient considerations.

By impounding a large volume of water, dams profoundly influence the hydrological regime, both upstream and particularly downstream. In the Indian context, rivers originating in Himalayan region such as the Ganges, Yamuna and Kosi have significant snowmelt flow from March through June, which increases with the advent of monsoon until late October. The monsoon flow in these rivers on the average is around 70 per cent of the annual flow. In contrast, the peninsular rivers such as Narmada, Tapi, Godavari, Mahanadi, Krishna and Cauvery are monsoon-fed rivers and convey 90 per cent of their annual flow between June through October. This is also true of Yangtze and Mekong rivers. Large impoundments in these rivers have led to not only intra-season flow variation but also blockage of migratory path/loss of habitat, which has resulted in a large decrease in river fish species and population, particularly in Sutlej, Mahanadi, Krishna and Cauvery (Sugunan 1995) and in Yangtze (Chen et al. 2003).

Water Logging

Water logging and increase in salinity in the command area due to mismatch of crop water demand and supplementation.

Environmental Impact of Water Resources Projects

Hydropower dams affect flow regimes in rivers worldwide (McCully 1996) and pose major challenges relative to conservation of native riverine biota. These dams not only fragment river systems, but also alter downstream flow regimes and existing river fauna (Table 7.1). Hydropower dams operated for peak load power generation impose frequent (often daily) flow fluctuations equivalent to storm run-off events in natural systems, as well as altering seasonal flows. Ecologists apprehend that unless managed to mimic as closely as possible the pre-impact patterns of flow-regulated river will destroy in-stream habitat. One case study from South Asia is given below to illustrate some of these impacts.

The Indus river, originating from Mansarovar lake in Tibet, flows for about 2,880 km to the Arabian sea through India and Pakistan draining 1,081,718 km^2 of basin area. The Indus delta is the fifth

Table 7.1: Change of Flow Regime and its Impact on Selected Dams in Orissa, India

Project	Project benefits	Change of hydrological regime	Impact
Hirakud dam, (Mahanadi river)	Irrigation: 1,59,000 ha; power generation: 347 MW Flood control of 5000 km^2	Drainage problems in delta area	A 12-km-long straight cut (300 m wide and 2 m deep) to sea from one of the deltaic branches of Mahanadi called Bhargavi at its 40th km shortened the river by almost 40 km. It has been operational for the last 30 years and has diverted up to 2000 m^3 sec^{-1} of high flood (almost 70%) directly to the sea. The cut has been helpful in restoring and sustaining agriculture by improved drainage over 30,000 ha of totally unproductive ill-drained land in the tail reach. The flood level has lowered by 2 m thereby facilitating surface drainage of over-bank lands into the river (Das 2004, 2005). The straight cut however has deprived the tail reach of the environmental in-stream flow and has itself experienced degradation in its hydraulic regime. The river has also experienced aggradations below the cut.
Rengali dam (Brahmani river)	Irrigation: 2,10,000 ha Power generation: 250 MW; Flood control: 3000 km^2	The estuary region which is currently receiving an annual inflow of 17,380 mm^3 will be deprived of 3000 mm^3 due to consumptive use through irrigation. In dry years, the inflow is likely to reduce to 10,000 mm^3 after abstraction for irrigation.	The mangroves currently experience salinity in the range of 5 to 35 ppt close to sea mouth. In view of the substantial abstraction of freshwater for irrigation (20 to 25%) concern was felt to study the impact of such abstraction which might lead to increased salinity impacting the estuarine mangroves. There is a distinct correlation between luxuriant mangroves which experienced salinity of 5–15 ppt, in contrast to stunted and less vegetative species experiencing salinity of 25–30 ppt.

Source: Das (2004) and Das and Mishra (2007).

largest in the world covering an area of 30,000 km^2. Over the last 50 years, several large dams such as Bhakra in India, Mangla and Tarbela in Pakistan and several barrages have abstracted considerable freshwater for consumptive use through irrigation in this arid basin. In Pakistan, the upper riparian province of Punjab has been very largely benefited depriving the lower province of Sindh of freshwater leading to devastating consequence.

Freshwater flow down the Indus in the 1950s was about 150 Million Acre Feet (MAF) and silt flow to the delta was 400 Million ton (Mt). These have been currently reduced to 20 MAF and 36 Mt respectively, due to massive freshwater abstraction in the last 50 years (Memon 2005; Qureshi, 2000). The consequence of such massive abstraction has been devastating for the delta affecting mangroves and fish population, enhancing pollution, degradation of water quality and sea water intrusion, due to reduction of river level. Although the inter provincial water accord of 1991 stipulates a flow to the delta of 1,21,400 cusec in April, the actual flow was only 36,275 cusec because of diversion to the upper province of Punjab impacting the main Kharif cropping season for Sindh. The appropriate hydrologic regime for reversing the environmental degradation has been studied by the Centre for Ecology and Hydrology, Wallingford, United Kingdom (Acreman et. al. 2000).

Environmental Impact Assessment (EIA)

Objectives of EIA

EIA can be defined as the systematic identification and evaluation of the potential impacts of proposed projects, plans, programmes, or legislative actions relative to the physical–chemical, biological, cultural and socioeconomic component of the total environment (Canter 1996). By identifying, both positive and negative impacts through EIA at the initial stage, the decision makers remain aware of the environmental consequence and plan for appropriate mitigation measures or even enhance the quality of environment. EIA is a continuous process in the project cycle that avoids serious and irreversible damage to the environment, protects human health and safety, identifies key impacts and measures for mitigating them, ensures efficient resources use, enhances social aspects, modifies and improves design and ensures good decision making and condition-setting.

Process Leading to Establishment of EIA

Lack of environmental planning in development has led to not only irreversible damage to natural resources and the environment but also to disruption of communal and social harmony. In the early 1960s, the United States realised the environmental damage attributable to development. A popular environmental movement led to enactment of laws protecting land areas from the threat of development, which furthered to control the hazardous by-products of industrialisation covering air, and water pollution and the production of toxic or radioactive waters.

With growing concern for environmental pollution, the US Environmental Protection Agency was established in Washington D.C. on 2 December, 1970. Earlier, the National Environmental Policy Act, NEPA, 1969 of the United States of America came into being requiring environmental protection in large-scale projects (Canter 1995). The phrase EIA comes from Section 102(2) of the NEPA, 1969, USA, which created a new era of environment awareness, as federal agencies were required to ensure environmental protection in all their activities. NEPA not only required EIA to be carried out for projects by the agencies to build, finance or permit, but also provided an important tool to the affected public by making available the right to bring a case to court.

The EIA process has now spread to South Asia. In the 1992 Dublin International Conference on Water and Environment, Integrated Water Resources Management (IWRM) was considered as the approach that historically encompasses environmental issues and encourages the participation of stakeholders and government considering a basin as a unit. The '3E principle', as advocated for Mekong river by researches of Helsinki University of Technology, covering economy, equity and environment (Varis et al. 2008) is applicable in the South Asian context. The 3E principle in the context of IWRM means 'Waters should be used to provide economic well being to the people, without compromising social equity and environmental sustainability'.

In the document 'Making Sustainable Commitments — An Environment Strategy for the World Bank (2001)', it is stated that 'Significant natural resource conservation concerns in South Asia include water quality degradation and local and regional water scarcity, degraded forest, coastal wetlands, freshwater bodies, etc., arising from poorly managed water resources'. The environmental concerns are being dealt within the bank-assisted projects in South Asia and Southeast Asia by addressing the trans-boundary issues.

Most developing countries in Asia have an established framework for environmental protection and environmental management (Asian Development Bank 1992). These countries also have some form of environmental legislation. Invariably, it includes a statement of a policy, goals, objectives and priorities. EIA and environment policy were formulated in Bangladesh (1991), Cambodia (1996), Peoples Republic of China (1979), India (established the Administrative Law Bench, 1973 and E.P. Act, 1986), Indonesia (1978), Lao PDR (no legislation), Malaysia (1974), Philippines (1977), Sri Lanka (National Environment Act, 1980 with Guidelines, 1993/95), Thailand (1992) and Vietnam (1994). The evolution of EIA can be summarised sequentially as follows:

1. Initial development in early 1970s.
2. Increasing scope between 1970s and 1980s.
3. Process strengthening and policy integration between mid to late 1980s.
4. Towards sustainability and biodiversity conservation in South Asia and Southeast Asian countries in the 1990s.

Historical Perspective of EIA in India

The foundation of EIA in India was laid in 1976–1977 when the Planning Commission asked the then Department of Science and Technology (DST) to examine the river-valley projects from an environmental perspective. This was subsequently extended to cover those projects which required approval of the Public Investment Board. However, these were administrative decisions and lacked legislative support. To fill this gap, the Government of India enacted the Environment (Protection) Act (EPA) on 23 May 1986. To achieve the objectives of the act, one of the decisions was to make EIA statutory. On 27 January 1994, the Union Ministry of Environment and Forests (MoEF), Government of India, under the Environmental (Protection) Act 1986, promulgated an EIA notification making Environmental Clearance (EC) mandatory for expansion or modernisation of any activity or for setting up new projects listed in Schedule 1 of the notification. Since then, there have been about 12 amendments made in the EIA notification of 1994.

The MoEF recently notified a new EIA legislation in September 2006. The notification makes it mandatory for various projects such as mining, thermal power plants, river valley, infrastructure (road, highway, ports harbours and airports) and industries to get environmental clearance.

Conceptual Approach for EIA Studies

To provide a basis for addressing the EIA process, a 10-step procedure suggested by Canter in 1996 is used for planning and conducting of studies (see Canter and Clark 1997). This procedure is flexible and can be adapted to various project types by modification as needed, to facilitate the addressing of special concerns of specific projects in unique locations. Table 7.2 lists this 10-step procedure and is being widely adapted by different countries in South Asia.

Each country in South Asia has designed various guidelines and procedures to be followed in conducting the EIA process, which are specific to the respective country. Similarly, every country has specified certain guidelines for impact assessment through collection and interpretation of environment-related data. The principles set by the MoEF for Indian projects are detailed in the following section.

The study area in the Indian context is generally considered to cover a 10 km radius from the construction location such as a dam, a powerhouse, water conductor system or from the periphery of the command area in the case of irrigation projects. The baseline information to be collected in the study area is prescribed. Selected environmental factors related specifically to water sector projects are described in the section that follows.

Water Environment

Hydrometerology

Availability of water in the basin over a representative period of 25–30 years, pertaining to both surface and groundwater, is broadly examined by reference to: (a) rainfall, (b) temperature, (c) wind speed, (d) humidity, (e) drought events and minimum rainfall, (f) sunshine hours, (g) extreme storm events, (h) river discharge data, (i) yield of the basin, (j) groundwater table fluctuation in the monsoon and non-monsoon periods, and (k) estimation of water need for the proposed objective(s). Simulation studies done to justify satisfactory performance of the project to meet the demand

Table 7.2: Steps for Conducting an EIA

Step	Activity	Description
1	Project description	Describe the type of project and how it operates in a technical context
2	Pertinent institutional information	Refers to multitude of environmental laws related to environment in a given country
3	Identification of potential impacts	Identify the potential impacts of the project activities
4	Description of the affected environment	Enable the selective identification of pertinent baseline factors for the study in subsequent steps
5	Impact prediction	Quantification (or at least qualitative description) of the anticipated impacts of the proposed project on various environmental factors
6	Impact assessment	Interpretation of anticipated changes related to the proposed project (assess the significance of the change)
7	Impact mitigation	Associated with identification and evaluation of impact mitigation measures
8	Selection of best alternative	Associated with the selection of best options out of alternatives available
9	Documentation	Prepare the EIA report
10	Environmental monitoring	Planning and implementation of environmental monitoring plan

Source: Canter and Clark (1997).

is generally considered as 75 per cent dependability for irrigation projects and 90 per cent dependability for power projects. Detailed techno-economic viability studies are separately carried out based on which environmental clearance is sought for.

Water Quality

The selection of site for sampling water quality is carefully done. For the hydroelectric projects, reservoir dam site, head race site, tail race site and for irrigation project reservoir dam site, irrigation network site and major drains are chosen as sampling stations. A critical examination for the site from which drinking water will be lifted is done. Considering that rain water harvesting is an environment-friendly approach, quality of rain water is also studied periodically. Standard physical, chemical and biological water quality parameters are studied.

Land Environment

A large quantity of muck is generated at the initial stage of construction, which needs to be estimated for each component such as dam base, spillway, powerhouse, head race/tail race tunnels, canal networks, etc. Careful reconnaissance of the project area is made to identify suitable sites for their disposal from stability, aesthetic and sanitation angles.

Data is collected pertaining to broad geology, presence of economically important mineral deposits, land form, physiography and physical and physiochemical data for soil representative, both for the catchment and the command area. Important parameters such as infiltration rate, hydraulic conductivity along with soil quality such as texture, bulk density, particle density, porosity, water holding capacity, pH, EC of saturated soil, organic carbon, available Phosphorus, Potassium, Nitrogen, Copper, Zinc, Boron, are studied.

Catchment Behaviour: Erodibility Characteristics

An important study pertains to assessing the erodibility characteristics of the basin by preparing thematic maps using lithology, drainage pattern, soil characteristic, land slides/slips, physiography and land use/land cover, which can be broadly delineated to dense forest, open forest, degraded forest, agriculture and settlement, waste land, rivers and water bodies.

Slope forms an important parameter and is categorised accordingly as very gentle (0–20°), gentle (20–35°), moderate (35–50°) and steep (50–80°). The free draining catchment area which contributes to silt deposit in the impoundment is subdivided into sub-watersheds of different erosion intensity. The sediment flow characteristic is a product of erosivity (climatic factor, slope), erodibility (soil) and areal extent designated as Silt Yield Index (SYI), a parameter developed by All India Soil and Land Use Survey (AISLUS) from 1969. SYI is defined as yield per unit area leading to overall sediment yield assessment and is utilised to decide upon the area to receive priority of treatment categorised as low, moderate and high. For every project, monitoring stations are established for observation of discharge and sediment load at suitable locations adopting standard sampling procedure. The sampling result over a year can be interpreted to give the annual Sediment Production Rate (SPR) expressed in volume of sediment moved past the sampling point per unit catchment area over unit time. This aids in determining the silt storage as well as identifying the problematic area prone to high erosion.

Another important study related to land is the correct estimation of water requirement in the rains, summer and dry seasons for the proposed cropping in the command area. Evapo-transpiration approach adopted in Food and Agriculture Organisation of the UN (FAO), namely Penman-Montieth method (FAO 1998), linked to climatic parameters, is used to estimate the crop water need and the supplementation. Depending upon the soil characteristics, irrigation management, which will be environment-friendly, is recommended. For example, ill-drained soils belong to poor irrigability class and need controlled application. Development of appropriate surface and sub-surface drains is studied at the impact assessment stage. A major concern has been the rise of groundwater table in irrigation command area, particularly in flat moderately drained command areas in India. Impact of irrigation on soil is studied with reference to its physical and chemical proportion and vertical and longitudinal profile of pervious–impervious layers in the command area.

Biological Environment

Change in use of land from virgin forests or vegetation to submerged area under a reservoir results in the most profound change to the terrestrial biota in particular and aquatic biota in general. Serious concern is raised with regard to submersion and degradation of large forest area in developing countries, particularly India. It is how-ever estimated that only 4 per cent of forest area has been lost in India (CWC 2003) which is attributable to the construction of water resources projects.

Terrestrial Biota

Seasonal study is conducted by characterising the study area as major forest types — evergreen, semi-evergreen, deciduous — and density-wise as dense, open and degraded types. Vegetation profile, floristic diversity, phyto-sociological studies by the internationally accepted and widely used Shannon and Wiener Index (1963) is carried out. By distributing to different components such as trees, shrubs, herbs, climbers grasses, the total number of species are enumerated. An important issue is the estimation of loss of biomass accruing from submersion of forest. As regards the terrestrial fauna, enumeration of different classes conforming to Tetrapoda, Mammalia, Amphibia, Reptilia, etc., is done over the pre-monsoon, monsoon and post-monsoon season. Butterflies of ecological importance need special study.

The data of faunal species is collected by close coordination with the wildlife warden and the principal conservator, who are generally government functionaries in all countries. Migratory route of wild animals is specially studied to ensure uninterrupted movement or appropriate remedial intervention.

Aquatic Biota

Fish, both resident and migratory, bear the brunt of riverine change brought about by a barrier. The status of ichthyo-fauna, fish and fisheries in the entire riverine reach to be impacted by the project is studied. Observations show that substratum of a river consisting of bed rock, boulders, cobbles and gravels (coarse sand) mainly supports the survival, growth and reproduction of macro-benthic organism. A close relationship is noted between the abundance of fish presence and population density of macro-zoobenthos/phytobenthos in par-ticular river reaches. Modification of macrobenthos occurs upon flow regulation. The habitat study of unregulated and flow-regulated sites influencing composition of fish assemblage (Freeman et al. 2001) is therefore an important input for the baseline data. The migratory species are specially enumerated by suitable technique with respect to their breeding areas, the seasonal trend, the water depth, velocity and discharge prevailing in the pre-project stage under satisfactory migration condition. The downstream impacts are assessed to arrive at the environmental flow release by considering not only the parent stream but also streams joining the river downstream. Data of zooplankton and phytoplankton in the riverine stretches, crucial for fish sustenance, are collected.

Socioeconomic Environment

Demographic profile of the population in the project area is collected to ascertain the number of people to be benefited or adversely affected. In India, scheduled caste and scheduled tribes are particularly identified for suitable rehabilitation. Information collected can be broadly categorised pertaining to socioeconomic profile of the area proposed to be submerged/benefited, relating to: (a) demography of people such as total household, male–female population, weaker sections and vulnerable persons, (b) occupational pattern, (c) educational profile, (d) land type pattern in submerged area, (e) other amenities, and (f) public perception regarding the proposed project.

Structured questionnaires are utilised for rapid village survey, household survey and attitude/psychology evaluation. The enumeration leads to estimating the Project Affected Families (PAF), categorising them as landless family, marginal farmers, small farmers and large farmers. The estimation of private land/dwellings to be acquired/displaced for rehabilitation is made relating to each administrative unit, such as a district or its subdivisions. Through a consultative process, new areas for rehabilitation and resettlement are identified, earmarking land for residence/agriculture/other occupation along with common community facilities.

Structural Impact

Two issues that relate to safety of the structures and their response to catastrophic natural calamities have to receive due attention.

Dam Break Study

The generation of a maximum probable flood consequent to the most intense storm event is likely at the project site. Although a structure such as a dam or barrage has a spillway to adequately and safely dispose of the flood, the possibility of a dam overtopping or breaching/breaking due to reasons beyond human control always arises. Dam break can be termed as a partial or catastrophic failure of a dam, causing uncontrolled releases. Land and communities downstream face incalculable loss of life and property as a flood wave several meters high travel downstream at a high velocity wiping out infrastructure such as road, bridges, railways and rural/urban habitation. A large amount of sediment also moves downstream along with the wave. A dam break model, which is an analytical model, is therefore utilised to predict the profile of the flood wave and the time it will take to reach a specified location. The information sought for are time of first arrival of the flood, its peak level and velocity, area and depth of inundation and duration. Detailed information on the topography of the downstream valley is collected. A suitable dam break model is used to prepare a disaster management plan for the downstream region to be affected in the unlikely event of dam break.

Seismic Study

From locational consideration, a project may be subjected to potential seismic hazard. Every country has delineated the entire geographical area to seismic zones considering the seismic–tectonic set-up, observed and reported earthquakes. The incidence of recent past earthquake of high intensity (above 5 in Richter scale) in the project vicinity is documented. Values of ground acceleration for different intensity of earthquake are then formulated. For smaller dams, it may be safe to adopt appropriate design indices. However, for large dams, underground structures such as tunnels, powerhouse, surge tanks, etc., site-specific seismo–tectonic investigations are mandatory requirement, which need to be conducted. For India, there is a National Committee for Seismic Design Parameters (NCSDP) for river valley projects involving members of several organisations drawn from Water Resources, Geological Survey, Meteorological Department, Geophysical Research, Institute of Remote Sensing, Water and Power Research Station and Department of Earthquake, University of Roorkee. The institutional mechanism for safe seismic design by involvement of experts thus exists.

The other issue is the possibility of Reservoir Triggered Seismicity (RTS), which represents the maximum level of ground motion consequent to filling or draw-down in a new reservoir. Small changes in the state of stress at seismogenic depths can occur if ongoing tectonic process have already caused near failure conditions. Instances of failure of hydraulic structures due to RTS are very few and no authentic information is available except in some Russian countries.

Environmental Clearance Procedure in India

In essence, the EIA is undertaken to ensure that the project on completion will be in compliance with the acceptable standards of environmental and social requirement, and within the sphere of review for a country's legislative and administrative framework as well as its relevant policy for the civil society.

A well-qualified multidisciplinary team conducts the EIA studies, which normally takes one to two years. Having identified the potential environmental impact due to activities envisaged during pre-construction, construction and operational phases of a project, the analytical model can be utilised to predict the overall benefit or damage in a quantitative way. The detailed methodologies for EIA are given by the Lohani et al. (1997) which need appropriate data collection and interpretation.

The Government of India has prescribed a maximum four-stage appraisal to accord environmental clearance for water resources projects.

- Stage 1: Screening for projects of command area less than 10,000 ha.
- Stage 2: Scoping involving preparation of a comprehensive Terms of Reference pertaining to relevant environment concerns for preparation of EIA report.
- Stage 3: Public consultation involving people affected by the project and other stakeholders.
- Stage 4: Appraisal of the project upon examination of the EIA, EMP reports and public.

Environmental clearance accorded by an Expert Committee (EC) is one of the statutory requirements for obtaining investment clearance to a project from a public investment bureau, which is the Planning Commission in case of India. Rehabilitation and resettlement, which is a crucial component of a project, cannot be initiated without the EC that stipulates a proper management plan. Impounding and diversion of water cannot start without proper environmental treatment of catchment of the intercepted basin and command area in an irrigation project. These considerations lend a significant importance to the EIA process. Along with public consultation which is mandatory in conducting the EIA, a definite trend of implementing measures to protect the environment — from conceptualisation through implementation phase — of a project has emerged. These mitigation measures are reflected in the environment management plan.

Environmental Management Plan (EMP)

An Environmental Management Plan (EMP) is formulated to manage/ mitigate the impacts based on the prediction of future environmental conditions, both quantitatively and qualitatively. The description given below details the nature of impact on the environment and the management action plan required. These plans incorporate year-wise physical and financial targets.

During the construction phase, air and noise quality monitoring programme is formulated to ensure acceptable levels. Muck disposal plan comprises identification of suitable site and detailing engineering measures for stabilising the site. Restoration of quarry sites by filling, levelling, grading and landscaping, along with protection by engineering and biological measures are also important. Development of a well-planned construction colony for staff with children parks, recreational facility, ornamental and flowering beds, avenue plantation with suitable irrigation arrangement, view point, watch and ward are also necessary. Health plan includes the assessment of technical/construction staff deployment, provision of a hospital and adequate and safe health measures, drinking water for construction and operation staff, awareness programme, etc. Sewage and Solid Waste Disposal Plan for residents including provision of free fuel for all categories of workers are also needed.

Catchment Area Treatment Plan comprises: (a) Engineering measures such as stream bank protection, contour trenching and bunding, gulley control measures, bench terracing and land slide control, and (b) Biological measures such as identifying suitable species and afforestation on forest land or government land for timber, fodder, horticulture, plantation and pasture development. Green Belt Development Plan provides plantation to insulate residential

area, and plantation along roads, dam site, powerhouse, large structures, reservoir periphery, etc., with suitable species. Reservoir
Rim Treatment Plan makes provision for arresting slides and slope
failure in vulnerable areas, and major streams, with retaining walls
if necessary.

Compensatory Afforestation Plan is the most important management measure to ensure minimal damage to biodiversity. The plant
species for afforestation, preferably indigenous, are chosen very
carefully after testing their ability to not only thrive in the new environment but also to ensure propagation of the species being affected.
In India, the Ministry of Environment and Forest accords separate
clearances for: (a) acquisition of forest land under the provision of
Forest Conservation Act, 1980 and (b) environmental angle of the
forest loss and mitigation.

Fisheries Development Plan is mandatory to provide for uninterrupted migration of fish species through low head barriers like
barrages/weirs by providing an appropriate fish ladder/fish pass. In
reservoirs, this is not possible. River flow change is required to be
minimised by ensuring that environmental flow is allowed downstream of the barrier by studying the habitat and migration need.
EFA is done with the specific objective of aiding fish migration by
appropriate methodologies currently in vogue but which are also
compatible with interpretation of the available data which is often
limited. Some of the EFA methodologies includes: (a) hydrological,
(b) hydraulic rating, (c) habitat simulation (or rating) and (d) holistic
methodologies (Tennant 1976; O'Brien 1987; Tharme 1996a, 1966b;
Arthington 2003).

The simplest, hydrological methodologies rely primarily on the
use of the hydrological data — usually in the form of naturalised, historical monthly or daily flow records — for making environmental flow
recommendations, where a set proportion of flow, often termed the
minimum flow, represents the Environmantal Flow Requirement (EFR)
intended to maintain the freshwater fishery and prominent ecological
features for an acceptable river health, usually on an annual, seasonal
or monthly basis. Extremely simplistic and the most widely used is
the 'Montana Method' (Tennant 1976), wherein environmental flow
regimes are prescribed on the basis of the average daily discharge
or the Mean Annual Flow (MAF). In general, 10 per cent of the MAF
is recommended as a minimum instantaneous flow to enable most
aquatic life to survive; 30 per cent MAF is recommended to sustain

good habitat; 60–100 per cent MAF provides excellent habitat; and 200 per cent MAF is recommended for 'flushing flows'. Particular attention is paid to protect spawning ground on pebbles, gravel and bank-side vegetation. For reservoirs, a fishery development and management plan is prepared to protect/promote indigenous species through construction/management of hatcheries, release of fingerlings, stocking of reservoir and appropriate reservoir regulation.

In a Biodiversity Conservation Plan, declaration of certain riverine stretch as restricted area is considered a necessity, particularly when the flow change can seriously affect the breeding pattern. A carrying capacity study for habitat of all aquatic species, if in abundance, is made in certain cases.

Command Area and Drainage Development Plan is drawn up to ensure good water management practices ensuring equitable distribution of irrigation water over the proposed command area by provision of field channels. The application over ill-drained (clayey) stretches needs to be carefully controlled by evaluation of crop water need and by good on-farm management. Equitable distribution is possible only by appropriate levelling and shaping of the command area. Both surface and sub-surface drains (primary, secondary and tertiary) need to be planned and constructed along with the distribution network. Drainage development has to reckon the consequence of topography, unfavourable subsoil and geology like existence of hard pan at shallow depths. As agricultural run-off have increased levels of nutrient and pesticide, a training plan for farmers to adopt optimum use of agro-chemicals have to be drawn along with regular testing of soil nutrient. In essence, a fertiliser and pesticide management plan is drawn up. As projects all over the world are faulted for inadequate rehabilitation and resettlement measures, every country — particularly India — has developed a well-formulated policy (Government of India 2006) which was amended in 2007 for rehabilitation and resettlement of Project Affected Families (PAF). Following identification of PAF at the EIA stage, compensation is estimated for the acquired agricultural land, and dwellings/trees of displaced families of different categories such as fully affected (losing home and land) and partially affected (losing land) and losing only homestead are estimated, which conform to the country policy and is in conformity with the desire of the ousted persons represented as a group. Provision of recreation/health/sanitation and education facilities in model villages is done through development of spatial and temporal maps. The plan must ensure completion of all facilities at least one year before the actual displacement takes place. Appropriate institutional arrangement, right from formulation through implementation and monitoring, must receive due attention.

Disaster Preparedness and Management Plan is also an essential component of the water resources development project which started with a preparation of an inundation map in the event of a dam break. If the water level in the reservoir rises above the stipulated maximum and even up to the top level, a status of emergency is reached and disaster thereafter occurs due to a breach. An action plan is drawn and manuals are prepared. These manuals include Emergency Action Plan which comprise Evacuation Plan and Setting of Control Centres and Preventive Action Plan with communication system set-up which describes equipments, materials, labour and expertise required at each location of deployment in an emergency along with the required personnel.

Upon completion of the EIA and EMP preparation, a public consultation/hearing, conforming to statutory regulations is held close to the project site well before the final clearance to the project is accorded (Khanna 2002). All the submissions and suggestions received in the consultation need to be incorporated in revising the EMP. Even collection of additional data may be needed as a part of EIA.

Effect of Implementing EIA

The enactment of Environment Protection Act (EPA) 1986 made a detailed EIA as a mandatory requirement for clearance of projects. Prior to 1986, every project clearance merely took note of the broad environmental-related issues and primarily, there was lack of evaluation of the pertinent status of any parameter and more important, correct estimate of protection measures was not prepared. However, the impacts on environment were substantially avoided in water resources development project after the introduction of EIA procedure. To illustrate the above statement, two examples of pre-EIA and post-EIA periods respectively are given below from Orissa, India.

Pre-EIA Project

The Upper Indravati Multipurpose Project with 600 MW installed capacity and with irrigation facilities for 128,000 ha was cleared for implementation in 1970 and completed in 1996. Techno-economic clearance and administrative clearances were accorded in 1979. This is a project with 4 large dams and 8 dykes totalling 5,000 m in length and varying in height from 10 to 73 m. The project intercepts a catchment area of 2,630 km^2 and has an average yield of 2,870 mm^3. The reservoir of 110 km^2 submerged 99 villages affecting a population of 16,000. Although half of the reservoir area was covered with forest, no compensation was paid for the loss of trees and a stipulation was made that the forest department would remove all the trees prior to submergence. Compensatory afforestation over area equal to the forest area lost was however done over non-forest government land. No biological studies on the forest species were carried out.

For rehabilitation of displaced families, more forest area was cleared for setting up the colonies with schooling, medical and community facilities. A model village was set-up first. Agricultural land with irrigation facility and home site was allocated for each family. In the absence of detailed demographic information, considerable escalation of cost occurred during displacement and rehabilitation.

The silt yield from the catchment was estimated from empirical formulae in the absence of landuse data, representative slope data, soil cover data and lack of remote sensing facility. Catchment area treatment measures were employed merely to stabilise the shifting cultivation patches and a few check dams were constructed as soil conservation measures. No biological or wildlife management plan was comprehensively done.

Post-EIA Project

A major earthen dam (30 m high) on river Ong intercepting 2,321 km^2 is being constructed in compliance with the environmental safeguards stipulated in revised guidelines (2006). The project is getting implemented as a plan scheme. The salient details of the project are given in Table 7.3. A comprehensive EIA has been carried out leading to an Environment Management Plan.

Management Measures Planned

A. Construction Phase, Involving Air, Land and Water

(a) Measures such as facilities for labour camp, water supply to residents/workers, sewage treatment, provision for free fuel, solid waste management, refuse storage, collection and transportation, restoration of construction sites.

Table 7.3: Detailed Information of the Project

Average annual rainfall	1,141 mm
75% dependable yield	340 mm^3
Live storage	289 mm^3
Length of earth dam	7,480 m
Central ogee concrete spillway	195 m
Gross command area	40,000 ha
Submergence area of reservoir	5,100 ha (51 km^2)
Private land	39.12 km^2
Forest land	2.13 km^2
Other government land	9.75 km^2

Source: Upper Indravati Project Report, Government of Orissa, Bhubaneswar, 1976.

(b) Sewage generated from various labour camps is proposed to be treated in septic tanks and disposed through absorption trenches. A total cost of ₹ 10 million (US $ 0.25 million) is earmarked for this purpose.

(c) Measures to control mosquitoes are to be adopted to destroy the habitat and interrupt the life-cycle by mechanical/biological or chemical means. Adequate arrangements are planned for the drainage of all borrow pits into the nearest natural drainage channel. Adequate vaccination and immunisation facilities shall be provided for workers at the construction site. The total expenditure for implementation of various public health measures shall be about ₹ 17.5 million (US $ 0.4 million).

(d) During construction, plantation will be developed on both sides of canals, which would improve the climatological conditions and bio-esthetics of the area. A cost of ₹ 10 million (US $ 0.25 million) is earmarked for this.

(e) Air pollution basically occurs due to primary crushing and fugitive dust from the heap of crushed material. It is planned to regularly spray the stacks with water to prevent the entrainment of fugitive emissions.

B. Biological Environment Measures

(a) Improved agricultural practices such as biological control of pests and diseases and soil conservation.

(b) Compensatory afforestation for loss of forest land is a major intervention to improve ecology. The Forest Conservation Act (1980) has stipulated strict forest protection measures and outlined procedures for compensatory afforestation, when the forest department agrees to diversion of forest for non-forest purpose. A cost of ₹ 18 million (US$ 0.43 million) is allocated for this.

(c) Suitable compensatory measures are planned for better management of the wildlife in the adjoining areas. A cost of ₹ 17 million (US$ 0.38 million) is allocated for wildlife management plan.

(d) Some measures are specifically recommended for improvement of livestock in the command area such as: (i) improve availability and quality of feed, (ii) improvement in marketing system to facilitate the movement of livestock and livestock products and (iii) establishing grass-legumes in pasture areas.

(e) Aquatic weeds are those unwanted and undesirable vegetation which reproduce and grow in water and if unchecked, may choke the water body posing a serious menace to canal management. Control measures such as manual and mechanical control, chemical control and biological control are planned. A provision of ₹ 2 million (US$ 0.05 million) has been earmarked for control of aquatic weeds.

(f) Few pockets in the command area, where groundwater levels are high, are excluded from the command area and groundwater wells are to be developed in these pockets for irrigation purpose.

(g) To promote and sustain both riverine and reservoir fish stocking and construction of fish pond for fish culture are planned. A provision of ₹ 2 million (US$ 0.05 million) is earmarked for this purpose.

C. Catchment Area Treatment (CAT) Plan

CAT plan highlights the management techniques to control erosion in the catchment area of a water resources project. Adequate preventive measures are thus needed for the treatment for the catchment for its stabilisation against future erosion. The catchment area treatment involves:

1. Understanding of the erosion characteristics of the terrain and,
2. suggesting remedial measures to reduce the erosion rate.

The Silt Yield Index Model (SYI), considering sedimentation as a product of erosivity, erodibility and areal extent was conceptualised in the All India Soil and Land Use Survey (AISLUS) as early as 1969 and has been in operational use since then to meet the requirements of prioritisation of smaller hydrologic units. The various erosion category of catchment area as per SYI index is shown in Table 7.4.

Table 7.4: Area under Various Erosion Categories

Serial number	Category	Area (ha)
1	Very low	–
2	Low	49,454
3	Medium	84,105
4	High	77,906
5	Very high	–
Total		211,465

Source: Upper Indravati Project Report, Government of Orissa, Bhubaneswar, 1976.

The objective of SYI method is to prioritise sub-watershed in a catchment area for treatment. The area under very high and high erosion categories is to be treated at the project proponent cost. A provision ₹ 3000 million (US$ 72 million) has been earmarked for Catchment Treatment Plan.

D. Rehabilitation and Resettlement Plan

The various provisions in the National Rehabilitation Policy (2207) have been kept in mind while preparing the rehabilitation and resettlement plan for the PAF of Ong dam project. There are about 6,228 PAFs who are likely to get displaced/affected due to the proposed project. Of these, 6,228 PAFs, 5,989 are in Orissa state and 239 in Chhattisgarh state. The fully affected families will be allotted land as per the availability while the partially affected PAFs will be compensated in cash.

Summary of major benefits are: (a) homestead land at the rate of 0.10 acre in rehabilitation habitat or cash equivalent at the rate of ₹ 50,000 (US$ 1,200) for self-relocation, (b) house-building assistance at the rate of ₹ 150,000 (US$ 3,600) for each displaces person, (c) transportation allowance at the rate of ₹ 2,000 (US$ 45) and (d) special benefits to displaced indigenous families and primitive tribal groups. The budget estimate for implementing this plan is estimated at ₹ 863.19 million (US$ 20 million).

E. Environmental Monitoring Programme

Monitoring is an essential component for sustainability of any water resources project. From the monitoring point of view, the important parameters are water quality, erosion and siltation, ecology, public health, aquatic weeds, socioeconomics, air quality, noise, landuse, afforestation, etc. An attempt has been made to establish early warning of indicators of stress on the environment. The Water Resources Department, Government of Orissa, is developing an Environment Management Cell (EMC) at the project office. The task of the group is to coordinate specific studies to carry out environmental monitoring and to evaluate implementation of environmental mitigatory measures.

Effectiveness of EIA Process

A few typical examples of how EIA has actually helped with Indian examples are given in detail in the section given below.

Wildlife

The stipulation of environmental clearance quite often leads to tangible protection of the environment — whether it is of land, water, biotic or socioeconomic character. Rengali, a major project in India constructed over river Brahmani, has a command area of 220,000 ha. The Left Main Canal, which runs parallel to the Brahmani river about 500 m to 1,000 m away, obstructs the migratory path of elephants who habitate over 500 km² reserve forest on the left of the command area. The elephants cross the canal from left to right and walk to the river in herds for drinking water and bathing. The project was accorded EC of MoEF of Government of India in 1998 after a critical study of their movement pattern was conducted. It was stipulated in the EC to provide a 100 m wide corridor with a ramp in to the canal from the over bank with suitable flank protection. A financial layout of 250 million Indian rupees (US $ 6 million) is earmarked for this purpose. These ramps, partly completed and under construction by the state forest department, have ensured uninterrupted migration of elephants. Salt licks are also constructed for the wildlife as the area is drought prone and in many years, it has become dry from December through June. Such provision, reflected through Environmental Management Plan, resulted in environmental safeguard.

Rehabilitation of Indigenous People

Reservoir projects generally result in submersion of villages which are quite often occupied by a large tribal population. In the EIA stage, data is collected on their culture and livelihood-earning pattern.

To ensure integrity of their cultural ethos upon displacement, an Indigenous Peoples Development Plan (IPDP), specifically tailored to their lifestyle, have been drawn for projects in Orissa. Substantial allocation of rupees 300 million (US$ 8 million) in the World Bank assisted Orissa Water Resources Consolidation Project (OWRCP), implemented from 1995 through 2003, has resulted in displaced tribals following their native lifestyle and not migrating to urban areas.

Environmental Monitoring Committee

To examine the adequacy of the environmental management plan drawn for a project and its effective implementation, an environment committee is constituted with interdisciplinary experts. The committee, through periodic field visits and collection of the environmental data, oversees the implementation of the formulated environment management plan. For water resources projects in India, a multidisciplinary inter-ministerial National Environmental Monitory Committee (NEMC-RVP) has been functioning from 1990.

Institutional Constraints in Implementation of EIA

Though EIA is essential for environment protection, there are practical difficulties on many occasions in implementation of EIA. Conducting EIA sometimes gets delayed and is fraught with difficulty for the following constraints:

Unavailability of Adequate Authentic and Technical Data

1. Lack of experts in the relevant disciplines is particularly true for South Asian Countries.
2. Inadequate EIA reports needing upgradation.
3. Poorly drafted terms and condition.
4. Late commencement in the project cycle.
5. Not managing the EIA to schedule.
6. Managing the competing and conflicting needs of ecosystem and people.

The following few steps can be taken for smooth conducting of EIA:

1. Empanelment of consultants with expertise in preparation of EIA report.
2. Making the EIA report available for the public.

3. Publishing a list of screening and final decisions taken along with conditions for approval.

References and Select Bibliography

Acreman, M. C., F. A. K. Farquharson, M. P. McCartney, C. Sullivan, K. Campbell, N. Hogson, J. Morton, D. Smith, M. Birley, D. Knott, L. Lazenby, R. Wingfield and E. B. Barbier. 2000. *Guidelines for Artificial Flood Releases from Reservoirs to Restore and Maintain Downstream Wetland Ecosystems and their Dependent Livelihoods: Final Report to DFID and World Commission on Dams.* Wallingford: Centre for Ecology and Hydrology.

Allen, R. G., L. S. Pereira, D. Raes and M. Smith. 1998. 'Crop Evapotranspiration Guideline', FAO Irrigation and Drainage Paper No. 56. Rome: FAO.

Arthington, A. H. 2003. Arthington, A. H., R. Tharme, S.O. Brizga, B. J. Pusey and M. J. Kennard. 2004. 'Environmental Flow Assessment with Emphasis on Holostic Methodologies', in R. Welcomme and T. Petr (eds), *Proceedings of the Second International Symposium on the Management of Large Rivers for Fisheries, Sustaining Livelihoods and Biodiversity in the New Millennium*, vol 2 (RAP Publication 2004/17), pp. 37–59. Bangkok: FAO Office for Asia and the Pacific.

Asian Development Bank (ADB). 2002. *Annual Report 2002*. Manila: Asian Development Bank.

Canter, Larry W. 1995. 'Cumulative Effects and other Analytical Challenges in NEPA', draft paper, University of Oklahoma, Oklahoma.

Canter, L. and R. Clark. 1997. *Environmental Impact Assessment*, 2nd ed. New York: McGraw-Hill Publishing Company.

Central Water Commission (CWC). 2001. 'River Valley Projects and Environment: Concerns and Management', CWC, Delhi.

———. 2003. *Annual Report 2002–2003 of National Environmental Monitoring Committee for River Valley Projects*. Delhi: Government of India, Ministry of Water Resources, Central Water Commission, Environmental Management Directorate.

Chen, D., X. Duan, S. Liu and W. Shi. 2003. 'Status and Management of Fishery Resources of the Yangtze River', in R. Welcomme and T. Petr (eds), *Proceedings of the Second International Symposium on the Management of Large Rivers for Fisheries, Sustaining Livelihoods and Biodiversity in the New Millennium*, vol 2 (RAP Publication 2004/17), pp. 173–82. Bangkok: FAO Office for Asia and the Pacific.

Das, B. P. 2002. 'Environmental Flow and River Protection for Salandi, Orissa, India', in P. Greenfield and S. Ward (eds), *Proceedings of River*

Symposium 2002, The Scarcity of Water, The Future of Rivers, The Future of Waters. Brisbane: IWA Publishing.

Das, B. P. 2004. 'Drainage Problem in Mahanadi Delta, India and Case Study of a Remedial Direct Cut'. Paper presented at the Symposium on River Restoration and Urban Streams of the World Water and Environmental Congress, 27 June–1 July, Salt Lake City, Utah.

Das, B. P. 2005. 'Environmental Problem of Drainage Congestion in Mahanadi Delta, India: Case Study of a Remedial Direct Cut'. Paper presented at the Symposium on Impacts of Global Climate Change of the World Water and Environmental Congress, May 15–20, Anchorage, Alaska.

Das, B. P. and S. Mishra. 2007. 'Impact of Water Resources Development on Bhitarkanika Mangroves in Brahmani Estuary', *International Journal of Ecology and Environmental Sciences,* 33 (4): 243–53.

Food and Agriculture Organisation of the United Nations (FAO). 1998. *Crop Evapotranspiration — Guidelines for Computing Crop Water Requirements.* Rome: FAO.

Freeman, Mary C., Zachary H. Bowen, Ken D. Bovee and Elise R. Irwin. 'Flow and Habitat Effects on Juvenile Fish Abundance in Natural and Altered Flow Regimes', *Ecological Applications,* 11 (1): 179–90.

Gleick, P. 2002. *The World's Water 2002–2003: The Biennial Report on Freshwater* Resources. Washington, D.C.: Island Press.

Government of India. 2002. 'National Water Policy'. Delhi: Ministry of Water Resources, Government of India.

———. 2006. 'National Policy on Resettlement and Rehabilitation for Project Affected Families', Ministry of Rural Development, Delhi.

International Commission on Irrigation and Drainage (ICID). 2004. 'Country Policy Support Programme (CPSP), Report on Water Policy Issues: India Country Study (Basin Level Assessment and National Consultation)', ICID, Delhi.

———. 2005. 'Water Resources Assessment of Brahmani River Basin, India', ICID, Delhi.

Jansson, R., C. Nilsson, M. Dynesius and E. Anderson. 2000. 'Effects of River Regulation on River-margin Vegetation: A Comparison of Eight Boreal Rivers', *Ecological Applications,* 10: 203–24.

Jones, G., T. Hillman, R. Kingsford, T. McMahon, K. Walker, A. Arthington, J. Whittington and S. Cartwright. 2002. 'Independent Report of the Expert Reference Panel on Environmental Flows and Water Quality Requirements for the River Murray System', Report to the Murray–Darling Basin Ministerial Council, Cooperative Research Centre for Freshwater Ecology, Canberra.

Kapoor, A. S. 2000. 'Bio-Drainage Feasibility and Principles of Planning and Design, Role of Drainage and Challenges in 21st Century'. Paper presented at the 8th ICID International Drainage Workshop, 31 January–4 February, Delhi.

Koehn, J. D. and S. J. Nicol. 2004. 'A Strategy to Rehabilitate Native Fish in the Murray-Darling Basin, South-eastern Australia', in R. Welcomme and T. Petr (eds), *Proceedings of the Second International Symposium on the Management of Large Rivers for Fisheries, Sustaining Livelihoods and Biodiversity in the New Millennium*, vol. 2, pp. 141–62. Bangkok: FAO Office for Asia and Pacific.

Khanna, R. K. 2002. 'Environmental Impact Assessment and Clearance of River Valley Projects'. Paper presented at the IWRA Regional Symposium 'Water for Human Survival', organised by Central Board of Irrigation and Power, 26–29 November 2002, Delhi.

Lohani, B. N. and N. C. Thanh. 1980. Impacts of Rural Development and their Assessment in Southeastern Asia, Environmental Conservation 7 (3): 213–16.

Lohani, B., J. W. Evans, H. Ludwig, R. R. Everitt, Richard A. Carpenter and S. L. Tu. eds. 1997. *Environmental Impact Assessment for Developing Countries*. Manila: Asian Development Bank (ADB).

McCully, P. 1996. *Silenced Rivers*. London: Zed Books.

Memon, A. A. 2005. 'Devastation of the Indus River Delta'. Paper presented at Proceedings, World Water and Environmental Resources Congress 2005, American Society of Civil Engineers, Environmental and Water Resources Institute, 14–19 May, Anchorage, Alaska.

Murray-Darling Basin Ministerial Council. 2002. *The Living Murray: A Discussion Paper on Restoring the Health of the River Murray*. Canberra: Murray-Darling Basin Ministerial Council.

O' Brien, J. S. 1987. 'A Case Study of Minimum Stream Flow for Fishery Habitat in the Yampa River', in C. R. Thorne, J. C. Bathurst and R. D. Hey (eds), *Sediment Transport in Gravel-bed Rivers*, pp. 921–46 Chichester: John Wiley and Sons.

Qureshi, M. T. 2000. *Indus Case Study Report*. Karachi: IUCN.

Sugunan, V. V. 1995. *Reservoir Fisheries of India*. FAO Fisheries Technical Paper 345. Manila: FAO.

Tennant, D. L. 1976. 'In-stream Flow Regimens for Fish, Wildlife, Recreation and Related Environmental Resources', *Fisheries*, 1 (4): 6–10.

Tharme, R. E. 1996a. 'A Global Perspective on Environmental Flow Assessment: Emerging Trends in the Development and Application of Environmental Flow Methodologies for Rivers', *River research and Applications*, 19 (5/6): 397–442.

———. 1996b. 'Review of International Methodologies for the Quantification of the Instream Flow Requirements of Rivers'. Water Law Review. Final Report for Policy Development, for the Department of Water Affairs and Forestry, Pretoria, Freshwater Research Unit, University of Cape Town, p. 116.

Varis Olli, Marko Keskinen and Matti Kummu. 2008. 'Mekon at the Cross Road', *Ambio*, 37 (3): 146–49.

Zalinge, N. Van, P. Degen, C. Pongsri, S. Nuov, J. G. Jensen, V. H. Nguyen and X. Choulamany. 2003. 'The Mekong River System', in R. Welcomme and T. Petr (eds), *Proceedings of the Second International Symposium on the Management of Large Rivers for Fisheries, Sustaining Livelihoods and Biodiversity in the New Millennium*, vol. 2, pp. 335–57. Bangkok: FAO Office for Asia and the Pacific.

8

An Integrated Fire and Water Management Strategy Using the Ecosystem Approach: Tram Chim National Park, Vietnam

Peter-John Meynell, Nguyen Huu Thien,
Duong Van Ni, Tran Triet, Martin van der Schans,
Deanne Shulman, Julian Thompson,
Jeb Barzen and Gill Shepherd

This chapter describes the application of the ecosystem approach for restoring the diversity of wetland habitats within Tram Chim National Park (TCNP), a remnant wetland of the Plain of Reeds in the Vietnamese Mekong Delta. During 2005 and 2006, a series of multi-disciplinary studies and adaptive management strategies provided a deep understanding of the relationships between hydrology, soil and water chemistry, and the plant and bird populations in the park. These studies illustrated how changing hydrological management had altered the components of the ecosystem. The fire and water strategy was designed around a combination of appropriate water level management and fire risk management measures which would allow the recovery of some of the original habitats in the Plain of Reeds. Implementation of this strategy has resulted in a remarkable recovery of the ecosystem in the park and influenced national park management policies in Vietnam.

Tram Chim is an area of grassland and wetland forests in the upper part of the Mekong Delta, close to the Vietnamese/Cambodian border. It is a remnant of the original Plain of Reeds — a large depression that was seasonally flooded by the Mekong river, but which has now been largely converted to irrigated rice fields. The TCNP was established to conserve a portion of the Plain of Reeds wetland habitat, especially for the Eastern Sarus Crane (*Grus antigone sharpie*) populations that use the grasslands for feeding during the dry season. Development of the fire and water management strategy for Tram Chim was one of the principal initiatives of the Mekong Wetlands Biodiversity Programme (MWBP), a Global Environmental Facility (GEF) funded project aimed at conservation and sustainable use of wetlands in the four countries of the Lower Mekong — Cambodia, Lao PDR, Thailand and Vietnam. It was implemented by the United Nations Development Programme (UNDP), International Union for Conservation of Nature (IUCN) and the Mekong River Commission between 2004 and 2007.[1] (www.mekongwetlands.org).

History of Change at Tram Chim National Park

Site Description

The TCNP was established in one of the lowest areas of the Plain of Reeds, a 20,000 ha area that was devoid of permanent human settlement and was called Tram Chim, meaning rear mangrove and bird swamp (Map 8.1). About 42,000 people now live on the edges of

Map 8.1: Location of the Tram Chim National Park in the Mekong Delta

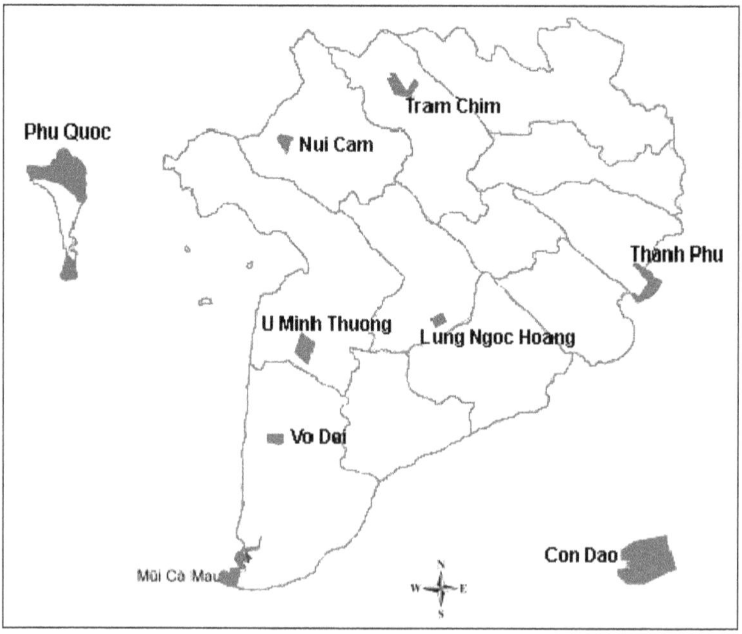

Source: BirdLife International and Forest Inventory and Planning Institute (2004).

the park, the majority of whom are poor farmers practicing intensive agriculture and aquaculture.

The park is a wetland complex comprising a mosaic of seasonally inundated grassland, regenerated *Melaleuca cajiputi* (rear mangrove) forest and open swamp (BirdLife 2004). There are six vegetation communities of which those dominated by *Eleocharis dulcis* and wild rice (*Oryza rufipogon*) are of highest conservation significance. The site supports 231 species of birds (MWBP 2005),[2] 130 species of plants (Pham Trong Thinh 1998), and 150 species of fish, one-third of the species known to occur in the Mekong Delta.[3] Of the 231 species of birds, 15 are globally endangered, threatened or near-threatened (Bao Hoa N. P. et al. 2006). The park's flagship species is the Eastern Sarus crane which is globally and nationally vulnerable.

The original Plain of Reeds was a floodplain of the Mekong river having no artificial canals and was covered with a thick mat of vegetation and scattered Melaleuca forest. At the onset of the high water season in the Mekong, water first penetrated slowly through small streams flowing up from the Mekong river. Later in the rainy season sheet flow, coming from the upper part of Prey Veng Province, Cambodia, floods across the Plain of Reeds until the basin fills and the remaining sheet flow continues to the sea. Through the dry season, water trapped in the depression is released slowly through small natural streams and by evapotranspiration. This floodplain recessional wetland supported tremendous biodiversity and regulated the 'heartbeat' of the hydrological regime of the lower Mekong River (ActionAid 2002).

During the conflicts that began in 1945, the Plain of Reeds was devastated primarily by napalm burning and drainage by the French, US and South Vietnamese military which aimed to remove vegetation that obscured activities of revolutionary forces (Le Dien Duc 1989; Tran Triet et al. 2004). Defoliants were used extensively in the Plain of Reeds (ibid.). Following reunification, approximately 700,000 hectares were converted to agriculture from 1975 to 1995 (Shulman 2002). In addition to changes in land use, seasonally flooded grasslands and forests have been fragmented by canals. Today, the hydrological regime of the Plain of Reeds is dramatically modified by the dense network of canals that abnormally lowers water levels during the dry season, desiccating major plant communities and increasing soil acidification through oxidation of acid-sulphate soils

(Le Quan Minh et al. 1998). In addition, major axial canals (Dong Tien, An Binh and Hong Ngu) connect the Mekong and Vam Co Tay rivers, enabling unnatural, rapid flow of water in and out of the basin (Beilfuss and Barzen 1994).

History of Management

Following reunification (1975), the lowest area of the Plain of Reeds, called Tram Chim, was identified as a place to remind people of the wetland environment that succoured resistance forces during the war. From the original extent of 20,000 ha., 7,000 ha. was set aside as a restoration area, while in remaining areas programmes to settle people and develop wetlands into active agricultural areas were initiated. Funds from the national government were used to dig more canals for irrigation and transportation as well as build bridges and other infrastructure, while provinces accepted the immigration of thousands of people. In this early stage, restoration activities had to be self-sufficient so the production of natural wetland products was encouraged. Fish, Melaleuca, grass (for grazing) and wild rice were the primary wetland products. Production, however, was hampered by residual impacts from the war. Drainage canals dug during the war, and new canals dug for development desiccated vegetation even more and extensive fires during the dry season destroyed plantations of Melaleuca.

In 1985, Tram Chim was established as an Agro-Forestry-Fishery enterprise aimed at preserving an image of the original Plain of Reeds while planting Melaleuca for commercial purposes (Map 8.2). New in this enterprise was the construction of small dykes that held water back from draining into canals. Though built initially to keep planted Melaleuca saplings from burning, these dykes expanded the restoration process by attempting to offset the desiccating effect of canals. Early dykes were relatively small (0.5–1.0 m high) so that rainy season sheet flows would still move through the basin. Efforts to re-establish more natural hydro-periods at Tram Chim not only helped protect Melaleuca but also allowed many other plant and animal species to recover. Sarus cranes first returned in 1983[4] but it wasn't until 1985 that scientists in Vietnam corroborated this recovery (Le Dien Duc 1991).

With the returning cranes, Dong Thap Province began to ban the hunting of the crane and established a district protected area.

Map 8.2: Management Zones of Tram Chim National Park

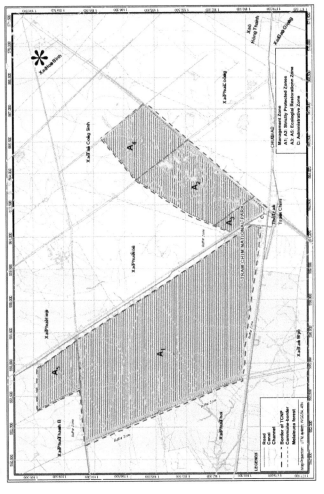

Source: BirdLife International and Forest Inventory and Planning Institute (2004).

Note:
1. 'Xa': commune (this is an administrative unit under a district which in turn, is under a province).
2. Names of communes: from lower corner (clockwise): Tan My commune, Phu Thanh commune, Phu Thanh B commune, Phu Hiep commune, Phu Duc commune, Tan Cong Sinh commune, Hoa Binh commune, Hung Thanh commune, Gao Giong commune, Tram Chim Town.

In 1991, the area became a Provincial Nature Reserve and in 1998 Tram Chim was recognised as a National Park by a decision by the prime minister (Nguyen Xuan Vinh and Wyatt 2006). Wetland management meetings were held from 1990 to 2005 to evaluate restoration actions and propose management objectives for Tram Chim. The meeting in 2004 led to the work which is presented in this chapter.

As a National Park, Tram Chim falls under the Special Use Forest (SUF) System. This has created a problem because the policy was designed for upland forest ecosystems, rather than for wetland ecosystems. The SUF legislation strictly prohibits fires and makes the managers of protected areas and provincial chairpersons personally responsible if they should occur (Van der Schans and Nguyen Huu Thien 2008).

Tram Chim National Park has a high incidence of unplanned fires. Some are accidental but park management believes that the majority are caused by local arsonists who resent the prohibition on local communities from utilising park resources (ibid.). Local people also light fires to improve resource extraction. For these reasons, the park management has aimed at maintaining high water levels all the year round to prevent fire. Water levels are maintained within dykes constructed around the four principal park management zones. In addition, more canals and ponds have been dug as fire breaks and water sources for fire fighting purposes. While the high water levels have reduced the immediate risk of fire, the long-term effectiveness of these measures is debatable as is the impact on species other than Melaleuca (Van der Schans and Nguyen Huu Thien 2008).

Pressures and Impacts of Management

BirdLife (2004) stated that the most important factor in maintaining suitable habitat for the Sarus Crane is appropriate management of water levels. In the year 2000, a partial draw-down of water levels was conducted followed by more extensive draw-downs in 2001 and 2002, with much evidence of recovery of natural vegetation. Unfortunately since 2002, due to the fear of fire, once again year-round high water levels were maintained in the park until 2005.

This inappropriate hydrological management at Tram Chim has led to degradation of grasslands, including the previously extensive Eleocharis beds, which are dry season habitats for the Eastern Sarus Crane and many other bird species. Crane populations declined precipitously from 1988 when over 1,000 birds (perhaps most of the known population) spent part of the dry season at Tram Chim to a low of only 13 per cent of the known population in 2005 (Nguyen Phuc Bao Hoa et al. 2008). The National Park was increasingly unable to fulfil the original objectives for which it was set-up and a new, less diverse and unnatural lacustrine wetland ecosystem started developing. In addition to abnormal inundation periods, dykes surrounding the Park disrupted the natural sheet flow of flood waters from the Mekong river across the Plain of Reeds. This resulted in more concentrated flood flows through the sluice gates built within the dykes. Canals constructed within the park have also affected sedimentation patterns.

Other impacts included deterioration of water quality as decomposition and the water-borne removal of organic matters were significantly reduced by the high dykes. Ironically, fire risk also increased with the accumulation of organic matter from seasonal grass cover and tree litter beneath Melaleuca trees. The growth of Melaleuca, the target of fire prevention action within the park, was also sluggish compared to areas outside the park which experienced more seasonally varying water levels. Fish populations and diversity were affected significantly as the tall dykes hindered the intake of fish fingerlings and eggs from the Mekong's water during the early flood season.

Since the inception of the park, conservation at Tram Chim began with the premise that local communities should be excluded. The park management tries to prevent people from entering the park by posting armed guards at stations on the periphery dykes to arrest those poaching resources. This has resulted in much resentment from the local community towards the park (Nguyen Huu Thien 1997). The majority of people who live near the park are poor and depend on the resources inside the park such as fish and wildlife for protein, wood for fuel and grass for fodder. Although the local community benefits from these resources, they tend not to appreciate the significance of the conservation efforts made.

Fire and Water Management Strategy (FWMS)

Development and Approach of FWMS

In early 2004, the (former) TCNP director reached an agreement with the MWBP and the International Crane Foundation (ICF) to develop an integrated FWMS for the park. Further building of drainage and canals were restricted. A monitoring programme encompassing hydrology, biological diversity (plankton, plants, fish and birds) and soil/water chemistry was initiated to evaluate experimental dry season draw-downs. To support these experiments, improvements were made to water control structures, damaged dykes and monitoring infrastructure, most notably staff gauges. The MWBP-funded studies were undertaken by a team consisting of local (e.g., Can Tho University) and international (e.g., University College, London) universities, fire ecology experts from the US Forest Service as well by staff of MWBP, ICF and TCNP.

The basic hypothesis of the FWMS was that management of water levels in the park, aimed at reducing the risk of catastrophic fire in the Melaleuca forests, was causing the degradation and loss of some of the key habitats, namely the grasslands. If dry season water levels could be returned to more natural hydroperiods described qualitatively in Beilfuss and Barzen (1994), whilst simultaneously managing fire risk more effectively, then the Tram Chim ecosystem might recover and the population of associated species, such as the Sarus cranes, rebound.

An ecosystem approach was fundamental to the development of the FWMS. It involved the application of a multidisciplinary team, considering ecosystem structure, function and interaction of species. The relationships between the key stakeholders, the park management and the surrounding communities, not always on easy terms, was considered as part of the solution to the issue of fires caused by humans, and the process of adaptive management, seeking practical solutions based upon the best available scientific and local information, was adopted.

The project started with an inception phase in March 2005 during which the team jointly developed a work plan that consisted of a series of research activities to guide immediate management actions of the park. The work plan incorporated inputs from the team as well as from provincial and national government representatives. Research activities addressed four major ecosystem components: hydrology, chemical and physical properties (soil and water chemistry), biology and fire, as well as their interactions. Two basic approaches were employed to guide data collection:

1. Survey and routine monitoring of key components of the ecosystem.
2. Monitoring of targeted management interventions designed to answer specific research questions and putting on trial potential management techniques.

Hydrology

As with wetlands worldwide (Mitsch and Gosselink 2000; Thompson and Finlayson 2001), hydrology plays a dominant role in influencing ecological conditions within the park (Beilfuss and Barzen 1994). This includes the distribution and extent of vegetation communities, habitat availability and quality for birds, fish migration and the incidence and severity of fire. Hydrological studies undertaken within the park aimed to provide insights into these relationships whilst experimentally implementing water management plans intended to restore historical hydroperiods following Beilfuss and Barzen (1994).

Temporary Water Level Control Measures

Two major improvements were made to the hydrological management of the park. First, the park management agreed to conduct a dry season draw-down of surface water in park management Zone A1 (Map 8.2). The goal was to measure water level responses to sluice gate operation throughout the largest section of the park. The recovery of vegetation as a result of the drier conditions was monitored alongside hydrological observations at the same location. The draw-down also facilitated monitoring of migratory fish movement into the park.

The ability to control water levels in other sections of the park, most notably Zones A4 and A5, had been limited since 2001 due to breaks in the surrounding dykes which were not repaired. As a result, these sections were permanently connected to the surrounding canal network, allowing them to desiccate in the dry season. Water management goals for these two sections was to have more extensive draw-down than planned for Zone A1 but not as extensive as what would occur with direct connection to the surrounding canals. As part of the interim strategy, simple water control structures (concrete spillways) were installed in the dykes surrounding Zones A4 and A5 to provide a means of providing planned water levels. Water level control was not possible in Zone A3 for this study.

Development of Target Water Levels

Wetland hydroloperiods generally exert a major control on vegetation species and their distribution (Wheeler and Shaw 1995; Wheeler et al. 2004). The effect of water levels upon both species growth and community composition can depend upon water elevation, duration, frequency and timing. Some plant species are sensitive to alterations in one or more of these parameters. Vegetation surveys from TCNP described distinct vegetation communities that corresponded to a soil elevation gradient. Hence flood depth and flood duration may be important. While *Panicum* had a wide tolerance to water level and grew well in both dry and wet conditions, the other graminoides such as *Eleocharis, Ischaemum* and wild rice had narrower ranges of tolerance to flood depth or duration. Melaleuca had a wide water level tolerance range within lower parts of the area whilst lotus was only found within deeper areas defined by permanent flooding. These plant surveys enabled the establishment of optimum hydrological conditions for the dominant plant communities within the park.

The combined result of a topographic survey undertaken for Tram Chim and water level records obtained from this study and historical park records enabled the establishment of a proposed hydrological regime throughout the park. The spatial pattern of hydrological conditions (principally duration of inundation) was compared to current vegetation patterns obtained through field surveys and interpretation of remotely sensed imagery. When compared to optimum hydrological conditions for the dominant plant communities, much of the park, notably Zones A1 and A2, were too wet for plant communities to persist. Current hydroperiods, as represented by water levels in early 2004, differed markedly from hydroperiods characteristic of the original Plain of Reeds ecosystem (Beilfuss and Barzen 1994). Prior to 2004, water level management had been driven by the overriding priority to prevent fire. However, this also induced negative conditions for characteristic vegetation and fauna associated with the Plain of Reeds.

Target water levels, reflecting more natural hydroperiods, were developed based on: (i) the conceptual understanding of the hydrology of the park including a simple water balance model; (ii) the distribution of dominant vegetation communities, and (iii) an idealised feeding pattern for Sarus Cranes in which grassland in different management zones of the park become available at different times as they dry out. Target water levels for the four management zones of TCNP (Table 8.1) were specified as the water elevation on the first day of each month and were provided for only the dry season since wet season water levels are dictated by the height of the flooding which was beyond the control of park management. To aid the implementation of draw-downs, target water levels were also translated into sluice gates operation guidelines.

Table 8.1: Target Water Levels for Tram Chim Management Zones (Units = cm a.m.s.l.)

	Target water level			
Date	Zone A1	Zone A2	Zone A4	Zone A5
1 January	143 ± 5	161 ± 5	At least 110	At least 137
1 February	126 ± 5	144 ± 5	93 ± 5	120 ± 5
1 March	110 ± 5	128 ± 5	77 ± 5	104 ± 5
1 April	92 ± 5	110 ± 5	59 ± 5	86 ± 5
1 May	73 ± 5	91 ± 5	At least 40	At least 67
1 June	At least 60	At least 80	No target	No target
1 July	At least 70	At least 75	No target	No target
1 August	No target	No target	No target	No target
1 September	No target	No target	No target	No target
1 October	No target	No target	No target	No target
1 November	No target	No target	No target	No target
1 December	At least 156	At least 174	At least 123	At least 150

Source: Ni et al. (2006).

Water Control Structures and other Hydrological Infrastructure Modifications

Although some of the dykes which surround the different management zones of TCNP have water control structures that allow water to flow in and out, dykes still reduce the exchange of water, nutrients, organic matter and fish by obstructing sheet flow through the park. Increasing the height of these dykes from 3 m to 5 m in 2001 dramatically increased these barrier effects. Internal dykes from canals built to provide access for fire control have exacerbated this trend by further altering water flow. Internal canals act as preferential pathways for movement of water and induce higher flow velocities than those experienced under historic sheet flow conditions. In order to allow more frequent sheet flow and increased fish migration, new structures were designed for a number of locations in the surrounding dykes whilst the removal of the dykes alongside the internal canals was recommended to enhance water exchange between the canals and surrounding land. The planting of vegetation buffer zones around these control structures to trap sediments as they flow over the new structures was also recommended.

Impacts of Changes in Water Management During the Interim Period

At the TNCP, inappropriate hydroperiods existed at two extremes: in Zones A1 and A2, water levels were managed at levels much greater than natural hydroperiods whereas Zones A4 and A5 had much drier than natural hydroperiods. Management objectives for 2005 and 2006 were designed to approximate natural hydroperiods in Zones A1 and A2 while improving wet conditions, but still conducting earlier draw-down than natural hydroperiods would allow, was the objective for A4 and A5.

A remarkable recovery of Eleocharis followed the draw-down experiment within Zone A1 during 2005. This regrowth, however, was from seeds. Underground tubers had already been killed by high water levels in previous years. A return to high water levels in subsequent years would quickly remove Eleocharis as their roots and tubers would not be fully developed after only one year of recovery. Prolonged health of Eleocharis, including the production and sustained growth of tubers, requires that lower water levels are repeated in subsequent dry seasons.

Prior to 2005, breaches in dykes surrounding Zones A4 and A5 resulted in extremely dry conditions, resulting in high fire risk for Melaleuca and poor growth of some graminoid species, most notably Eleocharis. The Eleocharis beds were of particular concern because they were one of the last feeding grounds available for the Sarus cranes following the imposition of high water level management in Zones A1 and A2 which inhibited growth of Eleocharis. Repairs to the dykes and construction of spillways resulted in the return of wetter conditions during 2005 and 2006, although the late closure of the water control structures in Zone A5 created conditions that were slightly too dry. Fire risk also declined with wetter conditions.

Although the impacts of the spillway construction on fish species movement in and out of the park were not systematically assessed, local fishermen suggest that passage and survival rates of fish through the spillways are high compared to previous years when fish migration only occurred through existing sluice gates (Marttin and De Graff 2002).

Fire Management

All fires in the park are caused by humans. The number and distribution of ignitions are thus directly correlated with the presence of people, their activities and behaviour patterns, and the amount of dry accessible area within the park. Efforts to educate local people are paramount to reducing unwanted fires and ongoing programmes in local communities that address this issue are described in the Park Fire Management Plan. Controlling water levels in the park, as a management strategy, also directly determines the amount of dry land accessible to people and their use patterns.

The frequency and intensity of fire are determined by weather, fuels and topography. With respect to fire behaviour, topography in TCNP is flat and weather is independent of human control. So the focus of fire-related management actions is limited to manipulating fuels and monitoring weather parameters that can be linked with preparedness actions.

Under the MWBP, the team reviewed relevant fire literature, collected data on the fire environment for analysis and modelling. Concomitantly, the park staff was trained in all aspects of these activities. The objectives were to describe the fire ecology of key species, quantify the fire environment and analyse hazard and risk under varying management strategies. The larger goal was to evaluate potential management actions as they relate to fire risk, specifically from varying water levels and fuel manipulation.

Data on water level, factors influencing fire behaviour and fire occurrence were compared to fire frequency, fire behaviour in both grasslands and Melaleuca, and grass fire spreading into Melaleuca stands. Of particular interest was the amount of dry ground created by varying water levels as it relates to occurrence of fire. Once conditions are present for a successful ignition, elements of the fire environment determine the intensity of fire and its rate of spread.

The team also analysed how fuel manipulations such as prescribed burning in grass, cutting and removal of grass, thinning and pruning of Melaleuca and prescribed burning in Melaleuca might influence fire behaviour during subsequent unwanted fires.

The data collected and tools developed included: (a) historic water level data at TCNP, (b) historic weather data, (c) historic fire occurrence records (d) on-site weather data, (e) fuel moisture in both grassland and Melaleuca sites, (f) grassland biomass, (g) a review of the literature on fire effects on Melaleuca and key grassland species including site visits, (h) fire modelling computer software, (i) descriptors of standard fire behaviour fuel models, and (j) descriptive information from experienced fire fighters on fire behaviour characteristics such as flame length and rate of spread.

The association between water level, dry land area, and available fuels and fire occurrence was then modelled using above data as shown in Figure 8.1. Although an ignition source may be present, the probability of an unwanted fire developing into a wildfire where suppression is required depends upon weather information. Historic weather information was used to estimate the probability of sustained fire occurrence as an indicator during the dry season of the degree of preparedness and mitigation measures required.

Figure 8.1: Relationship between Water Level, Dry Land Area and Probability of Fire Occurrence

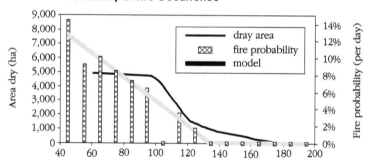

Source: Ni et al. (2006).

To describe the effectiveness of suppression activity for varying fuel manipulations, predicted fire behaviour was modelled for three grass models representing existing conditions, effects of prescribed fire and cutting and removing grass. Both prescribed fire and mechanical removal of grass reduced the severity of fire behaviour in grasslands and increased chances for fire suppression crews to control the fire. In a similar modelling analysis for Melaleuca fuels, similar results occurred: thinning and pruning as well as prescribed fire improved the probability of successful control of wildfire. Comparison of flame length and rate of spread under differing fuel load and structure within Melaleuca stands demonstrate two points: (a) tree mortality will be considerably lower in the event of a wildfire if the stand has been previously treated with biomass reduction though cutting or prescribed fire and (b) suppression crews will be more effective suppressing a fire where lighter fuel loads occur. The proposed management tools and interventions related to fire, advocated as a result of MWBP programme activities within Tram Chim, aimed to re-establish the important natural role of fire within the ecosystem (i.e., frequent, low intensity fires) while at the same time lowering the risk of high intensity fire which occurred less frequently but altered vegetation more dramatically.

The first recommended change in park management was the official recognition of the important and natural role that fire plays in shaping the ecosystem of the Plain of Reeds and that species native to this ecosystem, including Melaleuca, are well-adapted to fire as an abiotic process. Given that potential fire incidence will be higher with natural water level regimes, management tools and interventions to reduce the risk of uncontrolled fires were developed, primarily through manipulation of fuels.

1. The Park and Forest Protection Department (FPD) should establish routine practices to collect site-specific information on weather and fuel parameters that inform decisions related to fire suppression (e.g., preparedness levels, detection patrols, community awareness and timing of prescribed burning).
2. The FPD should continue to use prescribed fires within grassland areas as a means of reducing the numbers of wildfire ignitions and enhancing the ability of firefighters to suppress wildfires. Prescribed fire reduces the dead fuel component provided by the mat of dead grass in the following years resulting in decreased flame lengths. Conducting prescribed fires in corridors of human traffic and higher elevation areas (i.e., dryer) also eliminates fuels where fires have historically been started.
3. A prescribed fire programme in Melaleuca should be initiated, following experimentation, to develop optimal prescriptions for decreasing fuel loads and otherwise lowering risks associated with uncontrolled fire.

4. Reduction of fuel loads can also be achieved through the carefully controlled extraction of biomass by people. Within Melaleuca, simple silviculture practices such as thinning, pruning (e.g., cutting of lower branches) and the collection of dead woody material on the ground will reduce available fuels and intensity of fire. Similar positive impacts can be achieved through the removal of biomass from grassland areas by removing grass through cutting or grazing activities. When combined with prescribed burning, this non-fire removal of vegetation will reduce the severity of fire behaviour in grasslands and increase the ability of fire suppression crews to control the fire. Enabling people to extract resources from the park may also have the additional advantage of supporting rural incomes and improving relationships between the park and local communities.

5. Enhanced capabilities for fighting fire and use of prescribed burning as a fire management tool can be achieved through the adoption of new techniques and acquisition of additional equipment. The Fire Management Plan developed through collaboration with MWBP, the Forest Protection Department and Park Management details recommended management of fire within the National Park.

Ecosystem Recovery

Biomass Accumulation in Melaleuca Woodland under Waterlogged Condition

Melaleuca forest structure and above-ground dead biomass were measured at three forest plots representing 5 years, 10 years and 15 years under waterlogged condition. Under prolonged inundation, dead plant material increased on forest floor dramatically. The mean value of dried dead biomass on the ground in all plots was 15.4 kg m^2 and the average thickness of dead biomass layer was 12 cm. In some areas, the thickness of the dead biomass layer was as much as 80 cm. There are also many fallen Melaleuca trees in these forests. Long periods of permanent inundation cause Melaleuca to fall over (Pham Trong Thinh 1998). The fallen Melaleuca will eventually die, adding more dead biomass on the ground. Field observations confirmed that most Melaleuca forest areas in Zone A2, the most waterlogged area, were under a high risk of catastrophic fires because of the huge, accumulated fuel load.

Three factors contributed to the high rate of biomass accumulation in the Melaleuca forest of Zone A2 (the same conditions hold for some forest areas in A1):

1. Long period of inundation slows down the rate of organic matter decomposition,
2. The presence of high dykes and the absence of sheet flow during flood season reduces the amount of dead organic matters being carried away by sheet water outflow,
3. Melaleuca stems fall due to poor growth of roots.

Melaleuca Forest Regrowth after Fire

The structure (stem height, diameter and tree density) of Melaleuca forest regrowth after being burned by a wild fire in 1995 was compared to those of adjacent unburned forest. After 10 years of regrowth, the forest had regained the same tree density and stem diameter and were only slightly shorter than nearby unburned trees. This observation illustrates the ability of Melaleuca trees to quickly recover from fire and also the long-term negative impact of inundation by water where trees become stunted and eventually stop growing.

Response of Grasslands Following Prescribed Burns

Prescribed burning has been applied to grasslands that were contiguous to Melaleuca stands at Tram Chim to reduce fuel loads and risk of fire in the forests. It is important to understand how frequent fires will affect grassland communities. On 18 March 2005, Tram Chim Forest Protection Office conducted a prescribed burn on a large area of grassland in Zone A1. After the fire, the regrowth of five types of grassland communities were monitored: *Leersia hexandra*, *Ischaemum rugosum*, *Oryza rufipogon*, *Eleocharis dulcis* and *Panicum repens*. One year after burning, *Panicum* grassland had become more widespread. Areas of *Ischaemum*, wild rice and *Eleocharis* grasslands seemed to be reduced. If this result is repeatable, the expansion of *Panicum* grassland should be of concern as it could potentially reduce the overall diversity of Tram Chim grasslands. Confounding factors in this experiment include the maintenance of high water levels after 1995. In recent years, Panicum has become over-abundant in Zone A4, replacing most of Eleocharis grasslands, which was one of the main feeding areas of the Eastern Sarus crane. Panicum has strong rhizome growth and can recover quickly after being burned. Water management in Zone A4 has been drier than normal. Therefore hydrology can again be a confounding factor in this experiment. Though frequent fires may favour the reestablishment of some grassland species over others, the outcome of competition among plant species may further depend upon hydrology or other ecosystem processes currently unknown. Post-fire competition of grassland species should be studied further to assist the planning and design of prescribed burns.

Response of Eleocharis Grassland to Water Draw-down

The reduction of Eleocharis communities at Tram Chim has been one of the main management concerns and was thought to be correlated with water level management. Because cranes eat tubers of Eleocharis, concern focused not only on the area where Eleocharis was available but also the production of tubers which grow below the ground. In the 2005 dry season, the water level inside Zone A1 was drawn down. Coupled with low dry season precipitation that year, the soil of most of Zone A1 was exposed by March for the first time in a decade. Eleocharis reoccurred in large areas where mudflats had existed in previous years. Two transects were established to monitor the regrowth of Eleocharis. After one year of growth following the draw-down, above-ground biomass (aerial stems) no longer changed but below-ground biomass (roots and rhizomes) continued to increase. Importantly tuber production, after just one year of growth, did not occur. The average rhizome length between two adjacent Eleocharis clumps was significantly shorter in young Eleocharis of the draw-down area compared to old-growth Eleocharis clumps in Zones A5 and A4 where draw-downs still occur annually. This pattern of growth in Zone A1 is consistent with a plant community that is regenerated from seeds rather than from the underground rhizomes. It is likely that underground rhizomes and tubers may have died after being under permanent inundation for a long period. Annual draw-downs, as opposed to infrequent draw-downs, are more important for tuber production. Tuber production may be a useful indicator of the successful re-establishment of Eleocharis communities.

Water Levels and Cranes

At Tram Chim, most cranes associate with remnant wetland communities. In 20 years of observation at Tram Chim, no cranes have been seen in the surrounding rice paddy. Within remnant wetlands, the location of foraging cranes is easily predicted by their tendency to utilise grasslands dominated by *Eleocharis atropurpurea* or *Eleocharis ochrastachys*.[5] Though no food studies have been done, foraging behaviour suggests that Sarus cranes depend heavily on tubers produced by these two plant species. With the distinct relationship between Eleocharis abundance or Eleocharis tuber production and water levels, crane abundance was compared with water levels to examine long-term trends at a landscape level for Tram Chim. Crane numbers declined from a high of 1,052 birds in 1988 to 365 birds in 1992. Though water level data for this early period were not available, draw-downs occurred in 1988 and 1989, and were prevented in 1990–1992 because the surrounding dikes had been completed in Zone A1 that could impound water.

From the dry season of 1993 through the dry season of 2004, only one significant draw-down occurred (1995), where water levels dropped below the average soil elevation at Tram Chim for more than a week. The number of cranes declined from 1993 to 1995 but rebounded in 1996 following the draw-down of 1995. Presumably, Eleocharis below ground biomass had not been seriously affected during this period and could recover with just one significant draw-down. Following the initial recovery of crane numbers in the dry season of 1996, high dry season water levels resumed and crane numbers began to decline again, reaching 48 in 2001. The first draw-down, conducted in the current study, occurred in 2005. The number of cranes again recovered during this period but only to a flock size of 100–150 birds, still far below numbers in the 1990s.

Direct mortality caused by insufficient food during the years when water levels were kept high was the unlikely cause of lower crane numbers at Tram Chim. Overall number of birds in the Southeast Asia population (Barzen and Seal 2001) has not changed greatly during this period (Table 8.2). Instead, cranes probably moved to other wetland areas in the Mekong Delta or in Cambodia where suitable habitat persisted. Response of cranes to improved management at Tram Chim may therefore depend upon conditions at Tram Chim (which are now improving) as well as dry season conditions located in suitable wetlands elsewhere.

Grassland Bird Communities

Two studies of the grassland bird community at Tram Chim have been conducted in the dry season. In 1994, birds were surveyed in

Table 8.2: Population Estimates for the Southeast Asia Population of Sarus Cranes, 2001-2007

	Sarus crane numbers in the late dry season						
Location	2001	2002	2003	2004	2005	2006	2007
Mekong delta	411	527	494	417	366	391	317
ATT	228	345	339	365	334	373	394
Small wetlands	11	6	4	3	21	43	14
TOTAL	**650**	**878**	**837**	**785**	**721**	**807**	**737**
Per cent of immature birds					7.10	9.70	

Note: ATT is Ang Trapeang Thmaw in northwestern Cambodia while small wetlands refers to scattered wetlands of the dry dipterocarp forest in the northern third of Cambodia. Immature birds are those hatched the previous year and can be distinguished from adults using plumage characteristic.

three grassland communities that ranged from very wet (dominated by Oryza) to very dry (dominated by *Ishchaemum*; Larsen 1996). During dry seasons of 2006, bird communities were again observed but in wider variety of grasslands with locations spread among different types of grasslands at Tram Chim. Three of these areas were observed in 1994 using the same method. This work not only compares bird species composition among grassland types so that habitat associations can be formed, but they also discern the difference in bird communities from 1994, when water conditions at Tram Chim were closer to natural and 2006, after Tram Chim wetlands had been largely maintained under unnatural water level regimes.

In 1994, 30 bird species or bird groups were identified within the study area. This species diversity increased to 33 species in 2006. When only the three sites common to both studies were compared, the number of species was still higher in 2006 than in 1994. Interestingly, the species added in 2006 were the ones that use deeper water areas dominated by submerged aquatic macrophytes (*Dendrocygna javanica, Tringa totanus, Anas querquedula* and *Hydrophasianius chirurgus*) whereas the only species seen in 1994 but not in 2006 was a dry grassland species (*Tuto capensis*). Importantly, though the relationships between vegetation characteristics and bird species abundance remained similar in the two studies, abundant birds in 1994 were mostly associated with dryer grassland habitats (e.g., *Turnix* spp., *Cisticola juncidis, Alauda gulgula* and *Anthus richardi*) whereas abundant species in 2006 were mostly associated with substantially wetter habitats (e.g., *Porphyrio porphyrio, Egretta* spp., *Ardea* spp., and *Phalacrocorax niger*). These studies suggest that perennial high water levels have favoured bird species that prefer inundated systems, dominated by submerged aquatic and floating plant species, rather than flood plain recessional grasslands that are dominated by emergent grass species. The Plain of Reeds is typified by flood plain recessional grasslands though deeper water ecosystems did occur in the area as well.

The shift in bird species is a good example of how it is impossible to manage the ecosystem for the benefit of all species with a single management activity. Invariably, any management scenario will cause some species to increase in number, some to decrease and some to remain unaffected. The primary goal of Tram Chim, however, is to restore a portion of the Plain of Reeds ecosystem. The management of perennial high water levels during the dry season resulted in a decline in both plant and animal species that are adapted to grassland ecosystems which form a dominant characteristic of the Plain of Reeds. Typically, however, those bird species that increased in number were not of much conservation value, whereas the species that decreased in number were the ones which were most important.

Continued Recovery of the Ecosystem

The impacts of water draw-down on ecosystem were observed on several occasions. In 2006, after the first experimental draw-down during the research programme, the area of grassland increased from 800 hectares in 2007 to 2,700 hectares.[6] Nguyen Huu Thien et al. (2008) reported that from January to June 2007, IUCN–Vietnam with support from WWF–Greater Mekong Programme, ICF and the Hoa A Research Station of Can Tho University conducted a monitoring programme every two weeks on the impacts of the application of the recommended water levels. The results of the monitoring can be summarised as below:

1. The number of fire incidents is considered higher than in the previous years, but with less intensity. This is expected as a result of the accumulation of the dead biomass due to the prolonged inundation in previous years.
2. In Zone A1, where the actual water levels were appropriate to the recommended water levels, the soil and water quality conditions were found to be favourable for the recovery of the vegetation communities. With the recovery of tuber-producing

Eleocharis, the population of Sarus Crane visiting the park in 2007 was 125 individuals, higher than 89 in 2006 and 93 in 2005.

3. Although the actual water levels for the first three months of the dry season (January-March) were found to be similar to the recommended ones in Zone A2, there was no sign of recovery of the vegetation. This is due to the presence of a thick layer of dense biomass (up to 1.0 m) on the ground as a result of the prolonged inundation since 1996.[7] This presents a dilemma for management. On the one hand, if high water levels continue to be maintained for the purpose of prevention of fire, the biomass accumulation process would continue (due to low decomposition rate under water). This means fuels would continue to be accumulated and severe fires could happen in a particularly dry year in the future. On the other hand, if water is released from this zone, the immediate fire risk can be also very high. Under prolonged inundation, Melaleuca trees became tilted and developed thick aerial root systems that act as fire ladders enabling fires to climb to the tree crowns. This makes trees more likely to be killed when fire occurs.

4. Due to the damage of the periphery dyke, the soil became too dry (about 40 cm lower than the recommended levels) in Zone A4. The excessive dry conditions killed or withered vegetation and prevented regeneration in this zone.

5. In Zone A5, the actual water levels were lower than the target levels by 25–30 cm. The conditions in the early dry season, however, were conducive to the production of Eleocharis tubers in this area. Most of the cranes visiting Tram Chim in this year used this area as their main feeding ground in early months of the dry season, but were disturbed by the locals who were making claims on the land and were involved in a land dispute with the park. The disturbance caused the cranes to move to other parts of the park.

6. Although the area in Zone 3 used to be a good feeding area for the cranes, they have abandoned this area altogether. For many years, there have been no active management activities in this area resulting in the vegetation becoming poor and unhealthy. The whole area is severely invaded by Mimosa pigra.

Community and Livelihood Issues

After about 20 years of experience of park management, it has been recognised that there has to be some kind of cooperation with the local community in managing the resources of the park. A sub-project conducted by CARE International in Vietnam within the framework of MWBP was a pilot effort in promoting sharing of benefits with the local community. The approach taken by the sub-project was that poor, local resource users were organised into Resource Users Groups (RUGs). The RUGs would go into the park to conduct participatory resource assessments. Sustainable Resource Use Plans (SRUPs) would be formulated by the RUGs based on local knowledge and the information from the participatory assessments. The SRUPs are then presented to a panel of judges comprising park management personnel, knowledgeable local elders and local authorities. The RUGs defend and negotiate their SRUPs with the panel of judges to get endorsement. Once the SRUPs are endorsed, the RUGs enter into group contracts with the park for using resources according to the endorsed plans under the supervision of the park's guards.

The sub-project ended in March 2007. It created positive outcomes — the local users appreciated having access to the resources in the park and the number of arrests for poaching was reduced. Encouraged by the outcomes and experience of the sub-project, the WWF–Coca Cola Plain of Reeds Wetlands Restoration Project plans to continue the approach and upscale the work starting in 2008. It is hoped that conflict between the park and the local communities will be reduced while wetland resources will be used in a more sustainable way, improving the livelihoods of the local people. These would reduce the risk of fire caused by arsonists out of resentment, and would foster confidence in the management of the park to manage water appropriately for the recovery of habitat and thus biodiversity.

Future Management of TCNP

Since the MWBP ended at the end of 2006, the WWF–Coca Cola global partnership selected TCNP for the implementation of the project entitled 'The Plain of Reeds Wetlands Restoration Project', the design of which is aimed at continuing to address the issues at Tram Chim that the MWBP was addressing. This project continues to collaborate with the park management to take the fire and water strategy forward starting in the dry season of 2008. Draw-down of water was conducted following the target water levels recommended by MWBP. This has resulted in recovery of much of the grassland habitat in the park and the maintenance of the population of the Sarus Crane visiting the park in the dry season. There were 126 crane individuals recorded at the park in the dry season of 2008. Although water draw-down has been conducted for three years from 2006 to 2008 resulting in much reduction of the combustible thick layer of accumulated biomass, there was still much combustible biomass on the ground that led to a large fire in April 2008. A multidisciplinary team carried out a rapid assessment of the fire at 32 points with 12 points selected as permanent points for monitoring (Le Phat Quoi 2008). The fire covered an area of 469.6 ha, of which 166.29 ha were Melaleuca and 303.31 ha were grassland. Due to the water draw-down in the dry seasons of 2006, 2007 and 2008, much of the organic layers in the Melaleuca forest was exposed to oxygen and decomposed, leaving a 5–20 cm thick layer on the ground. The fire, however, did not affect much Melaleuca, except at the edges and the young Melaleuca stands where the fire climbed to the tree crowns. There are scatter areas of 0.1–0.3 hectares where all the trees were burnt off to the top, totalling 5–7 hectares. However, due to the thick bark, the core of the burnt trees were still found to be moist and were regarded as having a high chance of reviving. In the grassland area, the fire burnt off the layer on the ground but did not kill any grass. Evidence of total regeneration was seen on the area that was prescribed burnt 2 weeks before the fire. The fire also neither affected the water birds nor the cranes. In fact, 33 cranes were still seen feeding near the burnt area during and even two days after the fire. Some snakes and rats were found burnt.

Realising the management problems of a wetland such as Tram Chim in the legislation context of the Special Use Forest System, national and provincial authorities are seeking to address the limitations of the current protected area policy under the Special Use Forest System through the development of a pilot management statute for TCNP. This approach will allow management to be nuanced and rendered appropriate for the specific local circumstances at the park; the statute will allow the park to be managed in a way that is appropriate for the conservation of its species, habitats and wetland ecosystems and their sustainable use by local populations. A draft statute based on the body of scientific evidence and 3 years of piloting has been signed by the Ministry of Agriculture and Rural Development and is now pending ratification by the provincial government. Ratification by the provincial government has stalled due to lack of awareness of the supporting evidence by provincial departments; these departments still recommend that the park keep water levels artificially high in the wetland to suppress fire.

Ecosystem Management and Policy Change

The work at Tram Chim was used as one of a series of case studies conducted by IUCN's Commission on Ecosystem Management to identify some of the challenges to the implementation of the Ecosystem Approach (EA). Put very broadly, it is possible to summarise the EA in the following statements:

1. Ecosystems are not isolated: it follows that it is never enough just to consider protected areas, when planning conservation.
2. Human beings are ecosystem components.
3. Adaptive management is essential and over time, institutions will also need to adapt.

Tram Chim was a difficult case for the Ecosystem Approach from many points of view. It is currently fully implementing only two of the 12 ecosystem principles. It partially implements six others, but four principles could not be addressed at all (see Box 8.1).

The two principles being fully implemented are both part of traditional protected area management and so presented no new difficulties. Principles only partially implemented cluster into one or two themes. Principles 1 and 12 suggest that all sectors of society should have a say in the management of the area concerned, and be able to make an input into choices about its use. In Vietnam, as in many other countries, consultation of this kind is not the norm. Management has thus not 'considered all forms of relevant information' (Principle 11) — though this chapter shows that it has considered some and has only partially 'recognised that change is inevitable' (Principle 9). It is only partly clear what the limits to Tram Chim's functioning might be (Principle 6). If dykes were to be lowered and the other changes made that are suggested here, functionality would be improved and there would be better opportunities for the integration of conservation and use of biodiversity (Principle 10) than is currently the case.

Box 8.1: Principles of Ecosystem Management Approach

Principle 1: Objectives of management are a matter of societal choice **(partially implemented)**

Principle 2: Management should be decentralised to the lowest appropriate level **(not addressed)**

Principle 3: Ecosystem managers should consider the effects of their activities on adjacent and other ecosystems **(not addressed)**

Principle 4: There is a need to understand and manage the ecosystem in an economic context **(not addressed)**

Principle 5: Prioritising the conservation of ecosystem structure and function **(fully implemented)**

Principle 6: Ecosystems must be managed within the limits of their functioning **(partially implemented)**

Principle 7: The EA should be undertaken at the appropriate spatial scale **(not addressed)**

Principle 8: Objectives for ecosystem management should be set for the long term **(fully implemented)**

Principle 9: Management must recognise that change is inevitable **(partially implemented)**

Principle 10: Appropriate integration of conservation and use of biodiversity **(partially implemented)**

Principle 11: The EA should consider all forms of relevant information **(partially implemented)**

Principle 12: The EA should involve all relevant sectors of society **(partially implemented).**

Source: Shepherd (2004).

Finally, the most difficult of all principles that challenge the whole way in which Tram Chim and many other protected areas are managed, both at site level and more broadly, in the context of Vietnam's political structures. Tram Chim has been cut-off from the surrounding landscape by its dykes: a fortress holding it immune from a sea of rice cultivation all around it. From the land-use point of view, 'holding the line' in this way is quite understandable. But it has meant that Tram Chim is more cut-off from the broader flow of the Mekong river than it should be, and has much more limited relationships with the agricultural landscapes it is surrounded by than any real application of the EA would dictate. So there has been too little consideration of the park's impact on adjacent ecosystems (Principle 3), particularly in view of its important and neglected role as a fish nursery for wider areas. An appropriate spatial scale for the

Tram Chim ecosystem would involve going beyond the dyke boundaries and considering wider impacts and implications (Principle 7), but this cannot be dealt with by park managers alone.

Two other principles which Tram Chim could not meet highlighted failures in management issues at higher levels. On the one hand, certain aspects of management decision making ought to be located at site level but are currently located at provincial level, making for rigidity, and a frequent delay in decisions such as when to close and open sluices (Principle 2 concerns management devolution where possible). At the same time, issues which ought to have evoked a more coordinated management response at provincial level did not do so. So it is the case that more and more poor landless people have been encouraged to settle around Tram Chim by one provincial department, with no assessment by other departments of whether resources at the Tram Chim site could accommodate their needs. The high economic value of the site is thus relied on by one department while economic issues are systematically ignored by those responsible for Tram Chim as a protected area (Principle 4 concerns managing in an awareness of economic realities). IUCN chose to use the EA to investigate some of the difficulties the 12 principles highlight. In 2006, it held two workshops at which park managers from all over the Mekong Delta considered the main difficulties they were having in managing effectively. They used the EA to interrogate their own situation and identified four areas they regarded as problematic:

1. The impact on local livelihoods of parks and protected areas.
2. The lack of policy guidelines for sustainable use and sharing of benefits.
3. The lack of a forum for the resolution of land-use conflicts at provincial level.
4. The lack of a sectoral niche for wetlands at the national level.

A national level workshop to consider these issues was finally held in early 2008. Here, all the issues raised by these park managers and from other parts of Vietnam were considered from the policy point of view. Policy gaps and overlaps where wetlands are concerned were highlighted and revisions proposed, as were aspects of management devolution, benefit-sharing and co-management. Regional planning was also discussed, and it was decided that a Provincial Wetlands

Committee on a trial basis will be set-up in one province. It was felt that the EA has special value at this particular analytical and decision-making level. Several policy processes which were ongoing during 2008 are now receiving inputs as a result of this workshop, and it is expected that various sorts of new guidance will be issued.

Notes

1. www.mekongwetlands.org (accessed 10 October 2006).
2. Bird Survey Consultancy Report submitted by Nguyen Bao Hoa to Mekong Wetlands Biodiversity Programme (MWBP) in 2005.
3. WWF-Greater Mekong Plain of Reeds Wetlands Restoration Project at Tram Chim National Park, unpublished data, 2007.
4. Personal communication, Le Van Thoi, 1990.
5. J. Barzen, unpublished data.
6. Unpublished data, MWBP, 2007.
7. Personal communication, Nguyen Van Hung, 16 August 2007.

References and Select Bibliography

ActionAid. 2002. *Participatory Poverty Assessment (PPA), Plain of Reeds, Vietnam*. Vientiane: Mekong Wetlands Biodiversity Conservation and Sustainable Use Programme.

Barzen, J. and U. S. Seal. ed. 2001. *Eastern Sarus Crane PHVA Final Report*. Apple Valley: CBSG.

Beilfuss, R. D. and J. A. Barzen. 1994. 'Hydrological Wetland Restoration in the Mekong Delta, Vietnam', in W. J. Mitsch (ed.), *Global Wetlands: Old and New*, pp. 453–68. Amsterdam: Elsevier.

BirdLife International and the Forest Inventory and Planning Institute. 2004. *Sourcebook of Existing and Proposed Protected Areas in Vietnam*, 2nd ed. Hanoi: BirdLife International Vietnam Programme and the Forest Inventory and Planning Institute.

Buckton, S. T., Nguyen Cu, Ha Quy Quynh and Nguyen Duc Tu. 1999. *The Conservation of Key Wetland Sites in the Mekong Delta*. Hanoi: BirdLife International Vietnam Programme.

Larsen, B. 1996. 'The Avifauna of a Restored Wetland: Tram Chim National Reserve, Dong Thap, Vietnam'. Unpublished Ph.D. thesis, University of Minnesota.

Le Dien Duc. 1989. Inventory of Wetlands in Vietnam. Hanoi: CRES.

———. 1991. 'Eastern Sarus Cranes in Indochina', in J. T. Harris (ed.), *Proceedings 1987 International Crane Workshop*, pp. 317–318. Wisconsin: International Crane Foundation.

Le Phat Quoi. 2006. 'Report on Vegetation Mapping of Tram Chim National Park, Dong Thap Province, Vietnam', Consultant report, MWBP, Hanoi, Vietnam.

——. 2008. 'Report on Post-fire Assessment of the Fire in Tram Chim in April, 2008', WWF–Coca Cola Plain of Reeds Wetlands Restoration Project, Hanoi, Vietnam.

Minh, L. Q., T. P. Tuong, M. E. F. van Mensvoort and J. Bouma. 1998. 'Soil and Water Table Management Effects on Aluminum Dynamics in an Acid Sulphate Soil in Vietnam', *Agriculture, Ecosystem and Environment*, 68 (3): 255–62.

Marttin, F. and G. J. De Graff. 2002. 'The Effect of a Sluice Gate and its Mode of Operation on Mortality of Drifting Fish Larvae in Bangladesh', *Fisheries Management and Ecology*, 9 (2): 123–25.

Mitsch, W. J. and J. G. Gosselink. 2000. *Wetlands*. New York: Wiley.

Nguyen Huu Thien. 1997. 'Winning Support for Conservation at Tram Chim National Park', in R. J. Safford, D. V. Ni, E. B. Maltby and V. T. Xuan (eds), *Towards Sustainable Management of Tram Chim National Reserve, Vietnam: Proceedings of a Workshop on Balancing Economic Development with Environmental Conservation*, pp. 17–46. London: Royal Holloway Institute for Environmental Research.

Nguyen Huu Thien and Le Phat Quoi. 2008. 'Report on the 6-month Monitoring Programme at Tram Chim National park', Report to IUCN Vietnam, WWF–Greater Mekong Programme, ICF, and Hoa A Research Station of Can Tho University.

Nguyen Phuc Bao Hoa, Nguyen Van Hung, Doan Van Nhanh, Nguyen Hoang Minh Hai, Nguyen Nguyen Tan, Pham Quy Duong, Vo Thi Mong Tuyen and Truong Ngoc Dang. 2006. 'Grassland Birds Survey and Correlations between Grassland Birds and their Habitat Variables in Tram Chim National Park'. Unpublished report for Mekong Wetlands Biodiversity Programme, Cao Lanh.

Nguyen Phuc Bao Hoa, Tom Evans, Tran Triet, Hazel Watson, Hong Chamnan and Seng Kim Hout. 2008. 'Records of Non-breeding Sarus Cranes in Vietnam and Cambodia 2001–2007 Dry Seasons'. Unpublished report, International Crane Foundation, BirdLife International, Wildlife Conservation Society, n.p.

Nguyen Xuan Vinh and A. Wyatt. 2006. *Situation Analysis: Plain of Reeds, Vietnam*. Vientiane: Mekong Wetlands Biodiversity Conservation and Sustainable Use Programme.

Ni, D. V., D. Shulman, J. R. Thompson, T. Triet, and M. van der Shans. 2006. *Integrated Water and Fire Management Strategy — Tram Chim National Park*. Cau Lanh: UNDP/IUCN/MRC/GEF Mekong Wetlands Biodiversity Programme.

Pham Trong Thinh. 1998. 'Investment Plan for Upgrading Tram Chim Nature Reserve to National Park Status', Ho Chi Minh City Sub-FIPI, Ho Chi Minh City.

Safford, R. J., D. V. Ni, E. B. Maltby and V. T. Xuan. eds. 1997. *Towards Sustainable Management of Tram Chim National Reserve, Vietnam: Proceedings of a workshop on Balancing Economic Development with Environmental Conservation*. London: Royal Holloway Institute for Environmental Research.

Shepherd, Gill. 2004. *The Ecosystem Approach: Five Steps to Implementation*. Gland: IUCN.

Shulman, Deanne. 2002. 'Fire Management Assessment Tram Chim National Park, Dong Thap Province'. http://www.fire.uni-freiburg.de/iffn/country/vn/vn_2.htm (accessed 4 January 2002).

Thompson, J. R. and C. M. Finlayson. 2001. 'Freshwater wetlands', in A. Warren and J. R. French (eds), *Habitat Conservation: Managing the Physical Environment*, pp.147–78. Chichester: Wiley.

Tran Triet, J. Barzen, Le Cong Kiet and D. Moore. 2004. 'Wartime Herbicides in the Mekong Delta and their Implications for Post-war Wetland Conservation', in H. Furukawa, M. Nishibuchi, Y. Kono and Y. Kaida (eds), *Ecological Destruction, Health, and Development: Advancing Asian Paradigms*, pp. 199–211. Melbourne: Trans Pacific Press.

Van der Schans, M. and Nguyen Huu Thien. 2008. 'Case Study on the Application of IUCN's Ecosystem Approach Methodology to Tram Chim National Park, Mekong Delta, Vietnam'. Paper presented at IUCN's workshop on Ecosystem Approach, 9-11 January, Hanoi.

Wheeler, B. D., D. J. G. Gowing, S. C. Shaw, J. O. Mountford and R. P. Money. 2004. *Ecohydrological Guidelines for Lowland Plant Communities'*. Peterborough: Environment Agency.

Wheeler, B. D. and S. C. Shaw. 1995. 'Plants as Hydrologists? An assessment of the value of plants as indicators of water conditions in fens', in J. M. R. Hughes and A. L. Heathwaite (eds), *Hydrology and Hydrochemistry of British Wetlands*, pp. 63–82, Chichester: Wiley.

9

Valuation of Ecosystem Damages: A Case Study of Textile Pollution in Noyyal River Basin, South India

Prakash Nelliyat

Ecosystem services are the benefits people obtain from ecosystems, which include provision of services such as food and water; regulatory services such as regulation of floods, drought, land degradation and diseases; supporting services such as soil formation and nutrient cycling; and cultural services such as recreation, spiritual, religious and other non-material benefits (Millennium Ecosystem Assessment 2003). Ecosystem services or functions can be lost or diminished due to over-extraction or mismanagement of ecosystems.

In countries where environmental regulation is weak, the wastes or pollutants from different sectors are indiscriminately discharged into the ecosystems. When the volume of pollution exceeds the assimilative capacity of the ecosystems, the ecosystem functions of the receiving ecosystems are impaired. This ultimately impacts the well-being of people who depend on the ecosystems for their livelihood. Evidence suggests that water resources and the associated ecosystems have been seriously impaired due to pollution (Nelliyat 2002; Mukherjee and Nelliyat 2007). The demand for water from different sectors is increasing at a rapid rate in proportion with the growth of population and the economy (Lundqvist et al. 2003). Moreover, the structural transformation of the economy through diversification from agriculture to the industry and service sectors has significantly influenced the allocation of water resources. Since water in the industrial, service and domestic sectors is 'non-consumptive', 80 per cent or more of water used in these sectors is returned to the ecosystem as wastewater. Wastewater management strategies are often not adequate in many developing countries. Hence the impact of wastewater on the receiving ecosystems (such as land, rivers, canals, tanks, reservoirs and aquifers) is significant, particularly when there is accumulation of pollutants, resulting in loss of ecosystem services or functions.

The assessment and monetary valuation of damage to ecosystems due to pollution is useful in communicating the degree of damage due to pollution to the public and policy makers. Such valuations could facilitate decisions on investments to mitigate pollution. Assessment of damage can provide estimate of compensation that needs to be provided to pollution-affected people in case of litigation. This chapter presents a case study on the monetary valuation of ecosystem damage due to pollution from textile industry in the Noyyal river basin in the Tiruppur area in Tamil Nadu, a state situated in the southern part of India.

Methodology

Information related to industrial growth in Tiruppur and its ecological impacts were gathered from studies conducted by the State Pollution Control Board, other government departments and from scientific studies carried out by independent researchers. In order to study the impact of pollution on water and soil quality in the basin, the results from an extensive survey conducted by the Soil Survey and Land Use Organisation in Noyyal river basin was used. This study covered the entire pollution-affected area using a band with a width of 5–7 km on the north and south sides of the river. A grid pattern was used for selecting the samples. A total of 491 water and 466 soil samples were collected and analysed to understand the nature and area of impact of pollution on water and soil quality.

Monetary valuation of ecosystem damage costs due to pollution on the agricultural, domestic water supply and fisheries sectors was undertaken. Since the agricultural area affected by pollution was large, 13 villages were selected for a pilot survey and three villages for a detailed household survey. The monetary valuation of the damage on agriculture in the affected area was assessed through the 'productivity loss' method. Productivity loss in this case refers to the monetary loss due to the reduction in agricultural productivity because of pollution of the irrigation source.

The damage cost estimation on the drinking water sector covered the entire Tiruppur municipal area and villages affected by pollution. A study on social perception on impact of pollution was done in the Tiruppur municipality (52 wards), with the support of a local NGO (Centre for Environment Education). In some wards, the groundwater quality had deteriorated over the last 10 years and the residents now relied on public supply and/or purchase water from tankers. The cost incurred by the households for purchasing water (opportunity cost or defensive expenditure) is a way of estimating the loss due to unavailability of unpolluted supply of water through the natural ecosystem. Opportunity cost is a cost incurred on adopting the next best alternative option. To estimate the impact of pollution on rural areas, a large database provided by the Tamil Nadu Water Supply and Drainage (TWAD) Board was analysed. This database confirmed the high level of pollution of groundwater. The cost of various water supply schemes was estimated from the database. The impact on drinking water in the three villages chosen for was estimated through the 'replacement cost' method. Replacement cost is the cost of providing an alternative source of drinking water to replace water source affected by pollution.

To study the impact on fisheries, a biological study along the course of the river was conducted by the Zoological Research Department of Erode Arts College. The biological diversity including plankton, zooplankton and fish diversity, was studied and a bio-mapping exercise was done. Further, laboratory studies on respiration and biochemical alterations of biological entities due to pollution were studied. The valuation of damage to the fisheries sector was done based on the 'productivity loss' method with the data collected from the fisheries department.

A focus group discussion was also conducted in the pollution-affected areas to understand the magnitude of pollution and its socioeconomic implications. Information related to compensation strategy was collected from literature and government records. Detailed discussions were also conducted with the Loss of Ecology Authority (LEA) officials. LEA is the official agency formed by the state government based on a Supreme Court order to determine compensation packages for those affected by industrial pollution.

Water Resources and Ecosystem of the Noyyal River Basin

The Noyyal river is a tributary of the Cauvery, a major interstate river which passes through the states of Karnataka and Tamil Nadu and enters the Bay of Bengal. The Noyyal flows through the districts of Coimbatore, Erode and Karur and the urban centres of Coimbatore and Tiruppur, in western Tamil Nadu. A number of water intensive industries, particularly the textile industry located in Tiruppur area, discharge their untreated and partially treated effluents into the river. In addition, sewage from Coimbatore and Tiruppur cities is also being discharged into the river without treatment, making the Noyyal one of the highly polluted rivers in the country.

The Noyyal is a seasonal river which has substantial flow only for short periods during the north-east and south-west monsoons. Occasionally flash floods occur with heavy rain in the catchment areas. The river supplies water to several irrigation tanks. The Noyyal river basin covers an area of 3,510 km^2 and is located between north latitude 10°56′ and 11°19′ and east longitude 76°41′ and 77°56′ (Madras School of Economics 2002). The length of the Noyyal river is about 170 km from west to east. The average width of the basin is 25 km.

The annual rainfall varied between 600 mm to 3,000 mm in the Noyyal river basin (Central Ground Water Board 1983). Generally, Noyyal is a water-deficit basin since the average rainfall is only 617 mm with increasing water requirement to meet the needs of the rapidly growing population and industry. River water has been harvested for irrigation by the construction of anicuts (weirs) and tanks since the sixteenth century. Thirty-one tanks and 23 weirs have been used to irrigate about 16,250 acres of land. Hence the livelihood of a large number of people in the basin, particularly the poor farmers, depends on the river ecosystem.

Except the 23 anicuts, there were no major reservoirs in the Noyyal river basin until 1980. After 1953, when the Lower Bhavani Project (LBP) canal — a freshwater canal from the river Bhavani in the neighbouring basin which irrigates the downstream part of Noyyal — came into being, the seepage waters of the scheme drained into Noyyal at the location where the LBP canal crosses the river Noyyal. This seepage water (about 2,760 mcft or 78.45 mm^3 per year) flows into the Noyyal for many months every year. During the early 1980s, the government formulated the Noyyal Orathapalayam Reservoir Project (NORP) in two stages to utilise the heavy surplus water of the Noyyal that flows into the Cauvery.

Till 1980, water supply in the major portion of the basin was derived from deep dug wells reaching 35–45 metres below ground level. With the use of electric pumps, there has been considerable increase in groundwater withdrawals which in turn is depleting its level below 50–60 metres. In the industrial locations of the basin, particularly Tiruppur, groundwater pollution is significant. In brief, Noyyal basin was once a fertile landscape, supporting various crops like paddy, sugarcane, banana, turmeric, coconut, grapes, cotton, vegetables, cereals and pulses. All the surface water bodies were rich in biodiversity. Now the cropping pattern has changed primarily due to degradation of water quality. The industrial growth in Tiruppur, and the pollution problems have adversely affected the agricultural development in the lower basin.

Apart from agriculture, the water resources available in the basin are also used by the domestic and industrial sectors. Since Noyyal river is not a perennial river, it is not used as a source to supply drinking water. The two major cities, Coimbatore and Tiruppur, receive water from the Bhavani river in the neighbouring basin. However, the villages in the Noyyal basin generally depend on groundwater from public and private wells. The industries in Tiruppur also depend on groundwater.

The upper catchment of the basin is covered with forest. The natural forest in the basin consists of variety of flora. Around 34 species of plants have been recorded of which 3 are trees, 8 shrubs, 17 herbs, 3 climbers, 1 sedge and 2 grasses (Public Work Department 2001). The water bodies in the Noyyal river basin support a large number of aquatic organisms and a number of phytoplankton and zooplankton species are present. However, it was found that in Tiruppur, where the river is more contaminated, only one zooplankton species (*Daphnia*) was observed. Generally, phytoplankton also has become more or less rare in the river in Tiruppur (Public Work Department 2001).

Industrial Pollution and its Impact on Ecosystem

In recent decades, the water and land resources of Noyyal basin have been affected due to effluents discharged by textile industries in Tiruppur. The Tiruppur area is a rapidly growing textile (knitwear) industrial cluster located on the banks of the Noyyal river in Coimbatore district of Tamil Nadu. More than 9,000 small-scale knitwear related units are functioning here which provide employment to more than 200,000 people. The annual export value from Tiruppur during 2004 has been estimated as ₹ 45,000 million (Nelliyat 2007b).

The pollution problems of the knitwear industry in Tiruppur are associated with the bleaching and dyeing of textiles. During 1981, only 68 textile processing units were functioning in Tiruppur. However, due to the growth in exports of textiles the number of units has increased to 450 in 1991 and 700 in 2001. Correspondingly, the volume of effluents discharged by the processing units into the ecosystem has also increased considerably. During 1981, the total effluent discharged by the processing units was only 4.7 million litres (mld) per day. However, it increased to 39.5 mld in 1991 and 83.14 mld in 2001. Since the industries had not installed effluent treatment plants in the initial period, untreated effluent was discharged into the ecosystem till 1997. Even after the installation of effluent treatment plants most of the pollutants, particularly the Total Dissolved Solids (TDS), standard, was not met (Nelliyat 2005). The existing effluent treatment plants were not designed for reducing TDS, particularly the chloride and sulphate salts.

In Tiruppur, around 550 (78 per cent) textile processing units are functioning in non-industrial areas and discharge effluents (around 62 mld) on land. Hence the possibility for contaminating the groundwater in residential and agricultural areas is very high. Since 240 units are located at a distance of less than 300 metres from the Noyyal, the potential for polluting the river through direct discharge is also quite high (Nelliyat 2005). The river carries the effluents to downstream areas, especially to the system tanks and the Orathapalayam reservoir. Hence pollution is widespread even in distant areas. Pollution impacts are observed in different sectors such as agriculture, domestic water supply, fisheries, biodiversity and public health.

The damage to the ecosystem in Tiruppur area is due to the accumulation of effluents in groundwater and soil. Based on the information on quantity of effluent and its quality obtained from the State Pollution Control Board, the total pollution load generated by the textile processing units from 1980 to 2003 is estimated for the major parameters: TDS, Chloride, Sulphate, Total Suspended Solids (TSS), Chemical Oxygen Demand (COD), Biological Oxygen Demand (BOD) and Oil and Grease (Table 9.1).

Table 9.1: Pollution Load Generated by Tiruppur Textile Processing Units from 1980 to 2003 in Noyyal River Basin (Quantity in Tonnes/Year)

Period	TDS	Chloride	Sulphate	TSS	COD	BOD	Oil and grease
1980–2003	28,77,066	15,85,933	1,61,726	1,10,928	1,04,944	31,990	1,633

Source: Nelliyat (2005).
Note: TDS — Total dissolved solids; TSS — Total Suspended Solids; COD — Chemical Oxygen Demand; BOD — Biochemical Oxygen Demand.

Since none of the processing units treated effluents until 1997, untreated effluent quality is considered for estimating the load from 1980 to 1997. However, during 1998 and 1999 when the treatment plants were under construction, both untreated and treated effluents were considered for load estimation. From 2000 onwards, all the units discharged treated effluents. Hence pollution load estimation from 2000 to 2003 was limited to treated effluents. The pollution load, particularly the TDS discharged by Tiruppur industries, is large and is probably beyond the carrying capacity of the regional ecosystem. Around 80 per cent of the pollution load has accumulated in the Tiruppur area. Rainfall has only a marginal effect in reducing the severity of the impact (Nelliyat 2005).

Studies on water and soil quality proved the accumulation effect of pollution in Tiruppur area and downstream of the Noyyal basin. A number of studies have been carried out on groundwater quality by academic institutions, government agencies and researchers (Azeez 2001; Berglund and Johansson 2004; Central Ground Water Board 1983, 1993 and 1999; Centre for Environmental Studies 1996; Furn 2004; Jacks et al. 1995; Jacob 1998; Jayakumar et al. 1978; Public Works Department 2002, 2003; Rajaguru and Subbaram 2000; Ramasamy and Rajagopal 1991; Senthilnathan and Azeez 1999; Tamil Nadu Water Supply and Drainage Board 1999, 2001, and 1995–2001). The major conclusions of these studies are that open wells and bore wells in and around Tiruppur and the downstream stretch of Noyyal exhibit high levels of TDS (most areas > 3000 mg/l and some places even up to 11,000 mg/l) and chloride (generally > 2000 mg/l and certain areas up to 5000 mg/l) due to industrial pollution. Concentration of heavy metals including zinc, chromium, copper, and cadmium is also high in groundwater. The current values of pollution concentration are very much higher than the background levels for this region. Rainfall has only a marginal influence in reducing the concentration of TDS. The establishment of effluent treatment plants in Tiruppur has not had a positive impact on the quality of groundwater and the possibility of increased pollution concentration in groundwater in the future is high if effluent discharge by textile processing units continues. Studies also show that the available groundwater in Tiruppur areas and the surrounding places of Orathapalayam reservoir is not suitable for domestic, industrial or irrigation use.

The Surface Water Studies done by government departments and researchers (Central Water Commission 2000–2001; Public Works

Department 2001, Hydrology Research Station 1996; Jacob 1998; Palanivel and Rajaguru 1999; Soil Testing Laboratory 2000–2001; Tamil Nadu Pollution Control Board 1997–1999) indicate that the pollution (EC/TDS) concentration in Noyyal river is low till the river reaches Tiruppur. It increases considerably in the Tiruppur area and this continues up to the Orathapalayam reservoir. After Orathapalayam, there is some improvement in the quality of river water. Pollution concentration in the river is greater in summer than in winter and the existing moderate flow is not sufficient for diluting the pollutants. The Orathapalayam reservoir and system tanks have been badly affected by industrial pollution through high alkalinity, chloride, electrical conductivity, iron, phosphate and BOD. River and reservoir water is not suitable for aquatic organisms. Except during the rainy season when there is some dilution, the surface water is not suitable for irrigation.

Ecosystem Damage Assessment in Noyyal Basin

Identification and prediction of impacts on environment and natural conditions, quantification of direct and indirect impacts and monetisation of these impacts are the analytical steps in damage assessment. Valuation techniques like changes in productivity approach, loss of earning approach, opportunity cost, preventive expenditure approach, property value approach, travel cost method, replacement cost, relocation cost, etc., could be used to monetise environmental damages.

In Noyyal basin, the pollution load has exceeded the assimilative capacity of land and water, and has severely affected the natural environment in and around the Tiruppur area and the areas downstream of the Noyyal river. The level of pollution is high in the groundwater, in the surface water sources (river, tanks and reservoir) and to some extent in the soil, which has resulted in the degradation or damage of the ecosystem. According to Azeez (2001), the untreated textile effluent released to the environment is: aesthetically unpleasant; unsuitable for drinking and other human use leading to human health implications; and unsuitable for irrigation, unfit for livestock and not conducive for aquatic organisms such as plankton, invertebrates and fishes. The effluent has made the Orathapalayam check dam an environmental disaster, adding to the miseries of the local people.

The ecological consequences of the effluent are associated with degradation of the quality of the surface water body that receives effluents, sediment and soil, and groundwater.

Physical Estimation of Impact

Primary Water and Soil Quality Survey

A research study on water and soil quality was conducted by Soil Survey and Land Use Organisation (Soil Survey and Land Use Organisation 2002), to examine the physical impact of pollution in the Noyyal Basin. Since this study is focused on the ecosystem damage due to Tiruppur textile industrial pollution, data of above study pertaining to Tiruppur area and its downstream only has been considered for this analysis.

A total of 491 water and 466 soil samples were collected from both sides of Noyyal River (at 5–7 kms width) and analysed to understand the impact of pollution on water and soil quality. Water samples were taken from wells, streams, tanks, bore wells and canals. The water quality of samples were categorised in terms of the electrical conductivity: injurious (EC is >3 mmhos/cm) where water is not recommended for irrigation, critical (EC value of 1.1 to 3 mmhos/cm) where water quality is affected, but can be used for irrigation purposes and normal (EC value upto 1 mmhos/cm) where water quality is not affected. According to the results, as shown in Figure 9.1, water quality was highly degraded in Tiruppur area. There were significant improvements in water quality in the downstream of Noyyal for two reasons. First, because of the effluents collected at the Orathapalayam reservoir which was closed for most of the period, the possibility of water contamination downstream was less. Second, there was also dilution of the pollutants by canal water from the Lower Bhavani Project canal (Table 9.2).

The soil quality assessment was based on the pH, i.e., alkalinity and salinity based surface and sub-surface soil samples. An interesting finding here is that the soil degradation in terms of alkaline affected samples was much more in downstream of Noyyal compared to Tiruppur area (Table 9.3).

It was observed that the farmers in Tiruppur area cultivated only rainfed crops like maize and ragi and were not using the polluted water for irrigation. However, farmers downstream cultivated irrigated crops like paddy, banana, sugar cane, turmeric and tobacco

Map 9.1: Quality of Water (EC) in Noyyal River Basin

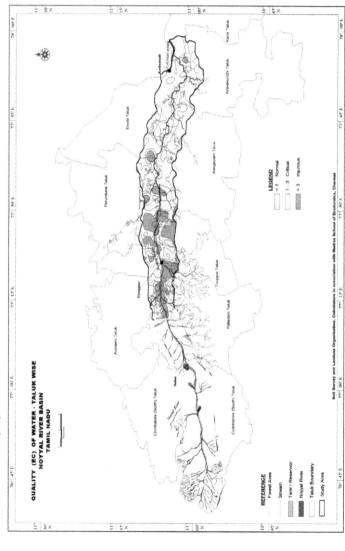

QUALITY (EC) OF WATER - TALUK WISE
NOYYAL RIVER BASIN
TAMIL NADU

REFERENCE
Forest Area
Stream
Tank / Reservoir
Noyyal River
Taluk Boundary
Study Area

LEGEND
< 1   Normal
1 - 3 Critical
> 3   Injurious

Soil Survey and Landuse Organisation, Coimbatore in association with Madras School of Economics, Chennai

Source: Soil Survey and Land Use Organisation (2002).

Table 9.2: Water Quality in Tiruppur Area and Downstream of Noyyal

Zone/water quality	Normal	Critical	Injurious	Total
Tiruppur area	62	106	82	250
	(25)	(42)	(33)	(100)
Downstream of Noyyal	136	84	21	241
	(56.0)	(35.0)	(9.0)	(100)
Total	198	190	103	491

Source: Computed from Soil Survey and Land Use Organisation (2002) Study.
Note: Numbers within parenthesis indicate the percentage of observations in each category over total samples.

Table 9.3: Soil Quality Status in Tiruppur Area and Downstream of Noyyal

Zone/soil quality	Category	Normal	Alkaline	Tending to alkaline	Total
Tiruppur area	Surface	62 (22)	65 (23)	152 (55)	279
	Sub-surface	147 (53)	75 (27)	57 (20)	279
Downstream of Noyyal	Surface	13 (7)	117 (63)	57 (30)	187
	Sub-surface	10 (5)	126 (68)	51 (27)	187
Total	Surface	75	182	209	466
	Sub-surface	157	201	108	466

Source: Computed from Soil Survey and Land Use Organization (2002) study.
Note: The numbers within parenthesis indicate the percentage of observations in each category over the total sample.

during early 1990s after the Orathapalayam project commenced. Occasionally, farmers were using the polluted water. Since the irrigation sources are contaminated, the possibilities of transfer of pollutants from water to soil is high.

Biological Studies

A detailed biological study was undertaken by the Research Department of Zoology (2002) to examine the impact of industrial effluents on the biodiversity of different surface water sources in the Noyyal river basin. The study revealed that water in the Noyyal river was highly polluted and was toxic to fish and other organisms in the stretch from Tiruppur to Orathapalayam. Consequently, there is a heavy loss of fisheries in the river, system tanks and the Orathapalayam reservoir. There was also loss of biodiversity and damage to the aquatic ecosystem.

Study on plankton diversity (phytoplankton and zooplankton) carried out at different stations revealed that greater number of species was observed in the river before and after Tiruppur. The maximum density was found in the downstream stretch of Noyyal in (3,678 individuals) while the least was recorded (726 individuals) in the Tiruppur area. Fish diversity in the river was also adversely affected in the industrial pollution affected areas, particularly in Tiruppur. However after Orathapalayam, the river supports seven species of fish.

Apart from the physical data on biodiversity, the biological studies covered toxicity studies on fish. A dose and time dependent decrease in the rate of respiration was observed in fishes when exposed to different concentrations of river water. The biochemical parameters such as muscle glycogen and liver glycogen were found to decrease while the blood glucose level increased due to stress, in the polluted stretch (Research Department of Zoology 2002).

The biological study revealed that the Noyyal river has very poor life supporting capacity in the industrial corridor of Tiruppur due to textile effluent discharge. However the river supports more life after Orathapalayam, since the dam is closed and there is influence of clean seepage water from the LBP canal. During the field survey, it was noted that livestock were also affected in some of the villages due to pollution.

FOCAL GROUP DISCUSSION

A focal group discussion was conducted in the pollution affected villages to understand the pollution impacts on ecosystem and associated socioeconomic problems and villagers' awareness on pollution. The members who participated in the discussion included the panchayat president, ward members, representatives of local non-governmental organisations, farmers and village elders. The major conclusions that emerged from the focal group discussions are: that before 1980, all local system tanks and groundwater were suitable for irrigation and farmers cultivated crops such as paddy and banana, but in recent times the irrigation sources have been highly polluted due to effluent discharge from Tiruppur, and farmers are cultivating only rainfed crops; compared to the earlier period, the soil quality had also deteriorated because of irrigating with polluted water; the area under cultivation and productivity of the crops had declined considerably primarily due to deterioration in both surface and groundwater quality and villagers were well aware that the major reason for the degradation of the ecosystem was industrial pollution caused by the textile units in Tiruppur.

Economic Value of Pollution Damage

The sectors experiencing ecosystem damage include agriculture, fisheries, domestic water supply, human health, livestock and bio-diversity (Nelliyat 2003). However in this case study, the economic valuation of damage was undertaken only in three major sectors: agriculture, domestic water supply and fisheries.

Agricultural Sector

The area located on both sides of the Noyyal between Tiruppur and Orathapalayam (located in Tiruppur, Perundurai and Kangayam taluks) was identified as areas affected by textile effluents. Based on the Electrical Conductivity (EC) values of water (from the primary water quality study), the areas were divided as severely affected, moderately affected and unaffected (see Figure 9.1) and output variations for different crops were estimated.

The average value of net output per acre was calculated using data from case studies of few villages. If the farmers were not able to raise a particular crop in an affected area, it was considered as a total loss of output. In the severely affected areas, farmers were not cultivating paddy at all, and the study estimated the loss due to this. In the severely affected irrigated area, the value of productivity loss per acre was estimated to be ₹ 7,362 per year while in the unirrigated area it was ₹ 2,910 per acre. In the moderately affected area, the damage was estimated based on the difference between output of unaffected and moderately affected area of paddy, i.e., ₹ 2,600 per acre in the irrigated area and ₹ 2,910 in the unirrigated area. The study considers that a single crop is cultivated per year.

Total cultivatable area in the pollution affected zone was estimated as 1,46,389 acres, of which 36,139 acres (24.7 per cent) could be classified as injurious, 53,938 acres (36.8 per cent) as critical and 56,312 acres (38.5 per cent) as normal for cultivation (see Figure 9.1). The taluk-wise damage cost in the agriculture sector is estimated based on the 'productivity loss' method (Table 9.4).

The total area was classified into irrigated and unirrigated in moderately and severely affected villages. In all taluks, the damage

Table 9.4: Taluk-wise Annual Damage Cost in Agriculture (Area Affected in Acres and Productivity Loss in ₹ Million)

Sl. No.	Name of taluk	Moderately affected				Severely affected				Total
		Irrigated		Unirrigated		Irrigated		Unirrigated		
		Area	Loss	Area	Loss	Area	Loss	Area	Loss	
1	Tiruppur	6,253	16.3	9,380	36.9	7,124	52.4	10,685	31.1	136.7
2	Perundurai	6,697	17.4	10,045	39.5	5,176	38.1	7,764	22.6	117.6
3	Kangayam	8,625	22.4	12,938	50.9	2,156	15.9	3,234	09.4	98.6
	Total	**21,575**	**56.1**	**3,236.3**	**127.3**	**14,456**	**106.4**	**21,683**	**63.1**	**352.9**

Source: Madras School of Economics (2002).

in severely affected area was more than the moderately affected area. The annual total damage in agriculture sector was estimated as ₹ 352.9 million.

Domestic Water Supply Sector

The domestic water supply sector is also affected by the textile pollution from Tiruppur. Domestic water supply includes the potable (drinking and cooking) and non-potable (bathing, washing, flushing and gardening) use of water.

LOSS TO TIRUPPUR MUNICIPALITY

Tiruppur municipality consists of an area of 27.20 km^2 and is divided into 52 wards (Tiruppur Municipality 1999). According to the 2001 Census, the total population of Tiruppur was nearly 300,000. Besides, the floating population of Tiruppur is very high and is estimated to be around 200,000 in addition to the resident population. According to Jacob et al. (1999), certain physico-chemical parameters like EC, TDS, Chloride, Sulphate Hardness, Sodium BOD and COD of the groundwater in Tiruppur exceeded the permissible level prescribed by the Bureau of Indian standards and World Health Organisation Standards for drinking water.

Tiruppur city had relied largely on groundwater to meet its domestic water requirement. With the growth of the textile industry, more people have migrated into the city while local water sources got polluted rapidly too. At present, Tiruppur municipality receives 33 million litres per day (mld) of water from the Bhavani river (Tiruppur Municipality 2001). However, this is not sufficient for the growing domestic and industrial water requirement of the city. Given the high pollution levels of the existing schemes, Tiruppur municipality (2000) has made the observation about the need for a third water supply scheme.

To estimate the magnitude of impact of pollution on households in Tiruppur municipal area, a detailed primary survey was conducted with the assistance of the Centre for Environmental Education (2002). Around 510 households living in 52 wards were surveyed. The major findings of the survey includes: that the quality of well water which was good before has deteriorated during the last 10 years; since the municipal water supply is insufficient to meet the household requirements, people are purchasing water and the amount spent for the water purchasing varies from ₹ 200 to ₹ 600 per month per household. It is clear that the households are purchasing water since they cannot rely on polluted local groundwater to supplement the public supply.

Based on the survey results, the damage cost in drinking water sector for Tiruppur municipality is estimated (Table 9.5). It is considered that Tiruppur has a population of about 60,000 households. Thus, the annual expenditure spent by the household ('opportunity cost/defensive expenditure') for purchasing water is ₹119.00 million.

Table 9.5: Damage Cost in Drinking Water Sector in Tiruppur Municipality

% of households	Total number of households	Average amount/ month (₹)	Total amount per year (₹ Million)
60	36,000	100	43.2
35	21,000	300	75.6
05	3,000	500	00.2
100	60,000	–	119.0

Source: Madras School of Economics (2002).

LOSS OF DRINKING WATER SERVICES IN RURAL AREAS

Apart from Tiruppur municipality, the drinking water supply in the villages located in and around Tiruppur and the downstream part of Noyyal are affected by textile pollution. To estimate the damage on drinking water, the cost of drinking water schemes as well as the cost of collection of freshwater was estimated. Most of the water supply schemes are executed by the TWAD Board and then transferred to the local government for maintenance. Generally, there are many reasons for the introduction of a water supply scheme such as population growth, urbanisation, industrialisation and pollution. The study showed that there was a positive relationship between the deterioration of water quality and cost incurred for water supply schemes in the villages affected by pollution.

So far, the TWAD Board has spent ₹ 271.6 million on drinking water supply schemes in the villages, which are located in the stretch between Tiruppur and Orathapalayam. The loss of drinking water services due to pollution was estimated by computing the total cost spent by the board for those habitations which have TDS values greater than 3000 mg/l. It was assumed that the board had to provide alternative drinking water schemes to the villages primarily due to the impact of pollution on local drinking water sources, i.e., the damage is the 'replacement cost' in terms of drinking water schemes. The damage cost to the affected villages was estimated to be ₹ 2.1 million.

The water supplied through the schemes is not sufficient for many of the affected villages. When the local sources were polluted, villagers were forced to bring drinking water from the neighbouring sources where there was no pollution. However, villagers have been using the locally available water for non-potable purposes. Based on the household survey information of pollution affected villages, the cost of collecting drinking water was estimated for all the households who live in the pollution affected area. From the survey, it is clear that in the affected area the households had to spend time for collecting water from more distant sources. The average quantities of water collected per household was 44 litres, for which they have to spend 1.20 hour per day, i.e., 36 hours per month. To value the cost of water, the total time spent for collecting water is multiplied with the rural average wage rate per hour for agriculture labourers (₹ 9). So the monthly cost is estimated as ₹ 324 per household. The study also estimated that 2,423 households were not having drinking water sources in their premises in and around Tiruppur and the downstream. The total annual cost for fetching water for all the textile pollution affected households was ₹ 9.4 million.

The overall annual damage cost of pollution in the domestic water supply sector was estimated to be about ₹ 130.5 million of which about ₹ 119 million is in Tiruppur town and ₹ 11.5 million in the affected villages, both in the form of drinking water supply schemes as well as the cost of fetching water from distant sources.

Fisheries Sector

Like the agriculture and domestic water supply sectors, the fisheries sector is also seriously affected by textile pollution. The impact on fisheries is mainly in the Noyyal river (between Tiruppur and Orathapalayam), system tanks (which are located near Tiruppur and downstream) and in the Orathapalayam reservoir.

LOSS OF FISHERIES IN THE NOYYAL RIVER AND SYSTEM TANKS

In the past, when the river was free from pollution, local fishermen depended heavily on the river for fishing. Different varieties of fish thrived in the past. However after the deterioration of the river water quality and ecology, the fish stock reduced considerably and fish mortality became a common phenomena. Generally, available fish are small, abnormally coloured (due to the effect of dyes) and are not edible, and hence are used only for manure (Nelliyat 2005). The biodiversity study revealed that the Tiruppur stretch of the Noyyal river did not support any fish species due to the discharge of industrial effluents (Research Department of Zoology 2002). However, seven varieties of fish species thrived after Orathapalayam.

The annual fish catch in the Noyyal river declined from 2,174 kg (1994–1995) to 540 kg (1997–1998) and the corresponding value reduction was ₹ 27,175 to ₹ 6,210 in current prices (Directorate of Fisheries 2000a). The fisheries activities in eight system tanks of the river (between Tiruppur to Orthapalayam) were highly affected. The panchayat used to auction the fish catch in five of the tanks which had high yield. The auction rate is reduced in all panchayats and the total annual reduction between 1999 and 2000 was ₹ 257,000.

FISHERIES LOSS IN ORATHAPALAYAM RESERVOIR

The fisheries department started fisheries activities at Orathapalayam reservoir in 1993. The fingerlings stock at Orthapalayam reservoir was 385,000 in 1993. Though the stock was increased to 700,000 during 1993–1994 and 801,000 in 1996–1997 (Directorate of Fisheries 2000b), no stocking took place in the reservoir in subsequent years. Towards the end of 1997, mass fish mortality occurred at Orathapalayam reservoir. In early December of 1997, dead fish were found floating near the dam site, which caused serious health hazards. Subsequently the district collector made arrangements to bury the dead fish amounting to several tonnes. The government also ordered the fisheries department to make immediate arrangements for removing all the fish in the reservoir to avoid further fish mortality and health hazards. The Public Works Department (PWD) also requested the fisheries department to stop the fisheries activities in Orathapalayam. Subsequently the director of fisheries ordered the Regional Fisheries Office at Erode to stop the fisheries at Orathapalayam (Directorate of Fisheries 2000c).

After realising the impact of pollution on Orthapalayam reservoir, the commissioner of fisheries conducted a detailed study (Hydrology Research Station 1996). The salient findings were: that the river water quality was unsuitable for aquatic organism; *Tilapia* is the main variety that was observed downstream of Tiruppur and Orthapalayam reservoir; plankton pollution was observed in huge quantity in reservoirs; bioassay test conducted revealed that the fish are bleached; although *Tilapia* may survive under present poor quality of water, it would not be conducive for the growth of other carps; fish caught from the reservoir get spoiled fast, and hence cannot be marketed to distant places; and high possibility of frequent fish mortality. Given these observations, the fisheries activities at Orathapalayam reservoir was completely stopped.

POSSIBLE TOXICITY IN FISHES

At present, even though the river, tanks and reservoir are completely affected by textile pollution, some informal fishing is still taking place in the water bodies and the fish caught is sold. Even the fish that survive in the Orthapalayam reservoir may be of doubtful quality. According to Azeez, (2001), Many dyes commonly used are known to have serious health implications. Azo dyes, which were used widely, are banned recently due to harmful properties including carcinogenesis. It is found that many species of fish survive in the Orthapalayam reservoir where all the chemical wastes from Tiruppur get collected. It is quite possible that the fishes accumulate these chemicals and are transferred to human beings while consuming fishes.

Based on the fish catch data, the estimated overall annual damage cost ('productivity loss') in fisheries sector as ₹ 1.5 million (₹ 14,000 in Noyyal river, ₹ 258,000 in system tanks and ₹ 1.2 million in Orathapalayam reservoir).

Total Economic Value of Ecosystem Damage

Table 9.6 provides the aggregate annual value of ecosystem damage in agriculture, domestic water supply and fisheries. The focal group discussions and survey conducted in the pollution affected area reveals the livelihood impacts, especially to the poor, due to degradation of ecosystem. Some of the villagers who live in the periphery of Tiruppur and even in the distant locations are getting employment opportunities in textile industries. However in the downstream pollution effected villages, which are not benefited by industries, the damage on ecosystem is a serious concern. The villagers in these areas have demanded compensation for the loss of ecosystem damage, particularly for the loss in agricultural productivity.

Table 9.6: Total Ecosystem Damage Cost due to Pollution (in ₹ Million)

Sectors	Annual
Agriculture	352.9
Domestic Water Supply	130.5
(a) Urban (Tiruppur municipality)	119.0
(b) Rural (water supply scheme)	2.1
(c) Rural households	9.4
Fisheries	1.5
Total	**484.9**

Source: Madras School of Economics (2002).

Compensation for Ecosystem Damage

Through the monetary valuation of damage cost of pollution, the magnitude of pollution can be better informed to the public as well as the policy makers. This might be helpful for suggesting or enforcing appropriate pollution control measures. Damage estimation is also necessary to decide compensation or for negotiation between polluters and affected parties.

In Tiruppur when the magnitude of textile pollution impact increased in the Noyyal river basin, a downstream farmers' organisation filed a court case during 1996. The writ petitions clearly expressed the downstream farmers' grievances against the bleaching and dyeing units situated in Tiruppur, the continuous efforts by the farmers for more than 30 years for the sanctioning of the Orthapalayam project and the current damage due to the textile pollution. But according to the petitioner (Kuppusamy 1996), at present the areas (Noyyal main canal command), which were starved for water are now objecting to release water from the reservoir for the reason that it is causing immense pollution in that area. The agriculturists felt that the water coming from the dam is very bad for the crops and the sludge which gets settled is toxic. Hence the entire investment for the Noyyal Orthapalayam Irrigation Project is a waste. Subsequently the farmers approached the LEA for compensation. The Supreme Court of India in a landmark judgement in the Vellore Citizens Forum case held that the 'Precautionary Principle' and the 'Polluter Pay Principle' as part of the environmental law of the country. The LEA, which is headed by a retired judge of the Madras High Court, has four members. The basic tasks of the LEA include estimating the monetary value of the ecological damage caused by industrial pollution (including reversing the damage); identifying the industries which have to pay compensation and fixing the amount for each industry; and identifying the farmers and the amount that they will receive as compensation.

In the case of the Noyyal river basin, the LEA sought the assistance of the Centre for Environmental Studies, Anna University, to assess the damage on ecosystem in the areas affected by the industries in Tiruppur. The university carried out water quality studies and identified the list of villages affected by the effluents. Each of the villages was classified based upon a weighted average of electrical conductivity in the sample wells in the village. Subsequently the monetary loss of agricultural productivity was estimated based on type of crop, average yield and procurement price. The compensation to the farmer was calculated on the basis of the cropping pattern as well as the average level of pollution in that village. In the case of villages which had access to canal water from other irrigation projects, the compensation was fixed at a lower level.

The amount to be paid by each industry was calculated based on its product, production level and the level of pollution control. The industries were ordered to pay the compensation amount in instalments to the district collector. The Collectors of the respective districts would distribute the compensation to the farmers as per the schedule provided by the LEA. The LEA awarded compensation of ₹ 248 million for the period 2002–2004 to a total of 28,596 farmers located in 68 villages in seven taluks in the Noyyal river basin. The industries have challenged the award on various grounds, and the case is being heard in the Madras High Court at the time this chapter was written.

Conclusion

In the Noyyal river basin, water ecosystem played a pivotal role in its overall socioeconomic development. Farmers depended on groundwater and surface water sources (Noyyal river, tank and reservoir) and cultivated different crops including paddy, sugarcane, banana, turmeric, coconut, cotton, cereals, pulses, grapes and vegetables. All the surface water bodies were rich in biodiversity and fisheries. Apart from agriculture, the water resource available in the basin was also used by the domestic and industrial sectors. Groundwater was the major source for domestic supply in the villages. All the water intensive industries, particularly the textile industries in Tiruppur, depend on groundwater for their processing. The water available at Tiruppur area was suitable for textile bleaching and dyeing and this is attributed as one of the reason for the growth of knitwear industry in Tiruppur. However in recent years, due to the textile effluent discharge from Tiruppur, the water ecosystem of Tiruppur area and downstream of Noyyal has been seriously affected. The groundwater and surface water sources are not suitable for domestic, agriculture or industrial purpose. The biodiversity and fisheries of surface water has also been seriously affected.

Since the groundwater and surface water sources are not suitable for cultivation, farmers incur heavy losses. Most of the saline sensitive crops have disappeared from the basin and the productivity of existing crops has reduced. The contamination of water has also affected the domestic water sector and households are purchasing or fetching freshwater from distances places. Since all the surface water sources are highly polluted, fisheries are affected and fishermen have lost their livelihood options. Industries are also bringing their required water from the non-polluted/distant areas through tankers or pipeline. The damage cost of pollution estimated was in agriculture, domestic water supply and fisheries. The aggregate annual damage cost was estimated as ₹ 484.9 million, of which ₹ 352.9 million is in agriculture sector, ₹ 130.5 million in domestic water supply sector and ₹ 1.5 million in fisheries sector.

The monetary valuation of damage cost of pollution is useful for framing ecosystem management policies and the payment of compensation for ecosystem damage. Recently the LEA estimated the monetary loss of agriculture productivity in industrial pollution affected villages in Noyyal river basin for providing compensation to the farmers. Moreover, the damage assessment provides an indication of pollution. If the damage cost of pollution on ecosystem is higher than cost of pollution abatement with serious socioeconomic implications and livelihood impacts, then this provides sufficient reasons to undertake pollution abatement measures and enforcement policies.

References and Select Bibliography

Azeez, P. A. 2001. 'Environmental Implications of Untreated Effluents from Bleaching and Dyeing', in S. Senthilnathan (ed.), *Eco-friendly Technology for Waste Minimisation in Textile Industry*, pp. 5–11. Coimbatore: Centre for Environment Education, Tiruppur and Public Works Department Water Resources Organisation.

Berglund, Per and Nina Johansson. 2004. 'A Ground Water Quality Assessment in Tiruppur'. Unpublished thesis, KTH Sweden.

Central Ground Water Board. 1983. 'Ground Water Resources of Noyyal, Ponnani and Vattamalai Karai Basin'. Chennai: Central Ground Water Board.

———. 1993. 'Ground Water Resources and Development Prospect in Ciombatore District, Tamil Nadu'. Hyderabad: Central Ground Water Board (CGWB) Southern Division.

———. 1999. 'A Note on Groundwater Pollution in Tiruppur Block'. Chennai: Central Ground Water Board.

Central Water Commission. 2000–2001. 'River Water Quality for Noyyal River'. Coimbatore: Central Water Commission.

Centre for Environmental Education. 2002. 'Report on Household Level Drinking Water Survey for 52 Wards in Tiruppur'. Tiruppur: Centre for Environmental Education.

Centre for Environmental Studies. 1996. 'Tiruppur Area Development Project: Environmental and Social Assessment'. Chennai: Centre for Environmental Studies, Anna University.

Directorate of Fisheries. 2000a. 'Fish Catch Data for Noyyal'. Chennai: Directorate of Fisheries.

———. 2000b. 'Fish Stocking and Catch for Orthapalayam Reservoir'. Chennai: Directorate of Fisheries.

———. 2000c. 'Fisheries Details on Orthapalayam Reservoir'. Chennai: Directorate of Fisheries.

Furn, Kristina. 2004. 'Effect of Dyeing and Bleaching Industries on the Area around the Orthapalayam Dam in South India'. Unpublished thesis, Uppsala, Sweden.

Hydrology Research Station. 1996. 'Pollution Studies in Noyyal — Orthapalayam Reservoir Project', Final Report. Chennai: Department of Fisheries.

Jacks, Gunnar, M. Kilhage, C. Magnusson and A. Selvaseelan. 1995. 'The Environmental Cost of T-Shirts', Report of the First Policy Advisory Committee Meeting on Sharing Common Water Resources, SIDA, Madras Institute of Development Studies, Chennai.

Jacob, Thomson. 1998. 'Impact of Industries on the Ground Water Quality of Tiruppur and its Ethical Implications'. Unpublished Ph.D. thesis, University of Madras.

Jacob, Thomson, J. Azariah and A. G. Viji Roy. 1999. 'Impact of Textile Industries on River Noyyal and Riverine Groundwater Quality of Tiruppur, India', *Pollution Research*, 18 (4): 359–68.

Jayakumar S. C., P. Loganathan and R. K. Sivanappan. 1978. 'A Study of Ground Water Pollution due to Human Agency around Coimbatore and Tiruppur', *Madras Agriculture Journal*, 65 (5): 329–33.

Kuppusamy. 1996. 'Affidavit of Mr. P. R. Kuppusamy', W.P. No. 1649 of 1996, Madras High Court, Chennai.

Lundqvist Jan, Paul Appasamy and Prakash Nelliyat. 2003. 'Dimensions and Approaches for Third World City Water Security', *Journal of Philosophical Transitions*, 358 (2440): 1985–96.

Madras School of Economics. 2002. 'Environmental Impact of Industrial Effluents in Noyyal River Basin', Project Report, Madras School of Economics, Chennai.

Millennium Ecosystem Assessment. 2003. *Ecosystem and Human Well-being: A Framework for Assessment*. Washington, D.C.: Island Press.

Mukherjee, Sacchidananda and Prakash Nelliyat. 2007. *Ground Water Pollution and Emerging Environmental Challenges of Industrial Effluent Irrigation in Mettupalayam Taluk, Tamil Nadu*, Comprehensive Assessment of Water Management in Agriculture, CA Discussion Paper 4. Colombo: International Water Management Institute.

Nelliyat, Prakash. 2002 'Environmental Cost of Industrial Water Consumption in the Lower Bhavani River Basin of Tamil Nadu, in N. Rajalakshmi (ed.), *Environmental Cost and Economic Valuation*, pp. 217–32. Delhi: Manak Publication Pvt. Ltd.

———. 2003. 'Industrial Growth and Water Pollution in Noyyal River Basin: India', in Stockholm International Water Institute (ed.), *Abstract Volume of The Thirteenth Stockholm Water Symposium*, pp. 365–68. Sweden: Stockholm International Water Institute.

———. 2005. 'Industrial Growth and Environmental Degradation: A Case Study of Industrial Pollution in Tiruppur'. Unpublished Ph.D. thesis, University of Madras.

———. 2007a. 'Public–Private Partnership in Urban Water Management: The Case of Tiruppur', in Barun Mitra, Kendra Okonski and Mohit Satyanand (eds), *'Keeping the Water Flowing: Understanding the Role of Institutions, Incentives, Economics and Entrepreneurship in Ensuring Access and Optimising Utilization of Water'*, pp. 149–60. Delhi: Academic Foundation.

———. 2007b. 'Trade Liberalisation and the Environment: With Special Reference to the Tiruppur Textile Cluster', *ENVISAGE*, 4 (2): 5–9.

Palanivel, M. and P. Rajaguru. 1999. 'The Present Status of the River Noyyal', in Bharathiyar University (ed.), *Proceedings of the Workshop on Environmental Status of Rivers in Tamil Nadu*. Coimbatore: Bharathiyar University.

Public Works Department (PWD). 2001. 'Environmental Status Report of the River Noyyal Basin', PWD Environment Cell, WRO Pollachi Region, Coimbatore.

———. 2002. 'Ground Water Prospective: A Profile of Erode District', State Ground and Surface Water Resources Data Centre, Water Resources Organisation, Chennai.

Public Works Department (PWD). 2003. 'Ground Water Prospective: A Profile of Coimbatore District', State Ground and Surface Water Resources Data Centre, Water Resources Organisation, Chennai.

Rajaguru P. and V. Subbaram. 2000. 'Ground Water Quality in Tiruppur', in Lakshmanaperumalsamy and Krishnaraj (eds), *Proceedings of the Workshop on Environmental Awareness on Quality Management of Irrigation Water.* Coimbatore: Department of Environmental Sciences, Bharathiyar University.

Ramasamy V. and D. Rajagopal. 1991. 'Ground Water Quality in Tiruppur', *Indian Journal of Environmental Health*, 33 (2): 187–91.

Research Department of Zoology. 2002. 'Biological Study of the Effects of Effluents in the Noyyal River', Final Report, Erode Arts College, Erode.

Senthilnathan, S. and P. A. Azeez. 1999. 'Water Quality of Effluents from Dyeing and Bleaching Industry in Tiruppur, Tamil Nadu India', *Journal of Industrial Pollution Control*, 15 (1): 79–88.

Soil Survey and Land Use Organisation (SS&LUO). 2002. 'Quality of Soil and Water for Agriculture in Noyyal River Basin, Tamil Nadu', Special Report-98, SS&LUO, Coimbatore.

Soil Testing Laboratory. 2000–2001. 'Water Quality Data for Orthapalayam Reservoir', Department of Agriculture, Erode.

Tamil Nadu Pollution Control Board (TNPCB). 1997–1999. 'Water Quality Data for Orthapalayam Reservoir', TNPCB Office, Tiruppur.

Tamil Nadu Water Supply and Drainage Board. 1995–2001. 'Regular Observation Well Water Quality Data', Tamil Nadu Water Supply and Drainage (TWAD) Board, Chennai.

———. 1999. 'Ground Water Quality Data for Tiruppur Region', Tamil Nadu Water Supply and Drainage (TWAD) Board, Coimbatore.

———. 2001. 'Hand Pump and Power Pump Water Quality Data for Tiruppur Block', Tamil Nadu Water Supply and Drainage (TWAD) Board, Chennai.

Tiruppur Municipality. 1999. 'Administrative Report of Tiruppur', Tiruppur Municipal Office, Tiruppur.

———. 2000. 'Water Supply and Street Light Report', Tiruppur Municipal Office, Tiruppur.

———. 2001. 'Water Supply of Tiruppur', Tiruppur Municipal Office, Tiruppur.

10

Recent Changes in Policy, Institutions and Utilisation of Ecosystem Services in Colombo Wetlands

Missaka Hettiarachchi, Kusum Athukorala, Ravi Peiris and Ajith De Alwis

Colombo is a city blessed with many rewards of nature. Principal among these is the vast 'necklace' of natural wetlands along the western boundary. Apart from the significance of these wetlands as natural ecosystems, they have been providing a range of water and environmental services for people living in the city and suburbs of Colombo for centuries. The rapid urbanisation, accompanied by weak urban planning, has caused a steady degradation of these wetlands. Some of the services provided by these wetlands have been severely threatened by the degradation. The frequency of urban disasters (such as floods) has also increased during the period. These events have increased the risk factors involved in development activities and aggravated the poverty of urban poor. This chapter presents an ongoing study carried out jointly by the University of Moratuwa and Network for Women Water Professionals — Sri Lanka, on the impacts of current environmental policy and wetland management strategies on the degradation of urban wetlands in Colombo.

Study Area

According to the Wetland Site Survey Report (1994) prepared by the Central Environmental Authority (CEA) of Sri Lanka, Colombo Flood Detention Area (CFDA) is mainly constituted by three wetlands, namely Kolonnawa Marsh, Heen Marsh and Kotte Marsh. Kolonnawa Marsh (the largest in area) was selected for the preliminary

* The study is separately funded by the International Foundation of Science (IFS), Women for Water Network (WFWF), and Senate Research Grant of University of Moratuwa.

investigation of environmental status as well as sociological survey. The geographical location of the CFDA is between Latitude 6° 52′ 55″–6° 52′ 45″ and Longitude 79° 52′ 35″–79° 55′ 15″. These wetlands are a combination of deep and shallow water marshes. The marshes drain into the Kelani river in North and South Colombo Canal System in the south (Map 10.1).

The average elevation of the CFDA wetland area is between 0.3–0.7 metres above sea level but the surrounding basin has hilly areas with significant height differences of 20–25 metres above sea level (CEA 1994). According to Cooray (1984), the present landform of the area can be categorised as 'Flood Plain', which would have evolved with the origin of Kelani river (second largest river basin in Sri Lanka). From the fifteenth to seventeenth century, the capital city of the Sri Lankan kings was situated in the Kotte, surrounded by these wetlands. A map prepared by the Dutch colonists in 1739 is the earliest graphical interpretation of the use of land of the area, which shows that around 70–50 per cent of the basin was then covered with marshes or paddy lands (Department of Irrigation 1947). The records of British colonial era state that the study area was inundated at least once a year with annual flooding of the Kelani river, mainly during the months of May or November.

Data and Analytical Methods

Analysis of policy and strategies on wetland management were mainly carried out as a literature survey. Regular site surveys were done on following aspects of the wetland: (a) Monthly variation of water quality at different stations for following parameters — BOD, pH, TDS, Temperature, DO, Nitrates, Phosphates, Fecal Coliform Count and Turbidity; (b) Study of the percentage cover of different vegetation in a set of selected plots in the marsh; (c) Study of diversity and abundance of fish and bird species of a set of selected plots in the marsh; (d) Study of wetland extent variation with aerial photographs of 1982, 1994, 1999 and 2000 using GIS.

Around 175 households from stakeholder communities within 200 metres from the boundary of the wetland were interviewed using questionnaires. Government officers, village leaders, local government politicians, religious leaders and senior citizens of the area were interviewed using structured questionnaires and informal discussions.

Map 10.1:   Map of the Study Area

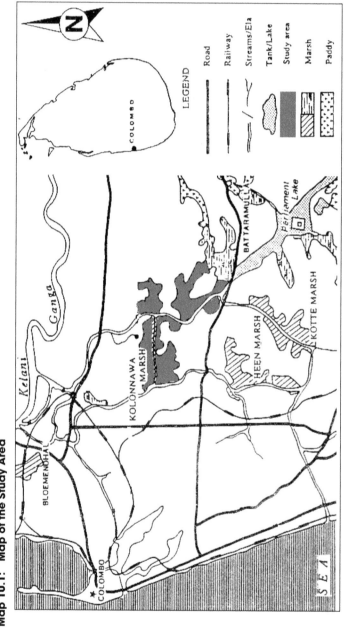

Source: Adapted from CEA (1994).

Results and Discussion

Recent Changes and Current Environmental Status

Large-scale encroachment of the marshes for housing and development started in the 1980s, when Sri Jayawardhana Pura Kotte was again declared the capital city of Sri Lanka in 1978. According to the Wetland Site Report (CEA 1994), 66 per cent of the families that have migrated to this area recently have settled after 1981. At present, the average population density in the basin is about 6,300 persons per square kilometre and the average housing density is about 1,500 units per square kilometre. According to key informants, a new pattern of sudden flash floods started during the 1990s. Floods that fully or partially inundated the areas within an average 100 metre distance form the marsh were recorded in the years 1992, 1999, 2004, 2005, 2006 and 2007.

Historically the major land use categories of the area were the marshlands, built-up areas and paddy lands. A trend of abandoning the paddy lands started in the mid-1960s which was gradually succeeded by wetland vegetation. The more recent trend was to convert these wetlands into built-up areas. Aerial photographs from 1980 to 2000 show that the average marshland area encroached per year was around 3.2 hectares Although this rate has continued at the same level in privately owned wetland areas, the recent government interventions have effectively minimised the illegal encroachment of state-owned portion of the CFDA wetlands.

Table 10.1 gives a summary of the water quality observations taken at four points in Kolonnawa marsh area between August 2007 and August 2008. The high values of Biochemical Oxygen Demand (BOD), Fecal Coliform, PO_4 and low Dissolved Oxygen indicate that there is a considerable pollution due to organic contaminants. The pH and NO_3 values do not significantly deviate from the range recommended for healthy ecosystems. Commonly found heavy metals such as Cd, Cr, Mn, Pb and Cu in industrially polluted streams were only found in levels below 0.1 mg/l. Therefore it could be safely assumed that the water pollution in the wetland is mainly caused by domestic discharges and solid waste.

Originally the marsh comprised characteristic freshwater marsh vegetation such as *Phragmites karka, Schoenoplectus grossus, Oryza sativa, Brachiaria mutica* and *Nymphaea lotus* (CEA 1994). However,

Table 10.1: Water Quality in the Kolonnawa Marsh

	BOD (mg/l)	pH	Fecal coli/(nos in 100ml)	NO_3 (mg/l)	PO_4 (mg/l)	DO (mg/l)	TDS (mg/l)
Average	26.5	6.42	4700	1.1	0.08	0.94	134
Maximum	70.0	7.91	23000	3.75	0.76	5.25	230
Minimum	5.0	5.5	40	0.0	0.0	0.07	90
Standard	4.0	6.0–8.5	1000	5.0	0.1	6.0	N/A
Reference	SL std	SL std	WHO (for any type of human contact)	SL std	USEPA	SL std	N/A

Source: Fieldwork conducted by authors.

Note: SL std.: Proposed Sri Lankan standard for surface water quality, USEPA: Environmental Protection Agency of USA standard, WHO: World Health Organisation standard for water for human consumption or contact.

the vegetation surveys carried out during this study revealed that such vegetation is being replaced by pollution indicative invasive species such as *Annona glabra*, *Eichhornia crassipes* and *Hydrilla verticillata*. *Eichhornia crassipes* cover in the open water bodies increased up to 70 per cent of the total water surface in October 2007. Tropical freshwater fish such as Murrel (*Channa striata*) and Stinging Catfish (*Heteropneustes fossilis*), said to have been found abundantly in the deep water areas, are now almost replaced by more pollution tolerant exotic invasive species such as Tilapia (*Saratherodon mossambicus*), Kissing Gourami (*Helostoma temminckii*) and Knife fish (*Chitala spp.*). During the ecological investigation, it was mainly found that the typical marshland habitats are increasingly being replaced by invasive vegetation and fish species (Table 10.2).

The socioeconomic data given in Table 10.3 reveals that a majority of households in the fringe area of the wetland are affected by floods and related events. They in turn contribute largely to further degradation of the wetland by discharging sewage, wastewater and solid waste to the wetland, directly or indirectly.

Table 10.2: Ecological Parameters in the Kolonnawa Marsh

	Chlorophyll content (mg/l)	Number of endangered macro-fauna species recorded in a station	% coverage of invasive plants species in a plot	No. of invasive fish species recorded in a station
Average	1.5	Not significant	55–45	2
Maximum	2.6	2	>90	5
Minimum	0.23	0	<20	1

Source: Fieldwork conducted by authors.

Table 10.3: Socioeconomic Data of Communities around the Kolonnawa Marshes

Parameter	Value %
Households effected by flood	60.0
Households effected by water-related diseases	35.5
Households with a sanitary latrine (septic tank, bio-filter or UDDT)	10.5
Households directly discharging wastewater into the wetland	52.5
Houses without proper solid waste collection or disposal method	39.5

Source: Fieldwork conducted by authors.

Currently Utilised Water Services of the Wetland

Wastewater Discharge

All the respondents interviewed had onsite sewage disposal methods (mainly soil soaking). However, only 10.5 per cent of the respondents had an accepted hygienic disposal method for their lavatories (i.e., septic and soakage pit, bio-filter). It should be noted that 52.5 per cent of the respondents directly disposed their wastewater (grey water) into the marshes or the nearest stream. Therefore it is clear that the CFDA wetlands are acting as the wastewater sink of the area.

Storm-water Drainage and Flood Attenuation

The storm-water drainage system in the area mainly consists of medium-sized rectangular open drains. Apart from the constructed drains, natural storm-water gullies and small streams are also present. All these drains, gullies and streams are ultimately connected to the marsh or a canal of the wetlands system. Therefore it should be noted that CFDA wetlands are the primary storm-water drainage infrastructure in the area. The importance of CFDA in flood attenuation in Colombo was well understood even by the first proposal for Colombo flood protection scheme in 1924, which was implemented by the irrigation department of Sri Lanka (Department of Irrigation 1947). According to a study done in 1966, Kolonnawa marsh alone had a flood retention capacity of 2.8 million cubic metres (Department of Irrigation 1966). Although a comprehensive environmental economic valuation of flood attenuation function has not been carried out in the CFDA wetlands, a similar study was done on the nearby Muthrajawela Marsh by L. Emerton and B. Kekulandala in 2002 (see Emerton 2005) where a value of USD 1,758 per hectare per year was estimated for the flood attenuation function.

Unutilised or Underutilised Water Services of the Wetland

It was revealed in key informant interviews that the streams and the pools in the wetlands were regularly used for bathing and washing until early 1980s. According to the survey, respondents are presently not using the streams or wetlands for their water needs. The main source of water in the area at present is piped water provided by the National Water Supply and Drainage Board and 31.6 per cent of the respondents in the area had dug wells. However, most of them (79.4 per cent) were not using water from the well for drinking. At present, the wetland is not used for bathing, swimming or pleasure boating. Therefore the current level of use of CFDA wetlands for recreational purposes is insignificant. According to key informant interviews, this canal network has served as a primary mode of transport in the past. Although the canals and streams within the CFDA wetlands area are still wide and deep enough to accommodate navigation, they are not used for any type of transportation at present. Restarting the use of the streams and canals within CFDA wetlands for transportation is an option worth investigating.

Current Environmental Policy and Wetland Management Strategies

Conservation of wetlands and public waterways such as CFDA wetlands were covered even under the earliest laws in Sri Lanka such as the Crown Land Ordinance (1947). The legal protection for marshlands was further strengthened with the Flood Protection Ordinance (1924), where flood prone areas or flood retention/detention areas were declared as 'Flood Areas' and were protected by the relevant 'Flood Authority' (District Secretary or Mayor). Under the provisions of this ordinance, the Colombo Flood Protection Scheme was implemented during the period 1924–1933, where the legal custodianship of the CFDA wetland was transferred to the irrigation department. However, the title of these lands was not transferred to the irrigation department.

The irrigation department developed a comprehensive flood protection plan with many engineering features such as bunds, lock gates and pumping stations and managed the marshlands in a flood protection approach. None of the reports or standing orders of the irrigation department view these wetlands as ecological features (Department of Irrigation 1993). With the estab-lishment of the Colombo District Low-lying Areas Reclamation and Development Board (1968) which later became Sri Lanka Land Reclamation and Development Corporation (SLLRDC), not only the custodianship but also the legal title of the public low-land areas were transferred by the SLLRDC Act of 1968. The SLLRDC carried the mandate for development of low-land areas for urban development purposes as well as protection of flood detention/retention areas, which were conflicting in nature. In the early period, the development activities were strong in the SLLRDC agenda and flood protection was addressed through passive measures such as maintenance of canals and streams. Privately owned marshy lands or fallowed paddy lands were only protected by the Agrarian Services Act of 1978, which prohibited the usage of any privately owned paddy land for any other purpose without permission of the Agrarian Services Commissioner [62 (1) of Agrarian Services Act of 1979]. The Urban Development Authority (UDA) was established in 1978 under the UDA Act of 1978. Its mandate was mainly to plan urban development in Sri Lanka. With the rapid urbanisation which began in 1980s in the CFDA area, defining the role of low lands in urban planning became a concern of the UDA. Major acquisitions of privately owned low lands within CFDA were made in early the 1980s for urban development purposes by the UDA instead of the SLLRDC [by provision 16 (1) of the UDA Act and the Land Acquisition Act]. An anomaly was created when the same wetland system came under the jurisdiction of different authorities with conflicting agendas.

The first evidence of interagency cooperation related to CFDA was seen in the formulation of Colombo Master Plan in 1978 (UDA 1998). Ecological importance of these marshlands was also appreciated in the Greater Colombo Development Plan and part of the Kotte Marsh (part of CFDA) was declared as a bird Sanctuary. In 1980, the landmark environmental legislation of the country, the National Environmental Act, was passed and the Central Environmental Authority (CEA) was established. As the apex body responsible for environmental management, natural features such as Colombo wetlands came within the subject area of CEA. The articles 16 and 17 of the National Environmental Act also entrusts CEA with an advisory role in land use and natural resources management in the country. Also, as the regulatory body handling environmental clearance procedure for environmental impact assess-ment of major investment projects and the environmental protection licensing of industries, CEA contributes largely in proper management of wetlands.

Although many landmark measures were taken in the area of wet-land protection during the 1980–1990 period, these were not strong enough to counter the effects of rapid urbanisation. The loosely defined mandates of new agencies (such as the CEA) and conflict of agendas among agencies SLLRDC, UDA and CEA) weakened the effectiveness of these new legislations and policies.

Severe floods which destabilised the city life in 1992, the alarming reports of degradation of the CFDA wetlands (CEA 1994) and the conditions of the Ramsar Convention on Wetlands (1973) instigated a new phase of environmental policies and wetland management strategies in Sri Lanka. During the same period, the regulatory role of the CEA was further strengthened by the National Environmental (Amendment) Act of 1988. The Colombo Flood Protection and Environmental Improvement Project was implemented during the same period where some private lands (especially in the Kolonnawa Marsh area) were acquired by the SLLRDC and demarcated as protected wetlands for flood retention. With the improvement of the Environmental Impact Assessment (EIA) and Environmental Protection Licence (EPL) procedures, the legal control over imprudent exploitation of wetlands got strengthened. Environmental monitoring was given a new dimension with the allocation of environmental officers for CEA at local government level.

This period distinctly marks a shift of government policy on wetlands. The functional values of wetlands (including the aesthetic values) within an urban context were identified in principle (UDA 1998). Formulation of urban development master plans with combined institutional roles (ibid.) was seen. The potential of developing bird-watching and ecotourism in CFDA wetlands was also evaluated and plans were prepared (CEA 1994). This phase of development of policy and strategy in wetland management successfully reduced the danger of mass encroachment of the most sensitive areas of the CFDA system due to haphazard urbanisation. However it cannot be denied that some decisions taken by the government itself seriously jeopardised sustainability of these wetlands. One such example was the selling of 70 hectares of low land owned by the UDA (acquired previously from private owners) to a private company to develop a sports and recreation centre. This decision was later declared as improper and unlawful by the Supreme Court of Sri Lanka [Fundamental Rights Petition S.C. (F/R)N0 352/2007]. Our investigation identified three public playgrounds and one multistorey housing complex in Kolonnawa Marsh area, which were constructed by filling the marsh or adjoining low lands. Another dimension to management of CFDA wetlands was added during the period 2000–2005 with the increasing concern over disasters. From 1994 onwards, Colombo district recorded a major flood each year up to 1999. A possible connection between the rapid degradation of low-lying areas in Colombo and some of these flood events was pointed out. The Disaster Management Centre of Sri Lanka (DMC) was established under the Disaster Management Act of 2005 (after the Asian Tsunami in 2004). The DMC functions as an independent body which coordinates among line ministries to achieve better disaster management. DMC was involved directly in clearing of unauthorised development in marshlands and widening of canals in Colombo following the floods of 2006. Although statutorily established, DMC's role in the complex web of managing the CFDA wetlands is yet to be defined.

Evolving through the above process, the National Environmental Policy and National Wetland Policy were introduced by the Ministry of Environment and Natural Resources in 2003 and 2006 respectively. Both the policies identified the importance of integration of stakeholder agencies, improved (equitable) public participation in decision making, environmental monitoring, environmental zoning and valuing of environmental services. However, the results of this study reveal that the environmental status of CFDA wetlands has been steadily degrading amid these new developments. Therefore the current policy and institutional framework should be critically reviewed.

Institutional Structure Related to Management of CFDA Wetlands

The institutional arrangement of the main stakeholder agencies relevant to CFDA wetlands prior to the Wetland Policy (2006) which still operative in practice is given by Figure 10.1. Their key roles with regard to wetland management are as follows: Irrigation Department — maintenance and management of the main head works of the Colombo flood control system; SLLRDC — protection of the wetlands and maintenance of streams/canals. It bears land ownership of nearly 80 per cent of CFDA wetlands; UDA — prime authority on urban development planning. It bears the land ownership of 10–20 per cent of the CFDA wetlands; CEA — enforcement of environmental regulations, issuing environmental clearance, environmental licensing of major industries and environmental monitoring; DMC — apex coordinating body on disaster management; Local Government Authorities (LGAs) — responsible for authorising all construction work, issuing permits and environmental license to small and medium industries. During the research, it was found that no NGO or CBO in the area were involved in wetland related work in the area.

Figure 10.1: Previous Institutional Arrangement

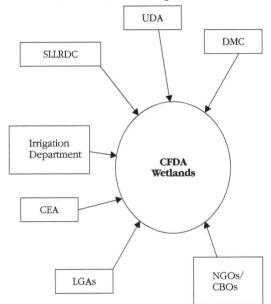

Source: Prepared by the authors.

There was no platform or provisions for these agencies for participatory decision making in this previous structure, which lead to sectoral and poorly-informed decision making on some occasions.

A major modification for the institutional structure in wetland management has been suggested by the National Wetland Policy (2006), but is not yet fully operational. The National Wetland Steering Committee (NWSC) was established and will function as the central advisory body in the wetland management sector in Sri Lanka. The Wetland Management Unit (WMU), which will be within CEA, is the main coordinating centre for wetland activities. Wetland Facilitating Committees (WFC) are to be established at different administrative levels (within divisional, district and provincial secretariats) as platforms for coordination of stakeholders. The lowest level unit in the suggested structure is the Wetland Management Committee (WMC). A WMC will be formed for each important wetland (or segment) which looks after its management by monitoring the activities of the government agencies and facilitating public participation in related decision making. The WMCs will be established under the provisions given in the National Environmental Act (1980) and will lead to formation of environmental committees to protect sensitive environments. Other line agencies, NGO and CBOs which are stakeholders of the wetland will act independently but in consonance with the proposed new institutions (NWSC, WMU, WFC and WMC). Figure 10.2 gives a graphical representation of the new institutional framework proposed by the National Wetland Policy (2006).

Analysis of the Current Policy and Institutional Roles

Historical development of the management process of CFDA wetlands initially had a focus on flood retention. It was on this basis that the Colombo wetlands were identified and demarcated by the irrigation department. However it can be seen that with the increasing urbanisation pressure, the focus of institutions which later took over the land tenure of CFDA (SLLRDC and UDA) shifted towards using these lands for urban expansion. However as described above, the repercussions (such as increasing urban floods) of this unsustainable approach has compelled the same agencies to consider integrated approaches in wetland management. Overlapping of institutional roles with regard to planning of use of land and land acquisition is also seen when legislative enactments such as Flood Protection Ordinance (1924), Land Reclamation and Development Corporation Ordinance (1968), Urban Development Authority Ordinance (1978) and National Environmental Act (1980) are compared. In the judgement against the sale of lands by UDA to a private company to develop a sports and recreation centre (Fundamental Rights Petition S.C.(F/R)N0 352/2007), the Supreme Court of Sri Lanka pointed out the weaknesses in definition of institutional roles in the current structure which has enabled UDA to overlook some of the environmental impacts of the project. The same impacts later affected the flood attenuation services of the wetland and created a crisis in urban services. Strengthening of the role of CEA in development decision making through improved environmental clearance and protection licensing procedures has also contributed towards this change.

The National Wetland Policy of 2006 could be identified as a milestone of the paradigm shift in the management of CFDA wetlands. It can be the backbone of a strengthened wetland management structure for CFDA. In principle, the policy identifies the importance

Figure 10.2: Institutional Arrangement Proposed by the National Wetland Policy

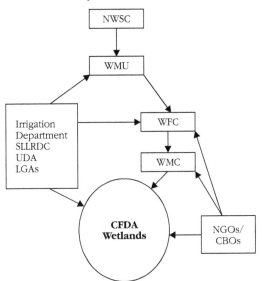

Source: Prepared by the authors.

of integrated decision making, regular environmental monitoring of wetlands, economic valuing of wetland services and zoning of wetlands according to sensitiveness. However many of the objectives of the National Wetland Policy are yet to be achieved. Some inherent weaknesses in the policy that may slowdown or weaken its proper implementation can be identified. Although the policy distinctly identifies the importance of services (especially water services) provided by the wetlands in an urban context, it does not make a distinction between urban and rural/natural wetlands. Management and restoration of urban wetlands is an emerging field of study at present and the importance of developing different management strategies for urban and rural/natural wetlands need to be recognised. A management approach based on restoration of the ecological functions of wetlands and the particular demands imposed by the urban environment should be followed in dealing with urban wetlands. The National Wetland Policy of 2006 is mainly based on a conservation approach rather than a function restoration approach. In an urban context (such as Colombo) where land value is soaring, reserving lands for nature conservation purposes can be less justifiable even among government agencies. A population growth rate of 2.2 per cent is projected for the period 1996–2010 in the urban areas surrounding CFDA wetlands (UDA 1998). However all the government stakeholder agencies agree in principle that CFDA wetlands provide essential urban services (CEA 1994; UDA 1994; UDA 1998). Therefore National Wetland Policy should adopt a strategy which highlights this aspect in protection or restoration of urban wetlands. Having a realistic estimation of the economic value of the services offered by the wetlands in urban areas is essential in strengthening the above strategy. Although the wetland policy recommends the use of environmental economic tools, this is merely indicated as a recommendation for the other stakeholder agencies (land use planners, UDA). It can be more effective if such valuation can be carried out by the WMU itself and then presented to other stakeholder agencies to support their decision making. Methods such as bio-economic modelling and cost-based valuation methods of environmental services of wetlands have been successfully used in truly integrated and informed decision making related to wetland management. Wetland protection policy analysis of wetlands in South Australia and wetland management in Hail Haor wetlands in Bangladesh are two such examples (Emerton and Bos 2004; Bennett and Whitten 2002).

The institutional arrangement given in the National Wetland Policy is highly centred on environment-related agencies (CEA, Ministry of Environment). Responsibility of initially setting up the proposed institutional framework lies with the WMU which will be a sub-entity of CEA or Ministry of Environment. How this entity can muster the support of strong line agencies with development agendas should be more clearly indicated in the strategies. The importance of disaster mitigation role of the urban wetlands is not properly indicated in the policy, which can be seen as a major drawback. Even at the strategy level, it does not specifically address the issue of disaster management. The results of this research clearly indicate that the CFDA basin is highly flood prone. Sixty per cent of the household investigated are threatened by flood. Therefore any management strategy developed for CFDA wetlands should have a disaster management perspective also.

A Framework to Integrate the Institutional Roles and Recommendations for Optimal Utilisation of Wetland Services

According to the results of this study, it can be reasonably surmised that CFDA wetlands provide the essential urban services of flood attenuation and drainage for its urban/suburban basins at present. This makes the CFDA wetlands an important water infrastructure of Colombo and and its suburbs. The economic value of these services (USD 1,758 per hectare per year as approximately calculated by Emerton and Bos (2004) is high and cannot be compensated easily under the given economic condition of the country. The flood attenuation service of the wetlands is also pivotal in disaster management planning. Therefore the management of CFDA wetland has the following three components: (a) Sustainable utilisation of the water services in the urban process; (b) Disaster mitigation or adaptation; and (c) Protection of the natural ecosystem. The authors contend that a successful management can only be achieved through an integration of these components (Figure 10.3).

Figure 10.3: Three Components of Urban Wetland Management

Source: Prepared by the authors.

At present, these three aspects are sufficiently covered by the mandates and statutory powers of the stakeholder agencies of CFDA. The importance of protecting the CFDA wetlands against the pressing land demand for the development needs of Colombo can only be justified by the urban services provided by them. Therefore the direct management role of CFDA should be borne by an agency which has the capacity to sustain these services. SLLRDC (which currently bears the tenure for the largest portion of CFDA wetlands) has ample statutory powers as well as technical capacity for this. However, the study shows that SLLRDC has had failures in preventing the degradation of the CFDA wetlands (water pollution, proliferation of invasive species) and managing flood-related disasters. National Environmental Act (1980) and Disaster Management Act (2005) provides CEA and DMC with necessary legal provisions to handle these issues. Their institutional capacity in these areas is also higher than SLLRDC. SLLRDC is not actively involved in use of land and planning of urban development. In CFDA areas UDA and LGAs are responsible for planning and authorisation of development work. Therefore a sustainable management of the CFDA wetlands can only be achieved through integration of the current roles of these agencies. This integration can only be achieved by providing a common platform for the stakeholder agencies to communicate specifically on wetland management. Such a platform should be established at the appropriate level, which is large enough in scale for effective decision making and small enough to focus on CFDA system. The divisional level Wetland Facilitation Committee (WFC), proposed in the National Wetland Policy, provides an excellent model for this. The Wetland Management Committee (WMC) may function mainly as a channel to facilitate public participation in wetland management. Public participation may be ensured by including the local CBOs in the WMC. However, the study revealed that there are no CBOs in the area which are involved in wetland related activities. Therefore a thorough campaign to improve the public awareness on the functional importance of the CFDA wetlands is required along with the suggested institutional reforms. The policy does not clearly delineate the way to make the decisions of these committees legally binding and how WFC is answerable to the WMC. These issues have to be sorted before actual implementation of the proposed reforms begin.

Conclusion

CFDA wetlands play an important functional role in ensuring the essential urban services of flood control and drainage in the city of Colombo and its suburbs. The economic value of these services can justify the need for protection and restoration of these wetlands against the development needs. The main threat to CFDA wetlands (and the sustainability of the services provided by them) at present is the degradation of its ecosystem caused by urban pollutants of domestic origin. At present, the disintegration among stakeholder institutions and the lack of stakeholder community participation is damaging the wetland management process. The existing stakeholder agencies have adequate statutory powers as well as technical capacity to handle their mandates. Therefore it is recommended to provide a common platform for stakeholder institutions and community participation. The institutional reforms suggested by the National Wetlands Policy of 2006 can be adopted effectively to achieve these objectives. Implementation of the above institutional and procedural changes will incur considerable transaction costs; therefore it is essential to have an indication of the economic value of wetland services to justify these costs.

References and Select Bibliography

Bennett, J. and S. M. Whitten. 2002. *The Private and Social Values of Wetland: An Overview*. Canberra: Land and Water Australia.

Central Environmental Authority (CEA). 1994. 'Wetland Site Report and Conservation Management Plan — Colombo Flood Detention Areas', Wetland Conservation Project. Colombo: Central Environmental Authority of Sri Lanka.

Cooray, P. G. 1984. *The Geology of Sri Lanka*. Colombo: National Museums Publications, Department of Government Printing.

Department of Irrigation. 1947. 'Kelani Ganga Flood Protection Scheme'. Colombo: Department of Irrigation.

———. 1966. 'Technical Report: Reclamation of Swamps in and around the City of Colombo', Division of Reclamation and Drainage Division of Reclamation and Drainage, Department of Irrigation Sri Lanka, Colombo.

———. 1993. 'Scheme of Organisation and Standing Orders to Safeguard City of Colombo from the Floods in Kelani Rive. Colombo: Department of Irrigation.

Emerton, L. ed. 2005. *Values and Rewards: Counting and Capturing Eco-system Water Services for Sustainable Development*, IUCN Water, Nature and Economics Technical Paper No. 1, IUCN — The World Conservation Union. http://data.iucn.org/dbtw-wpd/edocs/2005-047.pdf (accessed 2 March 2011).

Emerton L. and Bos L. 2004. *Value: Counting Ecosystems as Water Infra-structure*. http://data.iucn.org/dbtw-wpd/edocs/2004-046.pdf (accessed 2 March 2011).

Urban Development Authority (UDA). 1994. *Environmental Management Plan for Colombo Metropolitan Area*, vol. III. Colombo: Urban Development Authority of Sri Lanka.

————. 1998. *Colombo Metropolitan Regional Structure Plan*, vol. II. Colombo: Urban Development Authority of Sri Lanka.

WS Atkins International. 1985. 'Sri Lanka Water Supply and Sanitation Rehabilitation Project', Project Report Storm Water Drainage. Colombo: National Water Supply and Drainage Board.

11

Roles of Institutions and their Limitations in Integrated Management of Water, Forest and Land Resources

Dhurba Pant

Forests and water constitute two major resources wherein people's initiatives in management have made important contribution towards maintaining ecosystem services and livelihood enhancement in Nepal. This has contributed to the evolution of institutions that are central to conservation and utilisation of these resources. Water users' groups (WUGs) that have traditionally utilised and managed water resources at the local level are successful institutions established by local communities to derive collective benefits (Pradhan and Bandaragoda 1997; Pant 2000).

Likewise, the management of forests by local communities in the hills of Nepal is an example of sustainable resource management. Forest Users' Groups (FUGs) have evolved significantly over the last two decades (Soussan et al. 1995; Soussan 1998) as strong and formal local institutions that address not only the protection of forests but also various developmental activities in villages (Springate-Baginski et al. 2001). Therefore, water and forest provide examples of Community Based Natural Resource Management (CBNRM) approach at the local level. The successes of their management are confined to single resources for various reasons. Important among them is the government policy and legal provisions which promote a sectoral approach in the design and implementation of development programmes, which has resulted in the enactment of various laws for sectoral issues, and some of them contradict each other. As such, some of the policies and also laws have become redundant in implementation.

Although examples of multiple uses of the natural resources are found at the local level, integrated approaches in management, which contributes to incremental benefit from the use of resources are often difficult to find. Therefore, understanding institutional role and existing linkages or limitations for CBINRM is important and is the focus of this chapter. The presentation is based on an institutional study, which was part of an action research implemented since April 2005 in the Begnas catchment of western hills in Nepal, under the Challenge Programme on Water and Food (CPWF) of Consultative Group on International Agricultural Research (CGIAR). The focus of the institutional study was on understanding the role, existing linkages or limitations among institutions for integrated natural resource management at catchment level, which is believed to contribute to enhanced sustainable livelihood opportunities and reduced vulnerability for poor rural people in upper catchments. The study was intended to help government agencies in Nepal to develop and implement appropriate policy measures for Integrated Water and Resources Management Mechanism (IWRM), which is the major thrust of the Water Resources Strategy (HMGN 2002) and National Water Plan (HMGN 2005) of Government of Nepal.

This chapter sheds light on the role of forest and water-based institutions in promoting livelihoods opportunities, equity in benefit sharing among and across users and maintenance of ecosystem services, its sustainability and their limitations in actualising CBINRM. It also explores the dynamics of natural resource management through understanding of linkages among and between the local level and external institutions. An attempt is made to understand the role resource user groups and their institutions play in the management of natural resources in promoting livelihood opportunities and the constraints and opportunities for integrated management of natural resources through existing institutions.

Methodology

The study was undertaken in Begnas-Rupa basin in western Nepal. Altogether, 171 households from six communities — 86 households from upper watershed communities and 85 households from the valley floor — were selected for interviews with the household head. The communities in the upper watershed were Lamichhane gaon, Thapa gaon and Bhurtel gaon; and communities in the valley floor were Sainik Basti, Janata ko Chautara and Sat Muhane. Beside the use of a structured questionnaire, information was also collected through unstructured interviews using Participatory Rural Appraisal (PRA) tools and focus group discussions with the users from various resource user groups.

Socioeconomic Characteristics and Patterns of Livelihood

Physical Setting

The Begnas-Rupa basin lies in the western part of the middle mountain region of Nepal and occupies an area of 75.04 square kilometers. The Begnas basin is divided into the upper watershed area and the valley floor, with the Begnas Lake in the middle. The Begnas Lake was one of the natural lakes of the Pokhara valley with an area of 266 ha which was extended to 300 ha due to the construction of a 540-m-long and 6.9-m-high earth-fill dam in 1988. The average depth of the developed lake is about 8.5 meters.

The upper watershed area is mountainous with a slope ranging from 1,450 to 680 meters from north to south within a horizontal distance of about 5 km. Average hill slope of the watershed is about 15.4 per cent and the topography is steep and undulating. The upper watershed has several smaller natural gullies that provide irrigation to small patches of land and also drinking water to the people in valley floor areas. These gullies merge at the foothill of the upper watershed known as Shyangkhudi Khola that ultimately drains into the Begnas Lake, in which the Jalhari group is dependent for their livelihood. Therefore, the natural resources of the catchment are providing valuable ecosystem services to the livelihood of the people in the catchment. The valley floor area consists of flat lands with terraces with relatively high soil fertility and it receives irrigation water from the Begnas Lake through the Begnas Irrigation System (BIS).

Population

There were 550 households in the Begnas basin in 2005 and the population growth rate of the district where this basin lies is estimated at 2.62 per cent (CBS 2004). Based on the household survey, the average population of sampled household was 5.4. Male–female ratio was almost even with males slightly outnumbering the female population. The field survey also revealed that the out-migration from the upper watershed is higher (15 per cent) compared to the valley floor (7 per cent) which is largely attributed to the declining agricultural productivity and employment opportunity in agriculture.

Land holding

The cultivated/cropped area in the Begnas watershed covers about 60 per cent and 40 per cent in the valley floor and the upper watershed, respectively. The Jalhari has small homesteads in a small hillock adjoining the lake in the middle of the watershed. The land in terms of its use in the catchment could be categorised into three types. The first is the Khet, which receives perennial or seasonal irrigation and is used for cultivating rice, is highly valued and is an important source of livelihood for the households. The paddy in the summer and wheat, maize or potato in the winter and spring seasons are the major crops grown in this type of land. The availability of this land in the upper watershed area is limited. But most of the cultivated land in the valley floor is the Khet land. The second category of land is the Bari or Pakho land which does not receive irrigation and is primarily used for growing maize, millet, potato, soybean and black gram. Kharbari, the third type of land, is steeper than the Bari land and is used for growing thatch grass and some forest trees.

Almost all the households in the valley floor have some Khet land, whereas in the upper catchment about 90 per cent of the households possess Khet land and the size is less than 0.5 ha and they are considered small holders according to national definition. The percentage of households having Bari land is significantly more in the upper watershed indicating low agricultural production and productivity and more vulnerability to the climatic variability. Due to small holding, most of the households are owner cultivator and 15 per cent of the household have rented lands from others. This is an indication of the constraints in labour availability in some of the households, because of out-migration of able members.

Occupation

Based on the total population in the surveyed area, it was reported that over 85 per cent, including the students, are dependent on agriculture. The labour force available in the households generally includes the aging population or young school-going children, as many young educated people have moved away from the village in search of employment and a few study in the towns. At the same time, 10 per cent of the people are deriving income from off-farm activities that include service in the public sector, remittance, pension and small business. The Jalhari in the middle of the watershed, whose main occupation is fishing in the lake, derive almost all the income from fishing.

Food Sufficiency

The agricultural production and productivity is higher in the area which is 7.0 t/ha compared to the district average of 4.95 t/ha (DADO 2006) because of the availability of irrigation facility, fertilisers and other agricultural inputs with good extension services available because of close proximity to the district headquarter. The food sufficiency is higher in the valley floor as compared to the upper watershed. This substantiates information on the migration trend which is higher in the upper watershed. This also is indicative of the need of alternative livelihood opportunities for the people in the upper watershed compared to the people in the valley floor.

Annual Income

The household agricultural income in upper watershed is 15 per cent of the total income, which in average is less than $ 650 per household. The income from off-farm activities comprises almost 85 per cent in the upper catchment. This exemplifies declining importance of agriculture to the household livelihood in the upper catchment and their less dependence on it unlike in the past. The low level of income from agriculture could be attributed to the lack of irrigation facility and dependence on monsoon.

The proportion of agricultural income to total income for the households in the valley floor is nearly three times more compared to the upper watershed indicating the importance of irrigated agriculture. The average income from agriculture is also higher than the upper watershed because of the high cropping intensity due to availability of irrigation and poultry farming which is popular in the valley floor for household cash income. This also explains the inequity existing between the farmers at the upper watershed and valley floor due to differences in irrigated and rainfed agriculture.

The remittance is also a major contributor to the income of the households in the valley floor. This supports the argument that the young and educated are no longer interested in agricultural activities and tend to migrate to cities and abroad. One of the distinct features in the valley floor is the higher average income from other sources. This is largely due to the income from the pensions of the people who had earlier served in the ex-army, and who are present in large numbers in valley floor.

Project Intervention, Ecosystem Services and Institutions

Project Interventions

The period 1984–1996 was the turning point in the development of the upper watershed due to interventions by CARE Nepal and Department of Soil Conservation and Watershed Management of Government of Nepal. In coordination with local elected institutions at the village and district level, Begnas Tal Rupa Tal/Watershed Management Project (BTRT/WMP) was implemented in the study area in 1984 with a goal to stabilise the physical environment and increase the productivity of the project area through sustainable community management of its human and natural resources (Duijnhouwer 1997a).

With a focus on natural resource management, the project had tried to address the issues related to resource degradation, its availability and competition among multiple users. The project activities included the maintenance of ecosystem by preventing soil erosion and landslides through gully control, plantation in barren land, trail and canal construction, nursery and seedling production and conducting of environmental awareness programmes among local people. One of the major components it introduced was forest management through community by organising Forest Users Group (FUG), which included the conservation of existing forest and plantation in grazing land. After plantations, people started to protect the forest which has resulted in considerable increase in greenery with less erosion and landslides and that has had a positive effect on the lake downstream. The FUG activity was combined with agricultural and engineering activities to make soil conservation activities more effective (Adhikary et al., 1992). The Government of Nepal with support from the Japan International Co-operation Agency (JICA) introduced cage fish raising in the lake since 1982. This was one of the major interventions in the middle of the catchment designed to protect the lake environment and improving the livelihood opportunity for the Jalhari communities who are considered to be from a socially lower strata of the society.

The third project intervention with loan assistance from Asian Development Bank was the construction of the BIS with earth-fill dam and the canal network in 1984 and it was completed in 1988. It was implemented through a participatory approach, following the government policy of 1980s, which required users' involvement from the beginning of the project.

Overall, the project interventions in all the three reaches of the watershed have focused on ecosystem maintenance, livelihood improvement and development of institutions. Nevertheless, the effect of these interventions in the three reaches has been varied because of differences in focus on programme, intervention approaches, resource availability and institutional capability of users at local level.

Ecosystem Services

The activities during the project interventions were directed towards providing increased benefit to the users through the productive use of available natural resources and maintenance of these ecosystem services for sustained benefit. The major activities are discussed in the section that follows.

Community Forestry Programme

This was a successful programme and it contributed tremendously in regenerating and restocking the forests in the watershed. Villagers recall that the lands were degraded and the forest existed sparsely. The project strengthened the users group by supporting them in developing a constitution and operational plan, which was a requirement for the handover process of community forestry. This was a milestone in formalising the users group and preparing them for the ownership of the community forest, which the district office handed over to them in 1994. With the establishment of community forest management, the greenery in the area has increased along with household's ease of access to firewood and fodder.

Community Infrastructure Programme

The BTRT/WMP project in the upper watershed supported informal irrigation users group for the rehabilitation of the traditional irrigation systems. The support to the traditional irrigation systems was important in contributing to reduced erosion and increased water availability for drinking and irrigation. These irrigation systems are contributing to the food production in the upper watershed where availability of agricultural land is limited. As the livelihood opportunity in the upper catchment is limited, activities supporting the agricultural production and productivity received users' active involvement in the upkeep of these irrigation systems. These irrigation systems have been operating for a long time although it needs improvement to address users' need for increased irrigation water for crop. Beside, the project also supported village trail improvement through construction of stone pavement, which has stabilised the slopes and prevents slides. These schemes also provided seasonal employment to local people.

The Fishery Programme

The livelihood option available from the ecosystem has induced a great change in the livelihood of the group known as Jalhari who are involved in fishing activities. They were seasonally migrating to Begnas Lake since 1976 due to availability of adequate number of fish in Begnas Lake compared to other rivers in Pokhara valley. At that time, they were staying only for 6 months in small temporary cottages. They started to live permanently in that place from 1980, by constructing small houses in (*ailani*) uncultivated government-owned land. In the beginning, only 10 families settled there, which now has extended to 42 families due to separation of the adult family members and in-migration of 14 households from other areas. They are engaged in commercial fish farming in the lake with the technical support from the Fishery Centre of the government. This has contributed to the maintenance of lake environment. The activity has contributed to the improvement of livelihood of the fishermen group with regular income for the households. This improvement could be observed in their lifestyles and the amenities available in the households.

The BIS Project

The lake water is being used to irrigate 540 ha land in the valley floor. The irrigated area is said to be larger than that as claimed by some of the users. However, the WUA does not have records of the users and their land area. The new construction is an integration of the old network of canals with new canals and serves a larger area than what was served before by traditional canals. Therefore, the lake water is providing increased benefit through crop production to the farmers in the valley floor, although improvement in irrigation services could be instituted to enhance its impacts.

Upstream–downstream Linkage

From the foregoing discussion, it becomes clear that physically the upper watershed, the lake in the middle and irrigation system in the valley floor is linked. The action of the users in different reaches will affect each other. However, the linkage between the users in upper watershed and users in the middle and valley floor is one-way whereas the linkage between the fishermen group and the irrigation users are two-way. One of the direct linkages observed between upper watershed and the valley floor is the drinking water source at upper watershed for 200 households in one of the settlements in the valley floor. The drinking water is supplied through 15-km-long pipelines having 50 mm internal diameter, and a storage tank with a capacity of 100,000 liters. People in that settlement recognise the contribution of the people in upper watershed in conserving the water sources. As a result, they have also provided some construction materials like pipes and gabion wires for protection of source. A Users Committee (UC) manages the drinking water supply system. An operator has been appointed to operate the system on a daily basis. In all individual taps, water meters have been installed based on which water charge is collected, which is used for the operation and maintenance of drinking water system.

Institutions

Several local-level organisations — governmental, non-governmental, local elected institutions and Community Based Organisations (CBOs), and various user groups are engaged in the management of resources. There are organised Forest and Water Users' Groups having varied level of interest and capacity to respond to the need of the group members, both in the upper watershed and the valley floor. However, they carry out their activities independently except for some occasions when people affiliated to two or more institutions happen to interact, more in an informal way. A close look at the emergence of these institutions reveals that past development interventions in the watershed were instrumental in increasing their evolvement.

Forest Users' Group

The strengthening and continuation of Community Forests (CF) activities after the BTRT/WMP was facilitated by the policy shift of the government. The National Forestry Plan of 1976, the first major policy document, emphasised the participatory approach of forest management. The Master Plan for Forestry Sector (HMGN 1988) envisaged for integrated management of natural resources — mainly forest, land and water — to meet the people's basic needs through ecosystems maintenance. Following these plans, the government promulgated Forest Act 1993 and the Forest Regulations 1995 to facilitate participatory forest management. These have been the dominant legal documents for forest policy implementation in Nepal which standardised and institutionalised the process of forest handover and management to the Community Forest User Groups (CFUGs) by empowering and delegating authorities to the district-level offices and communities. The CF, the most focused programme, mainly in the middle-hill regions, are those parts of national forests that have been handed over to local communities to develop, conserve, use and manage the forest. The introduction of these acts paved the way for the institutionalisation of FUGs formed under BTRT/WMP project by registering them with the government authority and handing over of the forest from the government to these groups.

The FUGs have their own institutional arrangement with regard to appropriation, inventory of forest product, graduated sanctions, conflict-resolution mechanism and monitoring systems guided by collective-choice arrangements as agreed between the members. Types and availability of resources, resource use pattern, local socio-cultural, political–institutional arrangements and their linkages with them have mostly guided the activities of FUGs. It is due to the community-based management approaches and regulated by the provisions of the act that FUGs have evolved in diverse conditions over time and space. The BTRT/WMP final evaluation also acknowledged that compared to other user groups, the FUGs have higher sustainability (Kayastha et al. 1997) because of the direct benefit they receive from it. The FUGs are at different stages and are evolving as effective institutions in social development processes. However, lack of internal resource generation and dependence on government agencies for technical support are impeding their growth that could further contribute to the improvement of livelihood of its members.

Irrigation Users' Group

The Water Resources Act (WRA) 1992 and the Water Resources Regulation 1993 have protected the users' customary rights and have also enabled development of water resources for irrigation without requiring a license from the government. It has emphasised on participatory approach in irrigation development. Because of this, the users have mobilised resources both internally and from outside and have developed irrigation systems to the extent of their potential. Therefore the formation of the WUA was a prerequisite for implementing programmes at local level through outside support.

The WUA is conceived as an autonomous and corporate body having perpetual succession in case of its official registration and is mandated to discharge functions related to irrigation management. The users group formed under the BTRT/WMP project for the rehabilitation of the irrigation systems had informal status, as they were not registered. Therefore, they are not formally recognised and are functioning at present also in an informal way.

The WUA of BIS was the first formal water institution, as they are officially registered, in the valley floor having representation from the water users themselves and is maintaining linkage with Sub-Divisional Irrigation Office of the government since its formation. It is dependent on the annual grant from the government for operation and maintenance of irrigation system. However, according to the WUA chairperson the fund they receive is inadequate for the annual maintenance of the system because of the deteriorating condition of the canal networks. As a result, the users sometime collect cash for maintenance of branch canals on their own.

The WUA chairperson informed that since 2007, they have started collecting irrigation service fee but have not been able to enforce it strictly because of poor land records and inefficiency in the delivery of irrigation water. Therefore, the tendency is more on getting increased governmental support.

The WUA is weak in institutionalising and enforcing its functions as per Irrigation Policy (2003), which empowers it to collect Irrigation Service Fee (ISF) for the operation and maintenance of the irrigation system. One of the major constraints in the collection of ISF is the inequitable allocation and distribution of water among the users. The problems associated with it are both technical and managerial. The technical problem is the lack of updated land record in the command area served by the irrigation canal and extent of water availability in each of the branch command. The managerial problem is the inability of the WUA to draw and enforce new allocation rules due to lack of adequate resources to address some of the technical issues. In this respect, the WUA needs to begin collecting information on various management aspects — system development, water allocation and distribution, conflict among users and operation and maintenance issues — of the irrigation system.

The irrigation water that the BIS draws from the lake has its source in the upper watershed. However, the WUA of BIS is not involved in the protection of its water sources, both at the lake and upper watershed. This could be due to non-requirement of such activities as the FUGs in the upper watershed and fishermen's group in the middle are maintaining the forest and the lake environment respectively. This could be termed as institutional dependency or rent-seeking behaviour on the part of irrigation users.

Machha Byawasayi Sangh (Fisher's Group)

Having seen the formation of forest and the water user groups in the upper watershed and valley floor area during BTRT/WMP and BIS projects respectively, the Jalhari groups were also inspired to form a group of their own and established the Machha Byawasayi Sangh (Fishermen's organisation) in 1986 and were operating as an informal organisation until they registered it in the municipality in 1999. The formation of this group was the institutional response to the WUA of BIS in order to protect their right and collective interest of the group, as both (irrigation and fishermen) the groups are dependent on lake water for their livelihood. The formalisation of the group by registering with the competent authority was necessary for legal recognition. In that respect, the fishermens group was asserting their right in lake water, both organisationally and legally.

All the 40 households of Jalhari are the members of the group. The group activities are guided by a constitution. The members have to get a license from the fishery office near the lake by paying an annual registration fee, which signifies the government's approval for fish farming in common property resource. Fishermen have to pay a fee of NPR 2 (US$0.03) per kg to the District Development Committee (DDC) through the Fish Collection Centre (FCC), which buys fish from them. Also, they have to pay NPR 360 (US$ 5.5) annually to the municipality. This exemplifies the assertion of the right bestowed upon by the Local Self Governance Act 1999 to these local elected institutions. The act empowers these institutions to levy taxes for the use of common property resources under their jurisdiction.

Boaters' Group

Almost all the boaters come from the upper watershed and have formed a boaters association. It is a formal water user group formed in 1985 and was registered with the DDC in 1994. None of the boat operators are entirely dependent on the earning from the boat for their livelihood. The monthly income of NPR 1,000–NPR 1,500 (US $ 15–22) from the boat operation is not the major source of livelihood of these households. The earning is not satisfactory compared to the cost of making a boat which is NPR 17,000–NPR 20,000 (US $ 259.5–305.0). However, this is considered to be an additional job and also having an own boat makes it convenient to travel from their village to the other side of the lake for marketing and other purposes. The low income from boat operation could be one of the reasons as to why, compared to the fishermen's group, they are less concerned for the upkeep of the lake environment.

The foregoing discussions suggest that the evolution of institutions in the watershed area was necessitated for the implementation and continuation of the activities of development interventions. Beside, these institutions were also required to sustain the benefits from various development interventions accrued to the groups they belong to. Therefore, these institutions are playing an important role in managing the ecosystem services leading to the improved livelihood of their members as reflected in the activities of the FUGs, fishermen's and irrigation users' group in the valley floor. They have been protecting the environment and enabling their members' access to these services. However, the extent of the involvement of these institutions in ecosystem management is determined by their dependence on it, the benefit they receive from it, the resources they are able to mobilise for its upkeep and the support they receive from local and external institutions. There exists enough opportunity to expand and improve their performance for users' benefit.

To conclude, various projects in the watershed area helped in the formation of social capital as these programmes were implemented through various user groups like forest user groups, irrigation users groups, mothers group and drinking water groups. Groups other than forest user groups in the upper watershed have become dormant with the phasing out of the BTRT project, due to lack of activities which kept them active (Duijnhouwer 1997b; Kayastha et al. 1997). This happened to be the case in other parts of the country as well where traditional irrigation systems were rehabilitated. Most of these groups acted as construction committees due to dominance of local political leaders, vested interest of some of the committee members and lack of mobilisation of adequate resources for sustaining the rehabilitated work and became defunct after the completion of construction work (Pant 2000). Nevertheless, the informal traditional arrangement of management is continuing.

Government and Other Institutions

Besides the roles of water and forest institutions, the Fishery Centre of Government, local elected governments and International Non-Governmental Organisations (I/NGOs) are another set of institutions which are associated with resource management in the watershed. The Fishery Centre provides technical support to the Jalhari (fishermen's group) whose livelihood is dependent on raising fish in the lake. The Lekhnath municipality and the Kaski DDC are two locally elected bodies with legal entities that have a stake in the management of the natural resources at the local level, as provisioned in Local Self-governance Act (LSGA), 1999. The DDC/Village Development Committees (VDCs) and municipalities are empowered to manage natural resources in their respective areas and have the authority to levy taxes for use of natural resources. However, the services they should provide is not clear in the act. The major problem with them is that their roles and responsibilities are overlapping with those of the government line agencies and they also contradict the other legislative provisions such as the act regulating community forestry. Therefore, the role of these local elected institutions are not clear vis-a-vis users group formed for various activities, as their activities are regulated by the concerned act, rules and by the respective government functionaries. This has been one of the major gaps in formulating integrated approach for natural resources management in the watershed.

It should be noted here that the DDC and VDCs are without elected representatives as local elections were postponed in 2002 due to political unrest in the country. Till 2008, the elections were not held and thus their administrative functions are being discharged by the government officials. However, this has not affected the functioning of other users group indicating lack of functional linkage between local elected institutions and them.

Equity and Resource Management

One of the important aspects of resource management is the process through which users have the opportunity to participate in collective decision making by crafting rules (Ostrom 1992). These rules are important for the functioning of a group, as the users by becoming a member of the group, will have to adhere to these rules once agreed to by all the members of the group. Participation ensures access to benefit from resource management for which the users will have to invest their resources. Nevertheless, the access to benefit should be equitable among and across the resource users. Therefore, equity in common property resource management is important for the sustainability of resource users group. Therefore, it should be looked into from the perspective of equity within the resource users group and across the resource users group.

Equity within the Resource Users Group

Irrigation Users

Social relations play an ever-increasing important role in the individual and collective water rights to the irrigation water in the Begnas basin. Traditionally, water rights are tied to land in the irrigation command area and external interventions have effected changes by expanding the command area due to availability of increased irrigation water. However increasing competition in water allocation at the system level is becoming more socio-political in nature between users, both at the upper watershed and the valley floor, due to insufficient availability of water in relation to its demand. The irrigation users at the tail end of both the upper watershed and valley floor perceive that improvement in infrastructures and redefining the allocation rules could increase their access to water for irrigation. However, the users at the head reach do not agree to new allocation rules which are protected through their customary water right (Khadka 1997).

It could be observed that water allocation and distribution rules based on traditional water rights are contributing to maintaining status quo and promoting inequities. The existing inequities among the users at the head reach and at the tail end could be addressed with improvement in the infrastructure and by drawing new allocation rules. This is possible only when service delivery capacity of the irrigation infrastructures is improved without affecting the customary right of the users. For both these activities, external intervention is required and availability of funds is the main constraint. The current level of investment, which is mainly for minor maintenance, is not sufficient for major improvement. Therefore, insufficient mobilisation of cash and labour input from among the users and lack of external support is a barrier in improvement and drawing new allocation rules to increase water use efficiency and to address the need of new users. In the case of upper watershed, absence of their formal recognition by the District Irrigation Office (DIO) is a constraint in obtaining any governmental assistance. This is an evidence of the inadequate linkage or association that they have with the government agencies.

The use of water resource and struggles for access to water can be seen to operate at several levels. The claim of tail enders in the upper watershed for adequate share of water is legitimate since new irrigation development is not possible due to unavailability of water source. However, the users at the head end do not agree to this for fear of losing their traditional water right. The users at the tail end of the BIS in valley floor do not have adequate control of the irrigation system and also do not have adequate resources to implement any improvement that will help in reducing inequity. It could be argued that lack of financial resources and the existing institutional arrangement act as a barrier to technological improvement and introduction of innovative management practices.

Forest Users

FUGs have enforced rights, duties and punishments for its members who are involved in the management of CF. Their activities are guided by the broader framework of the CF Act in which roles and responsibilities of FUGs are specified and groups have the flexibility of drawing rules of their own. Therefore, the rules vary across the FUGs in the catchment. Although members are from various social strata, there is no discrimination in rights and duties among the members in the study area. FUGs harvest the forest products once a year from January to March in a plot-wise rotation and products are distributed among member households in an equal amount. However, the present practice cannot be termed equitable as the poorer households, especially the landless who are more dependent on the forest product for their livelihood, also receive equal amount of forest products. The gender aspects, in forest management are also weak in terms of representation of women users in the executive committees.

Since most of the forest products — fuel wood, fodder, agricultural implements and timber on request — are for domestic consumption, FUGs in study area are not financially strong to undertake other community activities due to lack of marketable products. This has been one of the constraints in providing direct financial benefit to the households. The valuation of the forest product based on the information provided by the users shows that the monetary value of the forest product to the households is NPR 10,500 (US $ 160) per annum that includes firewood, timber and agricultural implements. Likewise, the households contribute labour equivalent to NPR 5,400 (US $ 81). In this respect, the households are benefiting from their involvement in the CF activities. Some of the CF groups have tried the plantation of marketable species like Amriso (*T. maximus*) and other medicinal plants. However, they could not receive expert inputs for adequate planning to diversify the activities. This seems to be an important area where the concerned government agency, the District Forest Office (DFO), could provide expert guidance to diversify forest plantation. This is essential to contribute to the livelihood of user households through increase in household income from forest products.

Jalhari (Fishermen Group)

The equity issue is well addressed among the fishermen group as they have equal access to the benefit. The number of cages is decided among the group with the technical advice from the Fishery Development Centre. All 40 households of the Jalhari community keep equal number of cages (5–6) per household. Usually fish is harvested after one year, as it takes one year for the fingerling to weigh 1 kg. There is no restriction for collection of fish by user in one day, which is determined by the demand in the market and availability of harvestable product. One household earns NPR 6,000–NPR 8,000 ($90–122) per month and they are satisfied with their earnings and occupations. Women have formed their own group and are also actively involved in raising fish and selling it. This group has started cooperatives for saving money. They have also started a child care centre. They are aware of the importance of education and are able to afford the cost and are sending their children to school and university.

Due to increase in average household income (NPR 6,000–8,000) per month, they are able to purchase household amenities such as television, gas stove, electricity, telephone and cable television at present. This indicates that their economic status has increased. They lead a comfortable life compared to the people from higher caste in upper watershed, and this has become possible due to earnings from the lake. Due to the increase in their financial status, they have acquired higher social status by operating small restaurants where people from all social strata visit. Therefore, they are not socially discriminated as in the past and are not treated as lower class people.

Boat Operators

Everyone has equal chance to row the boat and each waits for his or her turn except in cases of emergency when someone has to be ferried immediately. Association regulates the activities of boaters through implementation of the rules, such as enforcement of queue system in boat operation and collection of fee from each boater on monthly basis. The female from each household are also actively involved in rowing the boat and are member of the association as that helps them to get a license to row boats.

Equity Across Resource Users Group

The equity across the resource users is difficult to assess as quantification of the cost to the maintenance and benefit from ecosystem services has not been made. The upper watershed is the source of water for irrigation, for the lake and also a source of drinking water for the users in the valley floor. From a catchment perspective, the forest and irrigation users in the upper watershed are mainly contributing to ecosystem maintenance. However, the value of benefit from ecosystem services generated from their activities to the downstream users — Fishermen's group who raise fish in the lake, the irrigation users who irrigate their land through lake water in the valley floor areas and drinking water users — is not assessed and there is gap in valuing ecosystem services. This flow of benefits is treated as externality and is uncompensated. Therefore this needs to be compensated by the downstream beneficiaries to the upstream users, which will help in addressing the equity issue. This would contribute to enhancement of the livelihood of people in the upstream who are poor compared to the downstream users. The irrigation users, due to lack of access to financial and technical resources, have not been able to derive the benefit from the irrigated agriculture. The increase in the efficiency of the irrigation through improvement in infrastructure and management could help in addressing the inequity among them. Likewise, the forest users group has not been able to derive adequate benefit compared to their contribution to the ecosystem services that they are providing to the users in the valley floor area. In this respect, a contribution from the fisher and irrigation users group would be useful for implementation of activities aimed at enhancing livelihood of the groups in upper watershed.

Theoretically, it is assumed that the activities of users in the upper watershed are detrimental to maintaining downstream environment. This has been acknowledged by the fishers group. They have observed changes in water quality, increase in water level and fish production due to check in flow of debris to the lake for water management. They have also seen that plantations have increased the crown cover at upper watershed. The changes that occurred were due to the linkage between forest and water is acknowledged both by the upstream and valley floor users. Therefore, one of the mechanisms to establish linkage among resource users from upper watershed and valley floor is to introduce Environmental Services Fee (ESF). However, benefit in valley floor due to actions of upstream users' needs to be established and the cost/benefit needs to be ascertained before its introduction. Beside, the enforcement of ESF is not possible without any intermediary who plays the role of a mediator between the resource users in upper watershed and valley floor area. Foremost of all, the users need to acknowledge and accept the concept of ESF, which is new to the users in the catchment.

Institutional Linkages and its Dynamics

From the foregoing discussions, it becomes clear that two sets of institutions have stakes in catchment management. The users' groups are the local-level stakeholders whereas the institutions that influence/control, facilitate/provide technical support and collect taxes from local stakeholders and also who claim ownership of local resources are the external stakeholders (Figure 11.1). The institutional linkage among the local stakeholders is horizontal whereas their linkage with the external stakeholders is vertical. The horizontal linkage is important for functional (working) relationship among resource user groups whereas the vertical linkage is important for structural (legal, policy and institutional support) relationship. The structural relationship is important as the activities of these resource users group are facilitated/controlled by sectoral agencies and their policies.

Though the WUG at valley floor (irrigation users, fishermen's group and Boaters Group) have linkages as they are at interface with each other, they still have to develop a working relationship among themselves. The irrigation users claim comes first over the use of the lake on the ground that the dam was constructed for diverting water to

Figure 11.1: Institutional Linkage between Local and External Stakeholders

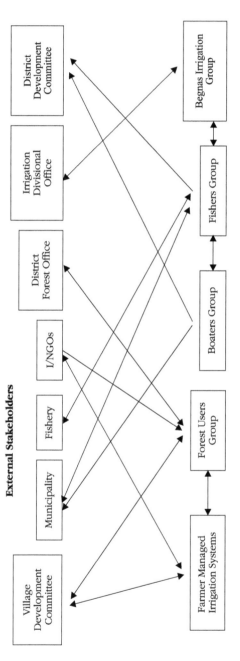

Source: Prepared by the authors.

irrigation. During the long dry spell, irrigation users release more water for irrigation whereas the fishermen's group wants to maintain water level in the reservoir that is required for the fish. This sort of conflicting interest needs to be addressed in order to avoid long-term dispute. This is possible through developing well functioning working relationship between them, which is not there at present. However, they do not have any formal relation with the BIS and are affected by the use of lake water which is controlled by irrigation users. In recent years, the fishermen's group has established a working relationship with the irrigation users' group and have persuaded them to obtain permission to plug in the irrigation outlet by placing nets in order to check the fish from flowing into the irrigation canal. This has been beneficial to the irrigation users as well because the maintenance cost of the canal, which used to be damaged by the fish catchers, has been reduced.

The figure 11.1 also shows that institutionally, the forest users groups in upper watershed are more isolated in terms of their linkage with other local resource users groups and government institutions, except the DFO. This is largely because their activities are regulated by the Community Forest Act (1993) which provides them autonomy and thus they can perform their activities independently without any linkage with other institutions at the local level. The lack of functional linkage with locally elected institutions has in fact restricted their access to local development funds as these are made available by the government through locally elected institutions. The synergy between the activities of two institutions would have facilitated the integrated management of the resources, which is lacking at the moment. It is interesting to note that the FMIS at upstream does not have vertical linkage with government agencies indicating that they do not have access to government resources, which are made available to the formal water user groups only. This has constrained improvement in the irrigation systems to address the equity issue as mentioned earlier.

The vertical linkage of users groups with external agencies, which happens to be sectoral, is also a barrier for integrated approach. For example, the BIS is linked to the Irrigation Sub-divisional office and the fishers group is linked to the Fishery Centre under the Agricultural Directorate at the district level and the DDC. Likewise, the FUGs at upstream are linked to DFO under the Ministry of Forest and Soil Conservation. These three district-level agencies represent three different ministries: Ministries of Water Resources, Agriculture and Cooperative and Forest and Soil Conservation respectively, and their work is guided by the policies, legal provisions and institutional arrangement of respective ministries. For example, it was revealed by the official of the fishery centre that they seldom interact with the BIS users group in the valley floor. Although they are one of the stakeholders of the lake, they do not have a say in the release of water for irrigation. In this context, it is important to note that none of the stakeholders at the local and at district level are clear on the ownership of the lake, though each of them claims ownership based on their own definitions of it. This reflects the policy constraint for integrated approach due to sectoral orientation and program implementation at local level. Therefore, it could be said that the integration at the policy level is more important in order to have integrated activities at the local level. This has also been a constraining factor for the establishment of institutional linkage between the upstream forest users and downstream water users and among resource user groups, which is fundamental to integrated resource management.

It could be concluded that precise policies for inter-sectoral and upstream–downstream resource management are still lacking in facilitating the integrated management of natural resources at local level. Beside, appropriate institutional provisions are required to be developed at the national, regional, basin, district and local levels.

Analysis and Conclusion

The Begnas watershed in the western hill of Nepal is endowed with natural resources, although the overall economic condition of the people, especially in the upper watershed, is poor. The development interventions in the past have initiated the users' involvement in the management of the natural resources through formation of user groups, which is being carried out at present as well. Over a period of time, they have transformed into formal organisations and have institutionalised the management of forest and water resources. This was largely influenced and facilitated by the shift in government policy at that time. FMIS at upper watershed are informally organised whereas forest users are formally organised. Households in the community have access to the forest product, mainly the firewood, but livelihood support through forest product is not happening due to absence of marketable forest products. Organisationally, they are strong but their linkage with other institutions downstream is non-existent, although they seem to be a major contributor for ecosystem services.

Existing inequity within and across the resource users, which is more prevalent among the irrigation users both in the upper watershed and valley floor, is not conducive to the integrated approach to management of natural resources. This is largely due to irrigation users' inability to mobilise adequate resources, both internally and externally, for the improvement of infrastructure. As far as water delivery at system level is concerned, it was assumed that users could manage the internal water distribution on their own in an equitable way without external input. This assumption is not turning into reality mainly because of existing traditional water right, lack of water use efficiency due to poor infrastructure and inadequate communication among users. Beside, the group of farmers who are benefiting from the existing allocation rules would like to maintain the status quo to retain their customary water rights. As a result, water distribution has remained inequitable leading to water use disputes. This is the major issue, as it is happening in many of the water use systems in the country. Therefore, the problem needs to be addressed both from functional and structural perspectives. From the functional perspective, improvement in infrastructure and redefining the allocation rules in consultation with users at the local level could help in this regard. Likewise, structural perspective requires legal/policy interventions that may enforce changes in existing water right provisions and allocation rules at local level. This includes, for example, crafting new allocation rules after interventions and ensuring its implementation, instituting mechanisms for access to external resources by the informal irrigation users and taking appropriate measures for ensuring ESF collection in government-supported irrigation systems. The users would assert their right to irrigation once they start payment of ESF. Most importantly, the lack of resources is not the only constraint. Redefining existing water right to improve water use efficiency is fundamental to addressing inequity issues.

The inequity across the resource users is exemplified by the eco-system services provided by the forest and irrigation users, which is neither quantified nor compensated. Activities of upper watershed users would have a detrimental effect on the environment of the lake for which they need to be compensated for by the valley floor users who are reaping the benefit from ecosystem services. The resource users at the valley floor recognise the ecosystem services of the upper watershed although it is not clearly visible and users are unable to identify and quantify it. The concept of ESF is quite new and because of this, the compensation mechanism is not in place. The stakeholders need to be made aware and convinced of this.

The role of an intermediary in introducing and collecting the ESF is important to establish linkage between upstream and downstream users. The collection of taxes by DDC/municipality from the fisher group and boat operators could be considered as ESF, but they are not investing it for the development of upper watershed and valley floor. They could act as an intermediary as they are the locally elected institutions having linkage with both the users in upper watershed and valley floor in other local activities, but not for natural resource management. The ESF thus collected could be compensated for by the activities which could contribute to improving the livelihood of people in the upper watershed, as they are poor compared to the users in the valley floor. In essence, the realisation of the ecosystem services provided by the users in upper watershed and developing compensation mechanism for it would help in building upstream–downstream institutional linkages at watershed level. This would help improve the governance at watershed level and would ensure equity across the resource users. Actualisation of such linkage at the watershed level would ensure improved governance as well as equitable benefit sharing at system level.

Though several water user groups including the WUA of Begnas Irrigation System are functioning downstream, the functional linkage among them is weak. Each of the user group is trying to maximise the benefit from the lake without making substantial contribution for its sustenance. In such a situation, quite often, conflicts arise within and among the institutions due to their divergent interests. The interest of various resource users group is reflected through their organisational undertaking. For example, during the long dry spell, the BIS wants to release more water from the lake by claiming first use rights of water from the lake, which is detrimental to raising of fish.

Therefore, proper delineation of ownership and management rights and duties of the users vis-à-vis external stakeholders would help in establishing functional linkage among them. This would ensure improved management and sustaining benefits from the lake. A holistic approach of lake management in consultation with relevant stakeholders can create a win-win situation for all. This is not happening due to lack of relevant policies that delineate the roles and responsibilities of both external and local organisations in common property resource management. This clearly indicates the need for developing appropriate policy, and legal and institutional provisions for integrated management of resources at the watershed level. This is because the activities of local stakeholders, who have vertical linkages with the sectoral line agencies, are guided by relevant sectoral policies.

With the realisation of existing institutional gap for CBINRM, this action research project initiated the process by facilitating the creation of platform, which 'is a venue where resource users and other stakeholders are brought together to discuss the issues related to resource management and it is believed that this process will contribute to the users and other stakeholders' understanding and thinking on CBINRM'.

The representatives from various resource users can present their views, interests and demands in the platform and the users, it is believed, could learn by sharing each others experiences for better resource management. For this, the users group could strengthen relationship with local institutions, government agencies and other external institutions for expanded and integrated activities on land and water management. Ultimately, the platform could facilitate in upscaling institutions for integrated natural resource management at catchment level.

Notes

This research was carried out for the CP-23 Project 'Linking Community-based Water and Forest Management for Sustainable Livelihoods of the Poor in Fragile Upper Catchments (Resource Management for Sustainable Livelihood)' of the Indo-Ganges Basin under Challenge Programme on Water and Food (CPWF) of Consultative Group on International Agricultural Research (CGIAR) and the funding received from CPWF for the research is duly acknowledged.

The author is grateful to Researchers — Dr. Umesh Nath Parajuli, Dr. Khem Raj Sharma, Dr. Binod Bhatta, Dr. Sabita Thapa, Mr. Joe Hill, Mr. Kamal Gautam and Ms. Pratima Shrestha — who were involved in individual components of the project studies.

References and Select Bibliography

Adhikari, J. and S. Ghimire. 2002. *A Bibliography of Environmental Justice in Nepal*. Kathmandu: Martin Chautari.

Adhikary Bharat, Denhoim Jeannette, Rajbhandari G., Fr. Ludwig Stiller, Sthapit Kesharman, Veerdig Henk. 1992. 'The Mid-term Evaluation of the Begnas Tal Rupa Tal (BTRT) Watershed Management Project, final report'. Kathmandu: Care Nepal.

Allan, A. 2004. *Report on Legal and Policy Background to WUAs in Nepal*. London: DFID.

Bajracharya, K. M. and S. M. Amatya. 1993. 'Policy, Legislation, Institutional and Implementation Problems', in K. H. Gautam, A. L. Joshi and S. Shrestha (eds), *Forest Resource Management in Nepal: Challenges and Need for Immediate Action*. The National Seminar 31 March–1 April, Kathmandu.

Bandargoda, D. J. 2000. *A Framework for Institutional Analysis for Water Resources Management in a River Basin Context*. Colombo: International Water Management Insitute.

Bhattarai, M., D. Pant, V. S. Mishra, H. Devkota, S. Pun, R. N. Kayastha and D. Molden. 2002. *Integrated Development and Management of Water Resources: A Case of Indrawati River Basin, Nepal*, IWMI Working Paper 41. Colombo: International Water Management Institute.

CARE. 1997. *The Begnas Tal Rupa Tal Watershed Management Project Document*. Kathmandu: Care Nepal.

Central Bureau of Statistics (CBS). 2002. *Population Census 2001*. Kathmandu: His Majesty's Government of Nepal (HMGN).

———. 2004. *Statistical Pocket Book of Nepal*. Kathmandu: National Planning Commission, His Majesty's Government of Nepal.

Chapagain, D. P., K. R. Kanel and D. C. Regmi. 1999. *Current Policy and Legal context of the Forestry Sector with Reference to the Community forestry Programme in Nepal*. Kathmandu: Seeport.

Chhetri, Rajendra Kishor. 2007. 'Ekikrit Jalsrot Vyabasthapan tatha Pani Sambandhi Adhikar'. Paper presented in Dialogue Forum on Water Right, Jalsrot Vikas Sanstha, 12 July 2007, Kathmandu.

Duijnhouwer, Jopie. 1997a. *Evaluation of Community Organisations in BTRT*. Kathmandu: Care Nepal.

———. 1997b. *Evaluation of Community Organizations in BTRT*, Part B. Kathmandu: Care Nepal.

Evan, Chris. 1997. *Qualitative Evaluation Report of The Begnas Tal Rupa Tal (BTRT) Watershed Management Project*. Kathmandu: Care Nepal.

Government of Nepal. 2006. 'Yearly Progress Report', District Agricultural Development Office (DADO), Kaski, Gandaki Zone, Nepal.

His Majesty's Government of Nepal (HMGN). 1963. 'The National Code', Ministry of Law, Justice and Parliamentary Affairs, Kathmandu.

——. 1976. 'National Forest Policy and Plan', Ministry of Forest and Soil Conservation, Kathmandu.

——. 1988. 'Master Plan for the Forestry Sector Nepal', Main Report, Ministry of Forest and Soil Conservation, Kathmandu.

——. 1992. 'The Irrigation Policy', Department of Irrigation, Ministry of Water Resources, Kathmandu.

——. 1992. 'The Water Resources Act', Ministry of Water Resources, Kathmandu.

——. 1993. 'The Water Resources Regulation', Ministry of Water Resources, Kathmandu.

——. 1993. 'The Forest Act', Ministry of Forest and Soil Conservation, Kathmandu.

——. 1995. 'Forest Regulations', Ministry of Forest and Soil Conservation, Kathmandu.

——. 1999. 'The Local Self-Governance Act', Ministry of Local Development, Kathmandu.

——. 1999. 'The Local Self-Governance Regulation', Ministry of Local Development, Kathmandu.

——. 2002. 'The National Water Resources Strategy', Water Energy Commission Secretariat, Kathmandu.

——. 2003. 'The Irrigation Policy', Ministry of Water Resources, Kathmandu.

——. 2004. 'The Irrigation Regulation', Ministry of Water Resources, Kathmandu.

——. 2005. 'The National Water Plan', Water Energy Commission Secretariat, Kathmandu.

International Water Management Institute (IWMI) and Water and Energy Commission Secretariat (WECS). 2001. *Integrated Development and Management of Nepal's Water Resources for Productive and Equitable Use in the Indrawati River Basin*. Colombo: IWMI.

Kayastha, Badri N., Kesharman Bajracharya and Ava Shrestha.1997. *Final Evaluation of Begnas Tal Rupa Tal (BTRT) Watershed Management Project*. Kathmandu: Care Nepal.

Khadka, Shantam S. 1997. 'Water Use and Water Rights in Nepal', in Rajendra Pradhan, Franz von Benda-Beckmann, Keebet von Benda-Beckmann, H.L.J. Spiertz, Shantam S. Khadka and K. Azharul Haq (eds), *Water Rights, Conflict and Policy: Proceedings of a Workshop Held in Kathmandu, Nepal, 22–24 January*, p. 24. Nepal. Colombo: International Irrigation Management Institute.

Ostrom, Elinor. 1992. *Crafting Institutions for Self-governing Irrigation Systems*. San Francisco: Institute for Contemporary Studies Press.

Pant, Dhruba. 2000. 'Intervention Processes and Irrigation Institutions: Sustainability of Farmer Managed Irrigation Systems in Nepal'. Unpublished Ph.D. thesis, Wageningen Agricultural University, Wageningen.

Pant, Dhruba, Sabita Thapa, Ashok Singh, Madhusudan Bhattarai and David Molden. 2002. 'Integrated Management of Water, Forest and Land Resources in Nepal: Opportunities for Improved Livelihood', CA Discussion Paper, 2005. Colombo: IWMI.

Pant, Dhruba, Madhusudan Bhattarai, K. Prashad, G. Rajkarnikar and David Molden. 2001. 'Interbasin Water Transfer and Irrigation Institutions: A case study from the Melamchi River Basin in Nepal', in Jerry Schaack and Susan S. Anderson (eds), *Transbasin Water Transfers: Proceedings of the 2001 USCID Water Management Conference, Denver, Colorado, June 27–30, 2001*, pp. 207–24. Denver: U.S. Committee on Irrigation and Drainage.

Pradhan, P. and D. J. Bandaragoda. 1997. 'Legal and Institutional Environment of Water Users Associations for Sustainable Irrigation Management Research Report'. Paper presented at the International Seminar on Irrigation Associations for Participatory Management, APO and Punjab Water Management Directorate, 6–11 October, Lahore.

Samad, M. 2001. 'Institutional Arrangements for River Basin Management: Some Emerging Issues', in R. N. Kayastha, U. Parajuli, Dhruba Pant and Chiranjivi Sharma (eds), *Integrated Development and Management of Water Resources: A case of Indrawati River Basin'*, Proceedings of a *workshop held in 25 April 2001 in Kathmandu, Nepal*, pp. 118–23. Colombo: Water Energy Commission Secretariat (WECS) and International Water Management Institute (IWMI).

Samad, M. and D. J. Bandaragoda. 1999. 'Methodological Guidelines for Five-country Regional Study on Development of Effective Water Management Institutions'. Colombo: IWMI.

Soussan J. 1998. 'Community Forestry in Nepal: Comparing Policies and Practice', Leeds University Working Paper.

Soussan, J., B. K. Shrestha and L. P. Uprety. 1995. *The Social Dynamics of Deforestation: A Case Study from Nepal*. London: The Partheon Publishing Group.

Springate-Baginski, O. 2000. 'Community Forestry Processes in Nepal: Progress and Potentials', Final report of the Leeds University/NUKCFP/NRI Collaborative Research Project, Leeds University.

Springate-Baginski, Oliver, Piers Blaikie, Om Prakash Dev, Nagendra Prashad Yadav and John Soussan. 2001. *Community Forestry in Nepal – A policy Review*. London: Department for International Development (DFID).

United Nations Economic Commission for Latin America and the Caribbean (UNECLAC). 1998. 'Network for Cooperation in Integrated Water Resource Management for Sustainable Development in Latin America and the Caribbean', Circular no. 6 and 7, ECLAC, Chile.

Upadhyaya, Surya Nath. 2007. 'National Water Plan and the Legal Regime on Water Resources'. Paper presented in Dialogue Forum on Water Right, Jalsrot Vikas Sanstha, 12 July, Kathmandu.

Water and Energy Commission Secretariat (WECS). 2002. *Water Resources Strategy*. Kathmandu: Government of Nepal.

12

Environmental Governance: Concept, Complexity and an Illustration

N. C. Narayanan and Jayati Chourey

Increasing environmental degradation threatening the sustainability of natural and thus livelihood systems and limits to hitherto tried technical and fragmented institutional solutions are prompting new ways of comprehending issues in natural resource management. A major related search in recent times is on innovative ways of environmental governance (hereafter EG), particularly with the hindsight of development thinking and practice in the last half century. EG as a concept got triggered from the need for approaching these cross-cutting issues with a more comprehensive understanding. Although attempts have been made along these directions recently (MEA 2005; IPCC 2007), there is an acute need for contextual understanding of such issues to decipher the 'sticking points' of change at the policy, programme and practice levels. The chapter attempts to explain the conceptual understanding of EG, complexity and contextual illustration through cases of three Indian wetlands.

This chapter is divided into four sections. The first section attempts to understand the conceptual evolution of EG and discusses various initiatives for governance in the context of environmental degradation. The second presents an overview of EG with particular emphasis on Indian wetlands. The key proposition is that the challenges to EG are basically triggered because of two reasons: (i) the complex nature of wetland ecosystems with competing uses and resulting multiple livelihood dependencies; and (ii) a complex governance structure due to the historical evolution hinged on a sectoral understanding of natural resources that led to a fragmented institutional structure (policies, laws and organisations). The third section describes the challenges to governance in three wetlands in India due to the environmental, socio-political peculiarities of each and illustrates the fragmented nature of institutions (policies, laws and organisations) of governance in these wetlands. The last section revisits the conceptual discussions on EG and reflects on the three Indian wetlands for a contextual understanding in order to decipher some general challenges at the conceptual and substantive levels.

Environmental Governance: Concept and Evolution

EG is an 'overarching principle' to regulate public and private behaviour towards greater accountability and responsibility to the environment. It operates from the most basic, i.e., the individual level to global and emphasises shared leadership and combined responsibility for ensuring environmental sustainability. EG is of irreplaceable importance for sustainable development as it provides specific instruments and tools for a comprehensive and multi-sectoral approach to environmental protection (Lead 2006). It is a complicated concept since it tries to club two different but related debates — one on environmental degradation and the other on the renewed concerns on governance ('good governance' as put forward from certain quarters). This section tries first to review these debates briefly to arrive at the genealogy of thinking that shaped debates in EG and then goes on to identify the initiatives on EG at different levels.

Changing Perspectives on Governance

The 'rediscovery' of governance has been done by mainstream International Financial Institutions (IFIs) like the World Bank as a rethinking of the pure neoliberal agenda that peaked development thinking in the 1980s. Many writers have been closely associated with this counter-revolution, along with policy makers like Bauer (1972), Balassa (1982), Lal (1983) and Bhagwati (1993). In subtly differing terms, these writers have all criticised what they consider as the damaging consensus on development economics that took shape soon after World War II that put state as the main designer, provider and catalyst of economic development. They suggested that such a consensus fails to see that most states are likely to be predatory, rent-seeking economic actors, and fails to recognise that state failures are at least as likely as market failures in the developing world, and far more damaging. They proclaimed 'trade not aid' and 'privatisation not nationalisation'. This argument found supporters on both sides of the Atlantic with Ronald Reagan and Margaret Thatcher as the torchbearers of this ideology. To some extent too, it has been associated with a shift in the policy direction of development institutions like the World Bank. This counter-revolution in development theory and policy has also been criticised for its simplistic accounts of the nature of 'real' markets in many developing countries and its one-dimensional accounts of what motivates apparently isolated economic actors (Stewart 1985: Killick 1986).

The process was hastened by the ending of the cold war, the dis-integration of the Soviet Union and fall of the Berlin wall. However, in the mid-1990s, with more than a decade of neoliberalism with a unipolar word, it was clarified that even for the market to work, there has to be strong institutions, particularly the need of a state. The World Bank emphasised on the importance for a 'scaled down' and 'tamable' version of state for the efficient functioning of the market. The World Development Report in 1997 argued against state's role as a direct provider of growth and favoured its role as a partner (World Bank 1997). As per this argument, complementary spheres of society like market and civil society have to be part of governance. Here, governance is the result of a social–political–administrative sharing process where state, market and civil society have their own role without the state having a central role since no single sphere has the sufficient knowledge to dominate a governing model (Rhodes 1997). In this liberal version of governance, institutions are created for bringing about coordination among different members of the society, thus preventing conflicts of interests by defining the rules of the game (Gorringe 1997). Such a version of institutions as rules of the game is celebrated by the new institutional economics school too (North 1990). The conditions for 'good governance' is the interactions and coordination among government, market and civil society under the framework of democracy and respect for human rights as described by the UN Resolution 2000, 64 (UN Commission for Human Rights). As per the United Nations Economic and Social Commission for Asia and the Pacific (UNESCAP), the governance concept is as old as human civilisation with the basic meaning being the process of decision making and also the process by which decisions get imple-mented or not. But of late, it has also been used in contexts of cor-porate governance, international governance, national governance and local governance.

A perspective of governance as defined by the United Nations System is shown in Figure 12.1. According to the United Nations, good governance has eight major characteristics. It is participatory,

Figure 12.1:   Characteristics of Good Governance

Source: UNESCAP. http://www.unescap.org/pdd/prs/ProjectActivities/Ongoing/gg/ governance.asp (accessed 15 March 2008).

consensus oriented, accountable, transparent, responsive, effective and efficient, equitable and inclusive and follows the rule of law. It assures that corruption is minimised, the views of minorities are taken into account and the voices of the most vulnerable in society are heard in decision making. It is also responsive to the present and future needs of society. Although this is a listing of normative ideas, it offers a prism to ray-out the concerns usually piled-up in discussions of 'good governance'.

In this framework, participation could be either direct or through legitimate intermediate institutions or representatives. However, we need to note that representative democracy does not necessarily mean that the concerns of the most vulnerable in society would be taken into consideration in decision making. This chapter argues for an inclusive idea of participation including the necessity for mobilising the weakest sections of the society to be part of it. The second requirement is fair legal frameworks with impartial enforcement of laws that requires an independent judiciary and an impartial and incorruptible police force. Transparency means that decisions taken and their enforcement are done in a manner that follows rules and regulations. It also means that information is freely available and directly accessible to those who will be affected by such decisions and their enforcement. It also requires free expression of views by all actors and mediation of these different interests in society to reach a broad consensus on what is in the best interest of the whole community and how this can be achieved. It also requires a broad and long-term perspective on what is needed for sustainable human development and how to achieve the goals of such development. This can only result from an understanding of the historical, cultural and social contexts of a given society or community. Accountability in this framework is not only for governmental institutions but also that of the private sector and civil society organisations. In general, an organisation or an institution is accountable to those who will be affected by its decisions or actions. Accountability cannot be enforced without transparency and the rule of law (UNESCAP). It is difficult to contest any of these ideas. However, the larger question is the political context under which EG decisions are taken, where fundamental questions like access to and control of resources and power relations are the key to such decision making. We conclude this discussion on EG with following observations.

EG needs to take the complementary role of the three societal spheres of state, market and civil society in decision making. Ideas of 'good governance' (like UNESCAP framework reviewed) with its normative concerns have to be contextualised in space and time with a political understanding of the concept. This is important since often these slip down to an uncritical managerial approach that advocates optimistically about the possibility of good governance without fully considering the political economy in which governance decisions are taken or not. We shall dwell upon this issue at the end of this chapter. EG needs to closely disaggregate the institutions (policies, laws and organisations) of governance.

Environmental Degradation and Initiatives for Governance

The ecological degradation and indications of an emerging crisis triggered the first global thinking on the subject with the publication of the *Limits to Growth* by the Club of Rome in 1972 (Meadows et al., 1972). However, the first attempts at such thinking came with the publication of *Silent Spring* by Carson (1962), who warned about the environmental consequences of the indiscriminate use of modern chemical pesticides, fungicides and herbicides. *The Tragedy of the Commons* (Hardin 1968) was part of the pioneer thinking about the relationship between property rights and environmental degradation, linking population pressure and unrestricted use of common property resources to degradation. The first global conference on environment, held in Stockholm in 1972, set in motion three decades of discussion, negotiation and ratification of a whole series of international environmental agreements. Schumacher (1974) was one of the earliest 'green' thinkers who put forward the idea that 'small is beautiful' and warned of the finiteness of resources. He argued to refrain from blind productivism that threatens the tolerance margin of nature and non-material needs of humans, and prescribed intermediate and appropriate technology, especially for developing countries. By the 1980s, the concept of 'sustainable development' was widely publicised by the World Conservation Strategy (IUCN 1980). It has since become central to the thinking on environment and development and a notable contribution is the report of the World Commission on Environment and Development (WCED 1987), also known as the 'Brundtland Report' mentioned earlier. The most important contemporary initiative is the United Nations Environment Programme (UNEP), which acts to centrally coordinate organisations and information. Major international conventions, held every decade or so (Table 12.1) guide the process of global governance, while

Table 12.1: Major Milestones of the Global Environmental Movements

Year	Milestone
1970	First Earth Day observed
1971	The Convention on Wetlands of International Importance Especially as Waterfowl Habitat (Ramsar Convention)
1971	This Endangered Planet, Richard Falk
1972	The UNESCO Convention Concerning the Protection of the World Cultural and Natural Heritage (World Heritage Convention)
1972	Stockholm Conference on the Human Environment
1973	The Convention on the International Trade in Endangered Species (CITES Convention)
1974	The Limits to Growth (Donella Meadows et al.)
1978	The Human Future Revisited, Harrison Brown
1978	The Twenty-Ninth Day, Lester Brown
1980	World Conservation Strategy, IUCN and UNEP
1987	Our Common Future, World Commission on Environment and Development (the Brundtland Commission Report)
1992	Rio Earth Summit
1992	Framework Convention on Climate Change
1992	The Convention on Biological Diversity
1994	Convention on Combating Desertification
2002	World Summit for Sustainable Development in Johannesburg
2005	Kyoto Protocol comes into force

Source: Kashwan (2005: 5).

a series of Multilateral Environmental Agreements (MEAs) simultaneously provides the basis of international environmental regulation (Goffman 2005).

Within the context of the evolution of global environmental politics and policy, the end goal of global environmental governance is to improve the state of the environment and to eventually lead to the broader goal of sustainable development (Najam et al. 2006). The profuse growth of global environmental governance system signifies the world's growing acceptance and appreciation of the scope and scale of the problem. However, this growth has also been unsustainable. Now, there is a general agreement that this system is more cumbersome and less effective than it must be if we are to confront the serious environmental challenges laid out in such international reports as those of the IPCC and the Millennium Ecosystem Assessment (Runnalls 2006). The major limitation to most of the global solutions till date is their lack of implementation. Although they are meant to complement each other, they often compete and even contradict to cancel out each other's impacts (Dach et al. 2005).

The environmental problems in Asia have grown in severity mainly due to the rapid and indiscriminate economic growth in the region. In most of the Asian countries, the state played a major role in initiating and formulating environmental policies to deal with these problems. The exception to this are Japan and India, where environmental groups and movements by citizens played a major role in introducing innovative policies and actions. The environmental NGOs are now emerging as promising actors in Asian countries (IGES 2001).

Although many bilateral agreements on environmental issues existed between South Asian countries, the South Asia Cooperative Environment Programme (SACEP) in 1982 was the first multilateral agreement, which was adopted by eight countries. It has the characteristics of a modest-sized regional environmental organisation, comprising three major organs — the governing council, the consultative committee and the secretariat. Besides providing secretariat and administrative services for implementing its own programmes, the SACEP also undertook other environmental initiatives such as the Malé Declaration on air pollution endorsed by UNEP (Shihab 1997 as quoted in Kato and Takahashi 2001). Also, the South Asian Association for Regional Cooperation (SAARC) has been concerned with trans-boundary and global environmental issues like natural disasters, climate change and trans-boundary movement of hazardous wastes and has developed an action plan. Besides these two major initiatives, several plans focusing on single issues have also been developed in South Asia. The launch of the Regional Seas Programme was called for by the SACEP member states at UNEP's Governing Council meet in 1982, resulting in the 'designation of the region as a part of UNEP's Programme' in 1983 (Abeyegunawardene 1997 as quoted in Kato and Takahashi 2001).

Efforts to curb pollution through legislation in India can be dated back to the mid-nineteenth century (Curmally 2007). Various legal acts and bodies relating to environmental protection are present in India such as on water pollution, air pollution, environment protection, National Environment Appellate Authority, National Environment Tribunal, animal welfare, wildlife, Forest Conservation and Biodiversity Acts (MoEF 2008). These acts and bodies have dealt with environmental regulation in a 'piecemeal manner' and have proved rather ineffective at reducing pollution levels (Curmally 2007). Despite the presence of such acts and environmental policies, environmental degradation continues to severely affect India more due to 'political and institutional failings' rather than 'technical failings' (Merrey et al. 2007).

Environmental Governance for Managing Indian Wetlands

Status of Wetlands in India

Wetlands nourish rich biodiversity and are also of great social, economic and cultural importance for communities. Despite their importance and value, wetlands are being modified or reclaimed. Wetlands can be ranked amongst the most highly threatened ecosystems on the planet. Worldwide, around 50 per cent of wetlands are estimated to have disappeared since 1900 (Wetland Internationals 2006). Global climate change is expected to aggravate the loss of wetland (MEA 2005).

India is no exception to the global scenario. Indian wetlands, the biodiversity hubs, provide ecological services to support livelihood of thousands of communities. However, they are still mismanaged and are often neglected. Wetlands suffer from unsustainable use and exploitation of their resources, drainage, alternative use and pollution. Like many other common property resources, everyone claims a stake in their use, but few are willing to pay for that use. As a result, the state of the country's wetlands is rapidly deteriorating (Parikh and Parikh 1999).

The high level of dependency on wetlands and their high vulnerability to degradation call for immediate actions. There is international alarm at the impending water crisis at the global level which is going to occur in the coming decades. Taking immediate actions towards sustainable management of existing wetlands is a global priority. While in the past two decades, there has been a significant expansion in international agreements, programmes and institutions pertaining to water and drivers of wetland change that contributed to broader awareness, there is a wide gap between formal policies and actual practices with lack of political will for implementation (MEA 2005).

India is a party to the Ramsar Convention, and has identified 25 sites (677,131 hectares) as wetlands of international importance under Ramsar Convention (Ramsar Convention 2006). However, without any clear policy and rules, the designation alone is not enough for the conservation and management of the wetlands. Apart from this list, the Ministry of Environment and Forest (MoEF) has identified wetlands of national importance. The number of such sites is increasing which is a positive sign but issues related to their management are also increasing (Panini 1998). India is also a party to the Convention on Biological Diversity (CBD), which addresses conservation of wetland through its international programme of work on inland waters biodiversity. There is no special legislation in India pertaining to wetlands, but several national policy documents and legislations make reference to them.

Existing Institutional Structure for Wetland Governance and Challenges

Institutions and institutional analysis have a wide and varying history in the past two decades. We follow the water institutional analysis provided by Saleth and Dinar (2004) to bring in some analytical clarity into the analysis of wetland institutions. Accordingly, institution of governance sets the rules and defines, thereby, the action sets for both individual and collective decision making in the realm of water resource development, allocation, use and management. Since these rules are often formalised in terms of three interrelated aspects, i.e., legal framework, policy environment and administrative arrangement, water institutions can be conceptualised as an entity defined inter-actively by its three main analytical components — water law, water policy and water administration (Saleth and Dinar 2004).

Box 12.1 provides an overview of the existing institutional struc-ture for wetland governance in India and also an account on three major components of a water institution — legal framework, policy environment and administrative arrangement. It is clearly visible that there are various environmental laws and policies which can be implemented to protect wetlands. However, the problem lies with the functioning of organisations responsible for execution of these laws and policies. Without the umbrella governance, wetland remains a threatened ecosystem due to factors such as jurisdictional overlapping and non-recognition as a separate ecosystem (Panini 1998).

The MoEF is primarily responsible for conservation and wise use of protected wetlands, mainly through the implementation of its National Wetland Conservation and Management Programme (NWCMP). Ninety-four wetlands have been identified under this programme (MoEF 2007). On the other hand, functioning of other central ministries such as water resources; agriculture; chemical and fertilisers; commerce and industries; science and technology; tourism; panchayati raj; and several state bodies (the departments of fisheries; agriculture; irrigation; environment, forests, science and technology; and tourism and pollution control board) have a direct bearing on the state of country's wetlands.

Lack of much needed coordination and conflicting aims of various government bodies have led to the exploitation of wetlands, and their consequent degradation (Parikh and Parikh 1999). The Irrigation Department, Fisheries Department and others maintain certain wet-lands, but do not maintain the multi-functional and multi-use aspects of these natural resources (Rao and Datye 2003). Another major drawback of the existing environmental governance system is that it is still state-centric. In the existing institutional structure, there is no provision for direct and active role for other actors of governance to play. Civil societies have been struggling for the reorganisation of their continued efforts in protecting the environment. On the other hand, financial organisations have yet to raise their consciousness to the level of willingness for conserving the environment.

Box 12.2 summarises the key players of environmental govern-ance in India across levels ranging from supra-national to local level organisations and provides causes for functional inefficiency

Box 12.1: Institutional Structure for Wetland Governance in India

Wetland Related Environmental Laws:

- The Indian Fisheries Act, 1857
- The Indian Forest Act, 1927
- Wildlife (Protection) Act, 1972
- Water (Prevention and Control of Pollution) Act, 1974
- Territorial Water, Continental Shelf, Exclusive Economic Zone and other Marine Zones Act, 1976.
- Water (Prevention and Control of Pollution) Cess Act, 1977
- Maritime Zone of India (Regulation and Fishing by Foreign Vessels) Act, 1980
- Forest (Conservation) Act, 1980
- Environment (Protection) Act, 1986
- Coastal Zone Regulation Notification, 1991
- Wildlife (Protection) Amendment Act, 1991
- National Conservation Strategy and Policy Statement on Environment and Development 1992
- Biological Diversity Act, 2002

Wetland Related Policies:

- National Environment Policy 2006
- National Water Policy 2002

Wetland Related Programme:

- National Wetland Conservation and Management Programme (NWCMP)

Wetland Organisations:

1) National Level

Legislative branch	**Executive branch**	**Judicial branch**
Parliament	Central Ministries Environment and Forests Water Resources, Agriculture, Chemical and Fertilisers, Commerce and Industries, Science and Technology, Tourism, Panchayati Raj …	Supreme Court

Environmental Protection Organisations: Central Pollution Control Board, National River Conservation Directorate, Central Water Commission, National Committee on Wetlands, Mangroves and Coral Reefs

2) State Level

Legislative branch	**Executive branch**	**Judicial branch**
State Legislative Assembly	Department of Irrigation, Agriculture, Fisheries, Forest, Tourism, Revenue, Rural Development, Panchayati Raj	High Court

Environmental Protection Organisations: State Pollution Control Board, State Biodiversity Board.

3) Local Level

District Administration and its line departments, namely Irrigation, Agriculture, Fisheries, Forest, Tourism, Revenue, Rural Development. District and Session Court.
Panchayats. Government initiated Cooperative Societies/Users' Associations.

Source: Framework adapted from Saleth (2004).

Box 12.2: Key Actors in the Environmental Governance Structure for Wetland Management in India

Environmental Governance	State			Civil Society	Market
Supra-National	United Nations, Ramsar Convention, GEF: Guidelines and Support to National Governments				World Bank
International	National Governments as international donors				MNCs
Regional	SACEP, SAARC (Mutual cooperation between Governments in South Asia)			WI, IWMI, SIDA, WWF etc. providing technical support	MNCs
National	Government of India			**Organised**	*National Commercial Organisations*
	Legislative branch	**Executive branch**	**Judicial branch**	***Civil Societies:***	
	Creates Environmental Laws (no direct laws or policy for wetlands)	***Ministries*** Environment and Forests Water resources, Agriculture, Chemical and Fertilizers, Commerce and Industries, Science and technology, Tourism, Panchayati Raj, etc. (Enforcement and administration of national law).	Supreme court *(PIL: Good Judgments in the context environmental issues)*	– Community-based organisations (Traditional Users' Associations) – Other social organisations (political, religious, educational, social welfare, etc.) – Environmental NGOs (National, International, Supranational) **Unorganised** – Marginalised user groups (e.g. landless labourers and other minor groups) *Least Participation*	

State	State Legislative Assembly (no direct Act or policy for wetlands)	CPCB, National River Conservation Directorate, Central Water Commission, National Committee on wetlands, mangroves and coral reefs; Irrigation, Agriculture, Fisheries, Forest, Tourism, Rural Development, Panchayati Raj (Enforcement and administration of state law)	High Court	State level Commercial Organisations
Local	District Administration	Line departments: Irrigation, Agriculture, Fisheries, Forest, Tourism, Rural Development	District and session court	Local market
	Panchayat			Local Market
Causes of inefficiency	Unclear Policies. Institutional Complexity. Conflicting aims. Lack of coordination and political will. Corruption.		Lack of awareness, empowerment and participation. Conflicts. Ownership issues. Corruption.	Unethical behavior, economic goals

Wetland Degradation

Source: Prepared by the authors.

of these players. Apart from the performance of the government, lack of public participation and unethical behavior of commercial organisations (market) are also equally responsible for the present situation. EG ideally needs all the actors — state, civil societies and market — together and facilitating co-management of wetlands under the umbrella of environmental governance as mentioned in section 2. The challenges to this will be clarified in the case studies.

Case of Three Indian Wetlands

Case Studies

Three wetlands, namely Kondakarla Ava in Andhra Pradesh, Vembanad in Kerala and Chilika of Orissa have been selected to examine contextual issues of environmental governance in Indian wetlands (Map 12.2). These wetlands pose different challenges to environmental governance and will provide insights into the ground realities of the process. Relevant features of Chilika, Vembanad-Kol and Kondakarla Ava Wetlands are presented in Box 12.3.

Like any other Indian wetland, Kondakarla Ava, Vembanad and Chilika suffer from ecological degradation and mismanagement due to conflicting interests. Chilika and Vembanad have been declared as Ramsar Sites and have received much wanted attention from national and international community whereas Kondakarla Ava, a lesser-known wetland, hinges on hopes based on local-level initiatives.

Chilika already has a formal wetland-level governance structure in place, known as the Chilika Development Authority (CDA). In case of Vembanad, there are various ongoing efforts by local, national and international organisations to initiate good governance practices. The government and government-initiated users' institutions have earmarked responsibilities for managing Kondakarla Ava.

Kondakarla Ava, Vembanad and Chilika have key actors from all three spheres, namely state, civil societies and the market, that are responsible for the present status of these wetlands. The governance structures are still state-centric and less democratic with civil societies having an insignificant role to play in the decision-making process. In such a scenario, there is hardly any scope for the voice of the unorganised and marginalised populace to be heard in any discussion on the aspect of environmental sustainability. These three cases are discussed in detail in the sections which follow.

Map 12.1: **Locations of Kondakarla Ava, Vembanad and Chilika Wetlands**

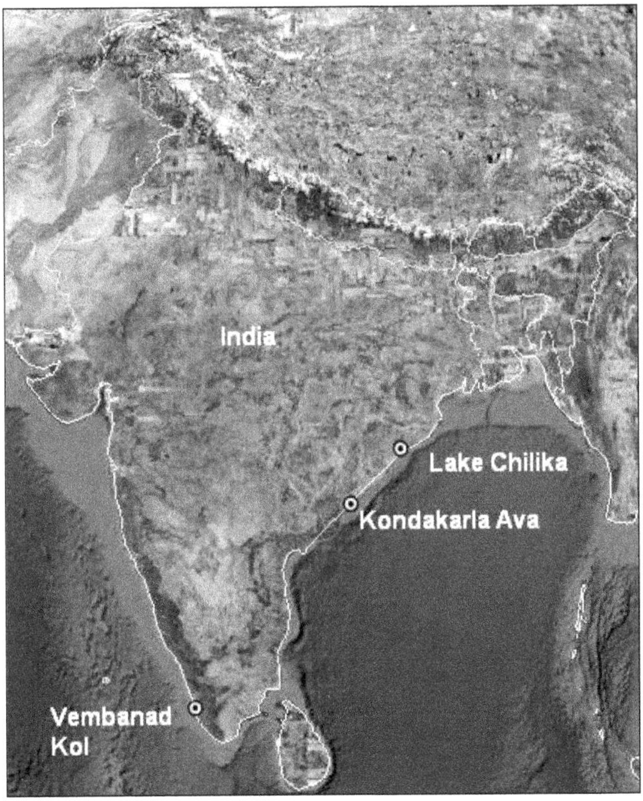

Source: Adapted by the authors from Google Earth.

Kondakarla Ava

Kondakarla Ava Wetland, the second largest freshwater lake in Andhra Pradesh (AP), is situated in Visakhapatnam district. This wetland, a part of the Sarada riverine system, is rich in its diversity of fauna and floral and is one of the major stop-over sites for many migratory birds flying through AP. The wetland serves as an economic backbone of its 17 surrounding villages. The economies of these villages to a larger extent revolve around it — for activities ranging from irrigation, fishing, washing, cattle rearing, etc. Kondkarla Ava is recognised as a priority site for Integrated Protected Area System (IPAS) by the Andhra Pradesh State Forest Department. It has been included as a

Box 12.3: Relevant Features of Kondakarla Ava, Vembanad-Kol and Chilika

	Kondakarla Ava	Vembanad-Kol	Chilika
Area (km²)	6.5	1,591	906–1,165
Location (figure 12.2)	Andhra Pradesh state; Visakhapatnam district	Kerala State (South India); Alleppey, Pathanamthitta and Kottayam districts	Orissa state (East India); Puri, Khurda Ganjam districts
Conservation status	Identified as a conservation site by the Asian Wetland Bureau and included under Integrated Protected Area System (IPAS) by the Andhra Pradesh State Forest Department	Ramsar Site since 2002	Ramsar Site since 1981
Ecosystem	Shallow freshwater lake	Estuarine system with freshwater and brackish water areas (direct connection to the sea)	Shallow estuarine lagoon; brackish water (connected to the sea by tidal channels)
Fauna and flora	14 species of Macrophytes. 25 species of fish including *Macrognathus aral* (listed under Data Deficient (DD) category in the IUCN Red List 106 species of local and migratory birds	Shrimp, finfish, shellfish. Spotbilled pelican, waterfowls, etc. Over 90 species of migratory birds Mangroves	Several hundreds of flowering plants. Hundreds of animal species including the green sea turtle, dugong, white bellied sea eagle, peregrine falcon, herrings and sardines
Livelihoods activities	Agriculture, open water capture fishery, washing and livestock rearing	Fish and shrimp farming, clams harvest, tourism	Fish and shrimp farming, clams harvest, tourism

Source: Prepared by the authors.

conservation site by the Asian Wetland Bureau and is recognised as one of the important sites for the development of ecotourism by the Tourism Development Corporation.

Despite all the importance attached to Kondakarla Ava, it has been subjected to severe environmental degradation in recent times. The surface area and level of water along with the local avian and fish populations have been decreasing and all this has adversely affected the livelihoods of the communities which are dependent on this lake. Conflicts and inequity in sharing Kondakarla Ava's resources is the major issue which forms an obstacle to its sustainability. Many government departments and community-based institutions have authority, i.e., decision-making powers over the wetland. Some of the roles and responsibilities of these decision makers overlap, leading to confusion.

The wetland is used by 10 Gram Panchayats (GP) which belong to two different Mandals (blocks) of two different electoral constituencies. The overall authority is vested in the district administration, which is supposed to coordinate all its line departments like revenue, agriculture, irrigation, fisheries, forest and tourism. The coordination between the irrigation department and the fisheries department is essential for wetland management. The irrigation department controls the water flows of the lake through the Water Users Association (WUA), which is formed with the farmers of the villages which receive waters from the Ava Irrigation Channel system. Ideally, the Kondakarla Ava WUA should be a body of farmers association, and associations for other water users like fishermen, washermen, livestock rearers, etc. Further, most of the members of the WUA belong to the downstream villages, while the villagers adjacent to the lake have least participation as they do not receive water through the irrigation channel.

The fisheries department should ensure that the minimum water levels (2 m) are maintained in the wetland as per prescriptions, but it has not taken that responsibility. Thus, during the periods of water scarcity, farmers' priorities prevailed at the cost of the lake ecology and the privileges of the fishermen. The fisheries department shrugs its responsibility after selling the seeds or nets required, and is not interested in protection of the rights of fishermen. On the other hand, all community-based institutions in the region aim to work for the welfare of the wetland users. However, it is ironical that despite the fact that subsistence livelihoods of user groups directly depend on the health of the wetland ecosystems, none of them has wetland conservation and management as one of their objectives. A very common perception here is that wetland degradation is a natural process.

With a plethora of 'stakeholders', it is almost difficult to distinctly define the rights and responsibilities for each one. There is no coordination between various government agencies and community-based institutions. There is no single body, which can coordinate and monitor all the wetland use practices. All these have contributed to the rise of several conflicts of very complex nature. Downstream farmers vehemently oppose lift irrigation scheme in the region. They have been agitating against the existing Vadrapalli and proposed Cheemalapalle lift irrigation projects. Their major grievance stems from the fact that there is inadequate water supply to irrigation channels as a result of excessive water usage through pumping of water with individual motors for lift irrigation by upstream villages. The downstream villages also state that they have the first right on the usage of water as they have been using the irrigation channels since a long time (this network of irrigation channels was built by the British), and pumping of water through motors.

The issue of encroachment in the marginal areas of the wetland, primarily for cultivation of crops, is another reason for conflicts among farmers. The WUA is demanding for a fresh mapping of the wetland. There are some points of contention between farmers and fishermen. Use of excessive water for irrigation by both upstream and downstream farmers results in failure in maintaining the permitted water level which affects fish production. Apart from these major issues, there are several other issues related to wetland use practices which have caused some minor conflicts among user groups. Conflicts and disputes that arise from these factors are not something that can be avoided, suppressed or ignored. A conflict over equitable sharing of wetland resources is the major obstacle to sustainability. Although various acts and policies for environment conservation and protection exist, they are not being implemented effectively for various reasons, namely lack of proper dissemination of information by the government machinery, general lack of awareness among various stakeholders, self-centered attitude and unhealthy politics. To facilitate the above, the National Water Policy 2002 was introduced to conserve water resources through participatory approach and provides clear guidelines towards institution building.

However, government apathy towards conservation and management of Kondakarla, despite the fact that it is the second largest freshwater lake in the state, is a cause of concern. In order to resolve all these problems, the wetland needs participatory governance structures comprising of all the stakeholders.

Kondakarla Ava has not received much needed attention. There are very few local level efforts to save this wetland. Shubhada, a youth action group, has been working since 2003 towards conservation of the Kondakarla Ava. It acts to achieve this objective through research, motivation meetings and awareness camps organised throughout the year in this region. As an initial step towards the formation of wetland governance, this group initiated village-level 'Wetland Development Committees' in 10 wetland user villages in 2006. However, sustainability and empowerment of these bodies, formation of a wetland level participatory institution and sustaining its own efforts are major challenges for this youth group.

Vembanad Kol

The Vembanad Kol and the adjacent landform is a large wetland eco-system of the southwest coast of India and is ecologically significant due to its mangrove patches and also being a habitat for resident and seasonal migratory waterfowl. A large proportion of the 1.6 million people living on the banks of Vembanad are directly or indirectly dependent upon this wetland ecosystem for their livelihoods. Major livelihood activities include agriculture, fishing, tourism, inland navigation, coir retting, lime shell collection, shrimp/crab farming and sand mining. The commercial nature of many of these activities leads to uncontrolled use of resource and this poses grave threats to the ecosystem (ATREE 2008). Based on the rich biodiversity and socio-economic importance, the Vembanad Kol was declared a Ramsar Site, a wetland of international importance, in 2002.

Vembanad Kol is a low-lying area with backwaters, canals and stream networks and falls partly in three districts. The rivers Achencoil, Pamba, Manimala and Meenachil, originating from the eastern hills, discharge their water and sediments into Vembanad Kol. The backwaters are connected to the Arabian Sea, which brings tidal influence and seasonal salinity into the system, a sizable area of which lies below sea level. A highly fertile tract of land replenished by silt brought by the river systems has contributed to the identification of this area as well suited land for rice cultivation from early days.

Reclamation of land for cultivation and flood control methods used to be undertaken by private farmers with assistance from the state. In Vembanad, the construction of a salinity barrier at Thanneer-mukkuom and its ineffective operations has affected the system. This barrier was constructed in a narrower part of the Vembanad Kol in order to prevent the ingress of salinity into the polders of Kuttanad (area for rice cultivation) during the summer season and also to retain the freshwater inflow from the rivers into the lake. Only two-thirds of the original number of gates is opened in July to release flood flow, but the gates are closed in mid-November. The structure has been relatively successful in keeping the waters in the Kuttanad free of salinity and has facilitated the addition of another crop in the dry season (SER 2007). Due to decrease in salinity, the area covered by invasive weed species has increased. The closure of salinity barrier has restricted the movement of juvenile prawns from the sea to the lagoon resulting in reduced area being available for breeding and spawning of some important fish varieties and the natural flushing of pollutants. Hence, biodiversity and fish catches have declined. The traditional fishermen are the most affected by these changes in the ecosystem.

The early 1990s saw the mobilisation of fishermen against the seasonal closing of the salinity barrier in Vembanad. The conflicts are the result of the twin processes of environmental degradation of the biophysical system and social marginalisation of sections of people directly dependent on the lakes for their livelihoods. The major trigger of conflicts is the capitalisation of resources by powerful societal groups. Users associations like Dheevara Sabha and Farmers'Association played a major role in mobilising communities.

The marginalised traditional fishermen use agitation as an overt tactic of protest and they also employ covert tactics like putting physical obstruction by fully closing the barrier in Vembanad. Such methods of agitation is the first stage. In Vembanad, the constant building of pressure by fishermen empowered through mobilisation led to the formation of a committee led by the district collector with representation from contesting sections. This was succeeded by a series of negotiations where the representatives of fishermen used the research results from the academia that suggested environmental degradation and the need for rationalisation of operations of the salinity barrier. They also mobilised the opinions of environmentalists to further their cause in the negotiation forums. The fishermen accelerated their protests and used a pressure-tactic of pressing for a permanent opening of the barrier. This led to the appointment of a commission of enquiry under a reputed scientist that suggested timely operations of the barrier as originally suggested in the design (with a closure of barrier restricted to three months). The government accepted this report and timely operations of the barrier were carried out. However, the sustainability of this resolution of the conflict depends on the continuity and institutionalisation of the process, including continued effective involvement by the state machinery.

Sectoral interests dominate the governance structure in Vembanad lake with division in the 'line' departments like departments of agriculture, fisheries, tourism and forest and environment, each trying to dominate with their own lists of priorities. In the Vembanad case, there is a conflict of interest between revenue administration on the two sides of the lake and also between the departments of agriculture and fisheries regarding the operations of salinity barrier. The respective district collectors, Members of Legislative Assembly (MLAs) and Member of Parliament (MPs) on different sides of the lake try to represent and pursue the interests of their dominant constituencies. The department of agriculture highlights the farmer interests, whilst the department of fisheries looks after the interests of fishermen (including the environmental externalities). The department of irrigation is caught in the crossfire between the farmers and fishermen interests and is always in a dilemma since they have to operate the closure and the opening of the salinity barrier. The Kerala Water Authority (KWA) requires the salinity barrier to be closed for freshwater availability. The Regional Agricultural Research Station (RARS) is responsible for monitoring the salinity intrusion in the lake. The closing date of the salinity barrier is supposed to be fixed based on objective scientific data. However, in practice, a political process described earlier guides the operation of the barrier. Thus horizontal sectoral divisions with the 'line department' approach of government system with different ministers heading their empires are a major problem. The departments function with narrow technically defined programmes, which may be counter-productive to the larger natural resource system. This horizontal differentiation within the state apparatus gets a spatial dimension when district collectors and local government institutions from different sides of the region take opposite stands in the conflict. In Vembanad, there is a coordination failure.

There is a need for intermediate level (between state level and local level) governance structures for management of Vembanad Kol. The district administration and panchayat raj institutions cannot take care of this since it involves multiple districts and panchayats. The involvement of panchayat raj bodies may deepen the conflicts because of their local focus and partisan attitude to the local resource use concerns with the risk of their being influenced by local elites.

Inclusion of the area in the Ramsar List in 2002 offers huge opportunities for development and implementation of a scientific management plan. Considering the fragile ecosystem of the wetlands, deterioration of water quality and consequent damage to aquatic organisms and the shrinkage of Vembanad Lake, Wetlands International-South Asia has formulated a management planning framework for this key Ramsar Site. The project was supported through the Dutch Embassy in India.

It has been proposed to establish a Vembanad Kol Authority to coordinate activities within river basins which will help in integrating coastal processes. The authority will have the responsibility for institutional and financial arrangements promoting communication, education, public awareness, and monitoring evaluation of activities within the overall policy framework to be developed and supported by appropriate legal framework (WISA 2007). Several local NGOs like the Kerala River Conservation Council and the Kuttanad Foundation have approached the government for implementing an integrated management action plan for this wetland (SER 2007). Some of the national level environmental NGOs like WWF-India and ATREE have initiated various projects towards creating awareness and institution building.

Efforts are on, but are fragmented and some times parallel. We may be optimistic that a governance structure will be established in a long run, but hope that the mistakes of Chilika, presented below, are not repeated.

Lake Chilika

Chilika lake is the largest lagoon in Asia spread over three coastal districts of Puri, Khurda and Ganjam. Hydrologically, Chilika is influenced by three major rivers — Mahanadi, Rushikulya and Bhargavi. Apart from the biodiversity and ecological peculiarities, Chilika lake sustains livelihoods of more than 200,000 people living in the 141 villages around it. Capture fisheries, the traditional livelihoods of traditional fishermen here, is organised through 92 primary fishery cooperatives. Six types of traditional fishing methods are in practice. The early 1990s saw the mobilisation of fishermen against enclosures for prawn culture in Chilika. The protests led to law and order problems that began to recur periodically.

A change in government policy regarding the lease of fishing rights that hampered the access of fishing grounds to traditional fishermen and rise of shrimp culture that enclosed portions of the lake were the major trigger points of conflict. Chilika Matsyajeebi Mahasangh and Khalamuha Fishermen's Cooperative Society led the campaign against the practice of culture fisheries.

In Chilika, the Government of Orissa created the Chilika Development Authority (CDA) in 1992 as a coordinating institution to look into the various claims of people dependent on the resource of the lake and for effective liaison with the government and other agencies involved in the management of Chilika. The management of Chilika was through the departments of revenue, fisheries, tourism and forests with their sectoral priorities. The CDA was created primarily to protect the unique lagoon ecosystem and to coordinate the activities of multiple institutions responsibilities for resource management. This was a response to the inclusion of Chilika in Ramsar's Montreux Record of degraded wetlands. The Ministry of Environment and Forests requested Ramsar Convention to place Chilika lake on the Montreux Record in June 1993. It identified various factors affecting the ecological character of the site. The purpose of the Montreux Record is 'to identify priority sites for positive national and international conservation attention' (Ramsar convention 2001).

The opening of a new lagoon mouth was the most important physical intervention by the CDA with technical studies and recommendations by the National Institute of Oceanography and Central Water and Power Research Station (CWPRS). According to the CDA, this has improved the hydrology, decreased invasive freshwater weeds, improved biodiversity, reduced silt loading from catchments, increased fish catches and thus income of fishermen. However, there are warnings from biological scientists and environmentalists regarding some unintended consequences of the technical intervention of opening of the new mouth in Chilika. They also highlight the livelihood crisis of fishermen from the nearby villages where the new mouth was opened and who cannot now engage in fishing due to the enhanced tidal effects.[1] They point to the unscientific nature of the CDA intervention done without proper modeling studies with the evidence of new deposition of sediments in the middle of the new lagoon mouth.

Since its genesis was related primarily to the environmental degradation faced by the lake, the forest and environment department was given administrative jurisdiction of the CDA with a governing body mainly composed of bureaucrats and people's representatives (see Box 12.4 for details). The CDA is supposed to work with flexible procedures, unlike a government department, to restore the ecosystem with the backing of academic studies, biophysical interventions, coordinating the work of departments and most importantly taking on board the aspirations of local people by working through NGOs and Community Based Organisations (CBOs). According to Ghosh and Pattnaik (2006), the institutional framework of Chilika is based on a principle of multi-sectoral collaboration with the CDA playing the role of a central coordinating agency. However, what is lacking is any mechanism for local people's voices to be represented in the CDA, which is evident from the structure of the governing body with representation mostly (17) by officials and (four) influential politicians.

In the Chilika case, there was a clear policy bias from 1991 to legitimise culture fisheries by the state, which is continuing with the new industrial policy and the latest bill drafted in 2002 that the government repeatedly tried to table in the state legislature. The class composition of people engaged in culture fisheries and their close links to state machinery is evident. This explains the failure of implementation of repeated recommendations made by a Commission of Inquiry set-up by the Orissa High Court, the Legislative Committee and the Supreme Court order to ban culture fisheries and the ineffectiveness of 'ceremonial' removal of *gherries* (enclosures) which came back within days of their removal by the revenue department.

The Chilika scene is less optimistic although the government has till now not been able to legitimise the illegal culture fisheries due to the mobilisation by traditional fishermen. Looking at the CDA's role, they adopted a 'participative' strategy and tried to play the newly fashionable role. It tried to enhance stakeholder participation and communication through a network of NGOs and CBOs to run an outreach programme for awareness building regarding the environment of Chilika. According to the former CEO of CDA, they have learnt from the conflict in 1999 that unilateral decisions could intensify conflicts and this was the key reason for the changed policy of participation and claimed that many activities, especially the new mouth restoration, was done through a consultative process with the community that enhanced the confidence of the local population.

There are two problems to such a participatory strategy. First, the NGOs are funded by the CDA and used merely for environmental educational activities (which might have a larger relevance) sidelining real livelihood issues and politics of marginalisation described earlier. Second is the 'benevolence' involved. One of the active NGO leaders who was interviewed complained of the clear policy shift of undermining NGO involvement by the new CEO who took charge from June 2006. During the interview, he was not apologetic about this shift since he found that the forest department has the capability to conduct the activities that the NGOs are presently doing. About the representation of the civil society in governance structure of CDA, he was adamant that since people are represented through their representatives like the MLA and MP, there was no need for any more involvement of NGOs. However, he vehemently argued for administrative control of officers in other government agencies in the jurisdiction of Chilika to be brought under the CDA for effective coordination. This raises a question mark to any possibility for a shift to a more democratic governance and clarifies the important role of the actor along with the structure needed for such a shift. Therefore, along with structural changes, sensitive and committed individuals are also needed for a shift in governance practices, which will then translate current benevolent notions of 'participation' to real practices of democratic governance.

The case of Chilika can provide a practical lesson for Vembanad and Kondakarla, which are at the earlier stages of evolution of environmental governance.

The three case studies reveal a fragmented institutional framework familiar to managers of natural resources in developing countries with: (i) several federal institutions (Ministries of Water Resources, Environment and Forests, Agriculture, and Rural Development, the Marine Products Export Development Authority, etc.); (ii) several state bodies (the departments of fisheries, agriculture, irrigation, environment, forests, science and technology, and tourism in each state, the Pollution Control Board, the Kerala Water Authority, several regional

Agricultural Research Stations, the Chilika Development Authority, etc.); (iii) several judiciary bodies (Supreme Court of India; High Courts of Kerala, Andhra Pradesh and Orissa); (iv) *panchayats* local institutions, (associations of farmers and fishermen, trade unions); (v) revenue departments in each district with representatives (collectors) at the district, *tahsildars* at the sub-district and village level; (vi) a multitude of regulations and legislations — the Wildlife Protection Act (1972), the Water Prevention and Control of Pollution Act (1974), the Forest (Conservation) Act (1980), the Environment Protection Act (1986), the Coastal Regulation Zone Notification (1991), the Indian Fisheries Act (1997), the National Water Policy (2002), the Biological Diversity Act (2003), the Coastal Aquaculture Authority Act (2005), etc. — for the most important one, and finally (vii) the supra-national bodies like the Ramsar convention that lists good practices for the sustainable management of wetlands and is a powerful tool to bring national and international conservation attention but do not have statutory powers. Box 12.4 illustrates the governance structures and actors that have varying levels and degrees of influence on the decision-making process in the three wetlands.

Challenges to Governance

The study attempted to conceptualise environmental governance and to examine the cases of three Indian wetlands with the major issues delineated. Environmental governance has gained currency at the level of rhetoric that led to formulation of many institutions (policies, laws and agencies for implementation). However, in practice, the implementation by state institutions (which are still largely responsible for it) is seen to be inadequate. We suggest that the shift from state to non-state (civil society and market) actors prescribed by good governance proponents is not happening on the ground. Both these processes lead to a situation where the normative gains form environmental governance is not at all achieved. The logic in wetland management still follows the age-old logic of access and control of resources by the powerful. Apart from issues of institutional and administrative fragmentation, the broader context in which policies evolve and the production process is organised is also important. Often, natural resources management is geared towards maximising economic returns with a thrust on techno-managerial efficiency and problems of social marginalisation and environmental degradation are overlooked (Scott 1998). Export market demands, for example,

Box 12.4: Wetland Organisation in the Study Area

	Chilika	Vembanad-Kol	Kondakarla Ava
Supra-national	Ramsar's Site of international importance (1981)	Ramsar's Site of international importance (2002)	–
International	World Bank, Wetlands International-SA, Ramsar Centre-Japan, India-Canada Environment Facility (ICEF)	Wetlands International-SA, WWF	–
National	Designated as a site of national importance	Designated as a site of national importance	Proposed Ramsar Site by SACON (2005), Proposed Conservation site by WWF and Asian Wetlands Bureau in Directory of Indian Wetlands (1993)
Government organisations	Ministry of Environment and Forests, Water Resources, Agriculture, Chemical and Fertilisers, Commerce and Industries, Science and Technology, Tourism, Panchayati Raj, National Bank for Agriculture and Rural Development, INS Chilika	Ministry of Environment and Forests, Water Resources, Agriculture, Chemical and Fertilisers, Commerce and Industries, Science and Technology, Tourism, Panchayati Raj	Ministry of Environment and Forests, Water Resources, Agriculture, Chemical and Fertilisers, Commerce and Industries, Science and technology, Tourism, Panchayati Raj

(Box 12.4 continued)

(*Box 12.4 continued*)

Environmental NGO	WWF, CEE, Whale and Dolphin Conservation Society (WDCS)	ATREE (The Vembanad Wetland Conservation Program), WWF, M. S. Swaminathan Commission	-
State level			
Government organisations	Water Resources Department, Fisheries and Animal Resources Development Department, Department of Agriculture, Forest Department, Revenue Department, Orissa Remote Sensing Application Center, Orissa Renewable Development Authority, Orissa High Court	Irrigation Department, Fisheries Department, Agriculture Department, Forest Department, Tourism Department, Revenue Department, The Kerala Water Authority (KWA)	Irrigation Department, Fisheries Department, Agriculture Department, Forest Department, Tourism Department, Revenue Department
Local Level Govt. Organisations	District administration with its line departments: irrigation, fisheries, agriculture, forest, revenue, tourism	District administration with its line departments: irrigation, fisheries, agriculture, forest and environment, revenue, tourism, PWD	District administration with its line departments: Mandal Revenue Office (MRO), Mandal Development Office (MDO), irrigation, fisheries, agriculture, forest, tourism

CBO	Watershed committees, Self-help groups, Bird Protection Committee, Campaign for Conservation of Chilika	Dheevara Sabha, Farmers' Association	Kondakarla Ava Water Users Association (WUA), Kondakarla Ava Inland Fishermen Cooperative Society, Vadrapalli Lift Irrigation Committee, Cheemalapalle Lift Irrigation Committee, Raitu Mitra Sangham, Washer-folk Society, Village Milk Producers Mutual Added Cooperative Societies, Women Self-help Groups (DWACRA), Youth Associations.
Environmental NGOs	Wild Orissa, Pallishree, Campaign for Conservation of Chilika Lagoon (CCCL)	Kerala River Conservation Council, the Kuttand Foundation	Shubhada-A Youth Action Group.
Wetland Level	Chilika Development Authority	–	–
Market	Shrimp culture	Rice/fishery	Sugarcane sold to nearby Anakapalli National Sugarcane Mandi.

Source: Prepared by the authors

often lead to environmental degradation and social marginalisation. This is very clear from the process of enhancement of culture fisheries in Chilika. Due to a high export-driven profitability, aquaculture in Chilika benefits from the strong support of the Orissa government. Aquaculture development policies aim at enhancing the overall economic situation of the states. The government scheme of supplying free electricity for irrigation in Andhra Pradesh has promoted excessive use of water for irrigation. The state, through regional economic development, has paid little attention to how the benefits are distributed: social marginalisation and environmental degradation are common and generally ignored. This is a characteristic of the recent Indian economic boom in the current era of liberalisation as also illustrated by land acquisition conflicts related to industrialisation (Balagopal 2007). Pressure over natural resources can also be triggered by strong state willingness towards achieving a given objective as illustrated by the rice-centric development observed in Vembanad and that find its roots in the willingness of the Kerala government to reduce the dependency of the state on cereal imports from other Indian states (Narayanan 2003).

In the above discussions, it is clear that there is threat to environmental sustainability of wetlands and there is an unequal impact of the process that is detrimental to vulnerable sections. EG is a generic term that includes a wide range of institutions, organisations, policies and actors, among which is the government, which shapes policies and the production of related outcomes. In the debates reviewed in section 2, the shift from 'government' to 'governance' has been associated with the dilution of the role and importance of the state: the state progressively lost some of its central functions to other spatial levels such as the supra-national and infra-national, and to non-state institutions such as private companies and voluntary organisations (Jessop 1994). The concept of governance incorporates a notion of public action where 'participation by the public in the process of social change' (Dreze and Sen 1989: 259) is demanded. It theoretically represents a rediscovery of civil society and of the role that institutions in that sphere can play in promoting collective, private and public ends (Mackintosh 1992). The dilemma of natural resources management and governance is about reconciling the many dimensions of sustainable development (the economic, socio-cultural, environmental dimensions and the politics encompassing them) that are overlapping but which are often conflicting in the long term.

Imbalance in meeting the requirements of those dimensions generally leads to conflicts and further degradation of the resource base.

Box 12.4 provides an overview of the multiple organisations at various levels in the three wetlands. The role and importance of local institutions have been diluted even as the legal and institutional framework of natural resources management broadened in line with the increasing recognition of the importance of environmental preservation for sustainable development. This institutional formalisation, aimed at better preservation of ecosystems, often led to further degradation of the environment. It is clear that sectoral interests dominate the landscape of natural resources management in these three wetlands: the different institutions generally pursue overlapping but conflicting objectives in the long run. The 'horizontal' segregation within the state apparatus between different government departments functioning with narrow and technically defined programmes is often paralleled by a spatial partition as populations of different regions may pursue different goals. Such fragmentation is generally counterproductive for the sustainable management of a larger natural resources system.

This does not mean that formalisation and planning are not needed. A strong institutional support is indeed essential to achieve governance systems supporting integrated resources management to ensure an environmentally and socially sustainable development (Falkenmark et al. 2007). But this reminds us of the needs for further integration and coordination among the different institutions involved in natural resources management (Merrey et al. 2007). Finally, inconsistencies between biophysical and political boundaries call for a meso-level body that integrates political/administrative divisions within given natural resource settings (a lake, a watershed, a river basin, etc.). However, lessons from CDA indicate the need for transforming such techno-managerial bodies into truly participating governance structures.

The institutions of governance reviewed in sections 2 and 3 are still state-centric and dysfunctional. The process on the ground as illustrated in section 4 shows the multiple and fragmented nature of governance institutions. It is also seen that there is rising dissent from the civil society with claims for more equitable share of resources. However, the relative power of these institutions vis-à-vis the combined power of state and market as in the case of Chilika is evident. In Kondakarla, too, there could not be any expression of dissent by a substantive section of landless agricultural labourers who were affected by the EG decisions of the powerful (and hence outside any 'civic' realm of action). In Vembanad, it is noted that the weaker realm of civil society is also fragmented by the conflict between farmers and fishermen unions and hence the concept of civil society is also not a monolith as is the case of the state.

For EG to become operational, a clear demarcation of institutions (policies, laws and clarity about role of organisations) is a necessity. Along with this, defining appropriate levels of governance and drawing boundaries to delineate the responsibilities (both in spatial and structural terms) of the different institutions involved in the use and management of natural resources is also needed. However, such coordination will not just be techno-managerial alone, but also political and hence the need for a democratic space to exercise citizen's rights. This is also because the actors in the various spheres wield unequal powers and hence the need for mobilisation of the marginalised sections who might be at the receiving end of 'development'. Here conflicts become signs of a healthy polity where there is latitude to mobilise and empower those who lose out. Thus democratic governance of natural resources also demands a conceptual understanding inclusive of mobilisation.

References and Select Bibliography

Ashoka Trust for Research in Ecology and the Environment (ATREE). 2007. 'Vembanad Wetland Conservation Program'. http://www.vembanad.org (accessed 5 June 2008).

Balagopal, K. 2007. 'Land Unrest in Andhra Pradesh', *Economic and Political Weekly* 42 (38–40): 3828–33, 3906–11, 4029–34.

Balassa, B. 1982. *Development Strategies in Semi-industrial Economies.* Baltimore: Johns Hopkins University Press.

Bauer Peter, 1972. *Dissent on Development: Studies and Debates in Development Economics.* Cambridge: Harvard University Press.

Bhagwati, J. 1993. *India in Transition: Freeing the Economy.* Oxford: Oxford University Press.

Carson, Rachel. 1962. *Silent Spring.* Boston: Houghton Mifflin.

Curmally, Atiyah. 2007. 'Environmental Governance and Regulation in India', Environment and Rehabilitation 97, Asia Link. http://www.archidev. org/article.php3?id_article=597 (accessed 5 April 2008).

von Dach, Susanne Wymann, Rosmarie Sommer and Ruth Wenger. 2005. 'Global Conventions and Environmental Governance', Info Resources Focus, Info Resources. http://www.inforesources.ch/pdf/focus_3_05_e.pdf (accessed 5 April 2008).

Dreze, J. and A. Sen. 1989. *Hunger and Public Action*. Oxford: Clarendon Press.

Falkenmark, M., M. C. Finlayson and L. J. Gordon. 2007. 'Agriculture, Water and Ecosystems: Avoiding the Costs of Going too Far', in D. Molden (ed.), *Water for Food, Water for Life: A Comprehensive Assessment of Water Management in Agriculture*, pp. 233–77. London: Earthscan.

Ghosh, A. K. and A. K. Pattnaik. 2006. 'Chilika Lagoon: Experience and Lessons Learned'. http://www.iwlearn.net/publications/ll/chilikalagoon_2005.pdf (accessed 1 November 2007).

Goffman, Ethan. 2005. 'Global Environmental Governance, Environmental Policy Issues, CSA'. http://www.csa.com/discoveryguides/ern/05aug/overview.php (accessed 18 June 2008).

Gorringe, P. 1997. 'The State and Institutions', The Treasury, Wellington, New Zealand. www.treasury.govt.nz/gorringe/papers/gp-1997.pdf (accessed 10 February 2004).

Hardin, G. 1968. 'The Tragedy of Commons', *Science*, 162 (3859): 1243–48.

IGES. 2001. 'Report of the First Phase Strategic Research. Environmental Governance Project', Institute for Global Environmental Strategies, Japan. http://www.iges.or.jp/en/pub/pdf/eg-e.pdf (accessed 7 April 2008).

Intergovernmental Panel on Climate Change (IPCC). 2007. 'IPCC Fourth Assessment Report: Climate Change 2007'. Geneva: IPCC.

International Union for the Conservation of Nature and Natural Resources (IUCN). 1980. 'World Conservation Strategy: Living Resource Conservation for Sustainable Development'. Switzerland: IUCN-UNEPWWF.

Jessop, B. 1994 'Post-Fordism and the State', in A. Amin (ed.), *Post-Fordism: A Reader*, pp. 251–79. Oxford: Blackwell Publishing.

Kashwan, Prakash. 2005. 'Environmental Governance in India: A Concept Note', Lead. www.leadindia.org/pdf/Environmental_Governance_in_India.pdf (accessed 5 April 2008).

Killick, T. 1986. 'Twenty-five years in Development: The Rise and Impending Decline of Market Solutions', *Development Policy Review*, 4 (2): 99–116.

Lal, D. 1983. *The Poverty of Development Economics*. London: IEA.

Lead. 2006. 'Stakeholder Participation in Environmental Governance — Introduction to Lead Global Training Session', Bhopal, India. http://www.leadindia.org/pdf/workbook-part-1.pdf (accessed 5 April 2008).

Mackintosh, M. 1992 'Introduction', in M. Wuyts, M. Mackintosh and T. Hewitt (eds), *Development Policy and Public Action*, pp. 1–12. Oxford: Oxford University Press.

Millennium Ecosystem Assessment. 2005b. *Ecosystems and Human Well-Being: Synthesis*. Washington, D.C.: Island Press.

Meadows, D. H., D. L. Meadows, J. Randers and W. Behrens. 1972. *The Limits to Growth*. London: Pan.

Ministry of Environment and Forests (MoEF). 2007. 'Conservation of Wetlands in India: A Profile (Approach and Guidelines). http://envfor.nic.in/divisions/csurv/WWD_Booklet.pdf (accessed 26 October 2010).

———. 2008. Ministry of Environment and Forests, Government of India. http://moef.nic.in/(accessed 15 June 2008).

Merrey Douglas J., Ruth Meinzen-Dick, Peter P. Mollinga, Eiman Karar, Walter Huppert, Judith Rees, Juana Vera, Kai Wegerich and Pieter van der Zaag. 2007. 'Policy and Institutional Reform: The Art of the Possible', Water for Food, Water for Life: A Comprehensive. Earth Scan. http://www.iwmi. cgiar.org/assessment/Water%20for%20Food%20Water%20for%20Life/Chapters/Chapter%205%20Policy.pdf (accessed in 5 June 2008).

Najam, Adil, Mihaela Papa and Nadaa Taiyab. 2006. *Global Environmental Governance: A Reform Agenda*. Denmark: IISD.

Narayanan, N. C. 2003. *Against the Grain: The Political Ecology of Land Use in a Kerala Region, India*. Maastricht: Shaker Publishers.

North, Doughlas C. 1990. *Institutions, Institutional Change and Economic Performance*. Cambridge: Cambridge University Press.

Panini, Devaki. 1998. 'The Ramsar Convention and National Laws and Policies for Wetlands in India', Case Study prepared for the Technical Consultation on Designing Methodologies to Review Laws and Institutions Relevant to Wetlands. Gland, Switzerland, 3–4 July 1998. http://www.ramsar. org/wurc/wurchbk3cs4.doc (accessed 18 April 2008).

Parikh. J. and K. Parikh. 1999. 'Sustainable Wetland. Environmental Governance-2'. Mumbai: Indira Gandhi Institute of Development Research.

Ramsar Convention. 2002. Ramsar Advisory Missions: No. 50, Chilika Lake, India (2001). 'Removal of Chilika Lake Ramsar Site, India, from the Montreux Record – 9–13', December 2001. http://www.ramsar.org/cda/en/ramsar-documents-rams-ram50/main/ramsar/1-31-112%5E22934_4000_0%20%20 (accessed 26 October 2011).

———. 2006. http://www.ramsar.org/cda/en/ramsar-news-archives-2006-india-s-latest-additions/main/ramsar/1-26-45-49%5E16874_4000_0__ (accessed 18 April 2008).

Rao, Y. N. and Hemant Datye. 2003. 'Overview of Indian Wetlands', in Jyoti Parikh and Hemant Datye (eds), *Sustainable Management of Wetlands: Biodiversity and Beyond*, pp. 41–98. Delhi: Sage Publications.

Rhodes, R. A. W. 1997. *Understanding Governance: Policy Networks, Governance, Reflexibility and Accountability*. Buckingham: Open University Press.

Runnalls, David. 2006. 'Preface', in Adil Najam, Mihaela Papa and Nadaa Taiyab (eds.), *Global Environmental Governance: A Reform Agenda*, pp. v–vi. Denmark: IISD.

Saleth, R. M. and A. Dinar. 2004. *The Institutional Economics of Water: A Cross-Country Analysis of Institutions and Performance.* Cheltenham: Edward Elgar.

Schumacher, E. F. 1999. *Small Is Beautiful: Economics As If People Mattered: 25 Years Later...With Commentaries.* Washington: Hartley and Marks.

Scott, J. C. 1998. *Seeing Like a State: How Certain Schemes to Improve the Human Condition Have Failed.* New Haven: Yale University Press.

State of Environment Report- Kerala (SER). 2007. Land Environment Wetlands of Kerala and Environmental Health, vol. I. Thiruvananthapuram: Kerala State Council for Science, Technology, and Environment.

Stewart, F. 1985. 'The Fragile Foundations of the Neo-classical Approach to Development', *Journal of Development Studies*, 21 (1): 282–92.

Wetland Internationals-South Asia (WISA). 2007. 'Conservation and Sustainable Development of Vembanad Kol backwaters, Kerala: WISA Outlines a Management Framework'. http://www.wetlands.org/southasia/En/news.aspx?ID=c2cea8ac-edce-41d7-8004-876bf5581ebc (accessed 18 April 2008).

World Commission on Environment and Development (WCED). 1987. *Our Common Future.* Oxford: Oxford University Press.

World Bank. 1997. *World Development Report: The State in a Changing World.* New York: Oxford University Press.

13

A Conceptual Framework for a National Policy on Financing Watershed Management in Sri Lanka

Hemesiri Kotagama, E. R. N. Gunawardena and K. A. I. D. Silva

Market economic theory distinguishes between demand and effective demand. An individual's desire for a commodity (demand) could be achieved in a market only if the individual is able to pay for it (effective demand). Metaphorically, sustainable development is universally desired, and to achieve sustainable development, environmental and natural resources need to be conserved. Policies, plans, programmes, projects and strategies to achieve sustainable development would be ineffective expressions of desires if not backed by mechanisms to finance (Panayotou 1993). Rees et al. (2008) has reported, quoting Biswas et al. (2005), that the World Summit on Sustainable Development held in 2002 had called for Integrated Water Resource Management (IWRM) and water efficiency plans and that:

> Regrettably, few any of the completed IWRM plans considered how the overarching water governance and management system is to be financed; indeed some did not mention financing at all. [...] The plans were typically silent on who should raise the funds, for what particular purpose and who should bear the pay back cost (Rees et al. 2008: 10).

The United Nations Summit on Financing for Development and the World Summit on Sustainable Development, both held in 2002, have examined potentials and strategies to finance sustainable development in relation to conservation of a wide array of natural and environment resources. The reports by Camdessus and Winpenny (2003), Van Hofwegen (2006) and Rees et al. (2008) are 'watersheds' on factually establishing: (a) the nexus between sustainable management of water and its financing, (b) that financing of the water sector has been neglected, and (c) conceptualizing mechanism to finance the conservation of the water sector. Camdessus and Winpenny (2003) has addressed the supply of finances to finance domestic water supply and sanitation whilst Van Hofwegen (2006) examined the demand for finances for water supplies by local governments. It is also reported that discussions on IWRM have not sufficiently deliberated the issue of financing the water sector (Rees et al. 2008).

An ideal conceptual framework to comprehensively examine the financing of the water sector would be to consider the demand and supply aspects of financing at each functional node (where water cycle interacts with people and provides a service) of the water cycle. This does not imply that examining financing of the water resource sector needs to be a single comprehensive study around the whole water cycle. What is implied is that, as expected by the concept of IWRM, water needs to be managed holistically around the water cycle, and thus finance it too holistically, integrated around the water cycle as a 'finance cycle'. This is succinctly expressed by Rees et al. (2008) as follows:

> When talking about a more holistic approach to water financing, we are not suggesting that financial allocations should occur through a major bureaucratic exercise which seeks and inevitably fails to coordinate every thing. Rather, it is seen as a process which considers the whole range of essential integrative functions and water services which require funding; examine financial sources that could be potentially be available for each function and service: attempts to ensure that most appropriate source is utilised for each purpose; and evaluate the institutional/reforms needed to increase the financial flows. In most developing countries (and some developed) many vital water resource management functions — such as catchments (watershed as referred here) management, systems analysis and planning, flood protection, research, hydrological and performance monitoring, public awareness, stakeholder consultation and institutional capacity building — are neglected and underfunded. Their continued neglect is unsustainable in environmental and socioeconomic development terms (ibid.: 9).

There is convergence of opinion that financing of watershed management has received the least priority among the functional nodes along the water cycle (Camdessus and Winpenny 2003; Van Hofwegen, 2006; Rees et al. 2008). Van Hofwegen (2006) reports that:

> The task force, however, is mindful of the equally important work of financing improvements in the management of river basins (watersheds

as referred here). River basins are the foundation of the water sector and a key component to ensuring water services are sustainable. Innovative financing at increased levels will be needed for resource management (Van Hofwegen 2006: vii).

The objective of this chapter is to propose a logical framework to conceptualise a policy on financing watershed management to achieve IWRM. A policy, which is an expression of expectation, should be pragmatic to guide action. Thus theoretical ideals of economics, experiences and opinions are presented and critically reviewed as guidance to formulating a pragmatic policy. The empirical basis is the experience in formulating a policy for financing watershed management as a part of formulating a watershed conservation policy for Sri Lanka (Kotagama 2001). Thus some of the data used dates back to 2001 and is used for illustrative purpose.

Empirical and Normative Aspects

A national policy is formulated recognising the social expectations of a nation and knowledge on possibilities to achieve expectations. Being explicit of such would facilitate dialogue and to arrive at public consensus about a policy.

The Need

The Second Water Forum that met in Hague in 2000 has indicated that additional annual investment of about US $ 100 billion was required for the water resource sector (Camdessus and Winpenny, 2003). Further, it has been estimated that annual funds required by the water sector would have to be doubled to achieve the millennium development goals by 2015 (Van Hofwegen 2006). In 1999–2001, the annual average commitment of aid to water supply and sanitation from all sources has fallen to US $ 3.1 billion from US $ 3.5 billion in 1996–1998. As expressed by Van Hofwegen (2006), the global water sector is experiencing 'decreased, static, or marginal' changes in finances, despite recommendation to double the finances to achieve the millennium development goals related to this sector.

The trend in water sector financing has been biased towards capital investments with neglence of operational and maintenance expenditures (Camdessus and Winpenny, 2003). As has been amply demonstrated in the irrigation sector, financing operational and main-tenance activities are essential to sustain satisfactory services (Small and Carruthers 1991). Global trend in financing the water sector suggest the impending and imperative requirement for a policy directed, planned process to secure finances for water sector investments.

National Sovereignty

Respecting national sovereignty is a prime requirement of a national policy. To respect national sovereignty and financing to be sustainable too, it is best that watershed conservation is nationally financed. However, in the short term, developing countries with budgetary difficulties may be required to depend on foreign finances to overcome cash flow problems and to meet some public costs of environmental conservation (the terms environmental conservation, watershed conservation/management are used as synonyms). Foreign finance can only play a temporary role with a catalytic or demonstration value and sustainable development cannot be achieved through indefinite dependence on foreign finances (Panayotou 1993). The availability of foreign finance, upon which Sri Lanka has up to recent times largely depended on for environmental conservation, has been gradually declining. This is due to increasing per capita income in Sri Lanka and other global political economic changes that have made Sri Lanka less eligible to receive foreign finances. Dependence on foreign financing as aid for environmental conservation have often been tied up with perceived loss of national sovereignty, such as the case on 'nature for debt swap' offer by the United States of America under the Tropical Rain Forest Conservation Act (1998) to Sri Lanka to preserve Sinharaja forest (EFL 2001). However, the international community derives substantial benefits from conservation of biodiversity and global climate conservation. This provides an economic justification for counties undertaking conservation to secure international finances. Water however is a resource that directly benefits a nation and local resource users. Van Hofwegen (2006) in assessing the issue of enhancing the access to finances by local governments for the water sector concludes by providing a compromising and realistic perspective that could guide a financing policy for the water sector. However, in the end, it is national and local governments that have to act (finance) and it is the responsibility of the international community to provide them with the support where needed.

Endogenous Financing

It is ideal if finance for environmental conservation is generated within the environmental sector (endogenous financing) through revenues from resource users. This is justified as the allocation of finance from general public budget for environmental conservation may decline in the future as demonstrated by the difficulties faced by the the Sri Lankan government in financing public investments. A sector that could be easily denied of public financial allocations would be the environmental sector, given that public financial allocations are ultimately done through a political process. Politicians would be interested in allocating finances to sectors that give direct, tangible and geographically identifiable benefits, whilst benefits from environmental conservation are indirect, non-tangible and geographicaly dispersed beyond the electorate. Therefore, to the extent possible, finances for environmental conservation must be endogenously (within environmental sector) generated. The possibility to achieve this in Ireland has been examined by Barrett et al. (1997).

Market Economics and IWRM

Among the four guiding principles of IWRM, one enunciates that 'water has an economic value in all its competing uses and should be recognised as an economic good'. This principle has been the most disputed and debated. Hence, clarification is required to build-up public consensus for policy formulation. What this IWRM principle implies, by recognising water as an economic good, is that water generates human utility in its use, water has alternative uses and water is scarce in relation to human wants. As water is scarce and has competing uses, it is rational to use it in the most utility-generating (socially valued) use. Such allocation of a resource to the most socially valued alternative is the role of an economic system. An economic system, which is a social institution, could be between the extremes of a centralised decision-making system, where an individual or a central bureau takes decisions on resource allocation or a decentralised decision system, such as the market that takes resource allocation decisions with the participation of individuals. Market decisions are guided by prices, and prices in turn are partly (the two parts being demand and supply) decided by preferences of individuals and their purchasing power (effective demand). If the markets are perfect, as defined in theory, the market would allocate resources to alternative uses of highest social value. Where conditions of a perfect market does not prevail, due to various reasons, and a resource is under priced, as often is the case with water uses, it leads to wasteful use of the resource. The market is a strong participatory decision-making institution. However, where the strength (purchasing power) of participating individuals varies, market decision on the allocation of a resource may not be considered equitable. Where demand, and thereby price, is largely influenced by the rich with high purchasing power leading to a high price, the poor with low purchasing power may not have sufficient access to water, which is essential for sustenance of life. Above is an intuitive explanation of the strength of the market economic system and its weakness in relation to IWRM principles. In essence, if the market economic system is used to allocate water where it could best be used, and where market failures are rectified, it would enable to conserve water within the principles of IWRM, such as allowing public participation in decision making and allocating water to the socially most valued use.

What has been explained in the foregoing discussion is the use of the market and prices to manage an allocate water. The market plays another role, and that is the financing of economic activities. A profitable price enables to recover costs and generate finance to sustain and further improve water services. Where prices do not prevail, the government would have to use alternative approaches to generate finance for water management.

Empirical observations have not been made on whether adoption of IWRM would increase the need for finances and change the processes of financing of the water sector. Conceptually, the need for finances would not drastically increase due to adoption of IWRM as it advocates only a paradigm shift on water resource management which may not require major additional investments, other than for institutional changes to facilitate a coordinating functions (management gum as referred to by Rees et al. 2008) of individual institutions that are responsible for a variety of management functions around the water cycle.

Stakeholder Participation

Stakeholder participation in decision making of water management, advocated by IWRM, could be realised through adopting markets to manage and allocate water by enabling stakeholders' to finance water resource management and development. The market economic system, as explained in the foregoing discussion, enables stakeholder participation through purchasing power, though participation may be inequitable as purchasing power could be inequitably distributed in society. The democratic political system enables stakeholder participation through voting processes and often the case is one-person-one-vote. National policies should be guided by formal institutions of decision making, such as the constitutionally accepted economic and political systems. Colloquially in the market system money decides, in the political system vote decides and with no system voice decides. Stakeholder financing does not mean that stakeholders need to fully finance water resource management or that every stakeholder participate equally in financing. Financing, least partial cost recovery, through charging some stakeholders who could afford is desirable than otherwise. As Camdessus and Winpenny (2003) quotes, urban households are able to pay up to 5 per cent of their income on water services. Further, of mostly the low income population, about 6 per cent in Delhi, 10 per cent in Dhaka, 19 per cent in Ho Chi Minh City and 44 per cent in Jakarta buy water from private water providers. As a guideline for policy formulation, it could be expected that public financing may continue to partially finance the water sector. Yet, private funding by stakeholders through payment for water services is an option to pursue.

Van Hofwegen (2006) recognises the recent trend towards decentralisation of governance, leading to local government institution becoming managers of water. This trend of decentralisation and bringing the management of water closer to its immediate stakeholders is in congruence with IWRM principles. However, Rees et al. (2008) opines that, 'decentralisation may not be highly desirable for attracting the level and type of funding necessary' in the water sector (ibid.: 11). Nevertheless, with the trend towards decentralisation of governance and given the facts that local governments are heavily dependent on central government finances and are often starved for finances; that they have less alternatives to raise funds due to control of central governments and other policy and legal reasons; and that there is the need to examine innovative ways to secure finance by local governments and strengthening the capacity of local governments to effectively utilise finances for the water sector (Van Hofwegen 2006), innovative financing mechanisms such as developing capital markets for water sector financing has been proposed and many case studies of alternative forms of financing the water sector at the local governmental level are provided by Van Hofwegen (ibid.).

Organisational Reform

The conceptual expectation of IWRM is that functions related to managing the water sector should be integrated. An ongoing debate is whether IWRM requires administratively integrating sub-sectors and institutions functioning in the water sector. The emerging consensus is that neither extremes of singularising nor fragmenting of sub-sectors and institutions in the water sector is optimal. The optimal organisational structure for IWRM would depend on history, culture, etc., of a society. From the point of view of financing IWRM, Van Hofwegen (Van Hofwegen 2006) observes that combining public services such as electricity and water, as was done in Morocco, would allow joining up commercially lucrative ones with those that are not, allowing cross-financing. Rees et al. (2008) too cautions that unbundling (fragmenting) organisations may lead to constraining financing options and increasing the cost of capital.

Sources and Mechanisms

Someone will bear the cost of watershed management. As Camdessus and Winpenny (2003) mentions, ultimately finances will be paid by one of the three parties: water users, tax payers and aid donors. In the mid-1990s, the breakdown of sources of finances for water and sanitation has been the domestic public sector 65–70 per cent, domestic private sector 5 per cent, international donors 10–15 per cent and international private companies 10–15 per cent. The mechanisms of financing through these sources (international donors, domestic public sector and domestic private sector) are many. International donors may provide multilateral and bilateral funding through international development banks or through international environmental financing tools such as the Global Environmental Facility (GEF), Clean Development Mechanism (CDM) and Debt for Relief. Domestic public sector financing could be through national budgets, local government budgets, capital markets, local development banks, micro-credit institutions, and domestic private sector financing could be through non-governmental organisations, corporate sector and finally by water users. The European Water Initiative maintains a website on international sources of finance for the water sector. The literature on use of alternative mechanisms of financing such as user charges, regulatory fees and levies, payments for ecosystems services, creation of pollution markets, pollution taxes and abstraction charges, etc., is replete.

Many experiments referred to as Payments for Watershed Services (PWS) have been conducted in financing watershed management that are conceptually compatible with theory of markets and IWRM concept (Echavarria et al. 2004; Kallesoe and De Alwis 2005; NIVA 2007; Dillaha et al. 2008). The structure of the PWS mechanism is that downstream beneficiaries of watershed services pay upstream land users to maintain optimal land uses for watershed conservation. Among the constraints identified as limiting the success of PWS are the inadequate specification of property rights to resources such as land and water, contradictions of payments in relation to existing laws on environmental and natural resource use, imperfect information on the 'commodity' that is being transacted (which could be partly overcome by ecosystem valuation) and more critically high transaction cost of the organisational and implementation process. It is too early to judge the effectiveness and sustainability of these experiments. Kallesoe and De Alwis (2005) have scoped the possibilities to adopt markets for environmental services in Sri Lanka and have concluded that such experiences in sustainable financing of environment conservation were unfound. However, recent changes in strengthening land use rights, and decentralisation of government and resource management which would reduce transaction costs, offers potential to adopt PWS like financing and resource conservation. The greatest potentials are perceived in relation to ecotourism, hydropower and green agriculture.

Privatisation of water services has been vigorously promoted as a general economic policy. As reported by Tecco (2008), the Second World Water Forum held in 2000 had stated that 95 per cent of the incremental investments in the water sector should come from private sources. However, private investments to financing the water sector have fallen than risen. The water sector has been the least attractive to the private sector. Only 3 per cent of the population in the poor or emerging countries is served by the private sector and private international companies are showing less willingness to invest in developing countries (Camdessus and Winpenny 2003). The general observation that could guide policy is that public funding would remain to be dominant in water sector financing and hence the need to strengthen public financing practices and institutions.

Some payments to mobilise resources in the water sector may be 'payment' in kind such as in voluntary participation in maintenance of irrigation systems and in watershed conservation activities.

In essence, the lesson for policy formulation is that no single source or mechanism would be appropriate and sufficient to finance watershed management. Use of a mix of source and mechanisms would be the most appropriate in financing watershed management in a sustainable manner.

Market and Financing Watershed Management

It has been proposed that IWRM should recognise that managing water and land is inseparable (Bandaragoda 2006). Watershed management involves land use management of natural areas as well as cultivated areas, other land uses such as for housing and roads, etc. Watershed management also involves functions of institutional coordination, data gathering, processing and planning, research, etc. Conservation of watersheds has implications on other environmental sectors such as biodiversity and global climate conservation, too. For simplicity of presentation, watershed management is limited to agricultural land use and its implications. Watershed management requires the adoption of optimal on-farm land use practices such as soil conservation and adoption of a range of conservation-oriented farming practices. Such decisions will be taken by individual farmers. Consider that the cost of watershed management is the costs on soil and water conservation activities undertaken on upstream farms. The private cost of watershed management undertaken by upstream farmers is considered to be equal to the social cost of watershed management. Marginal Social Cost (MSC) of watershed conservation is shown in Figure 13.1. MSC is the cost incurred in conservation of additional hectare of land, which increases with incremental land brought under conservation.

The benefits from watershed management could be classified as upstream and downstream benefits. The upstream benefits are maintenance of soil fertility and water holding capacity of soil, resulting in sustenance or improvement in agricultural productivity. The Upstream Marginal Benefit (UMB) of conserving land is indicated in Figure 13.1. The downstream benefits are sustenance of natural water supplies, reduced silting of waterways and reservoirs and reduced land slides, etc., resulting in sustenance of the supply of domestic and irrigation water, hydropower generation, reduced cost on flood and land slide damages, etc. The Downstream Marginal Benefit (DMB) is indicated in Figure 13.1. The Marginal Social

Figure 13.1: Social and Private Optimality of Watershed Conservation

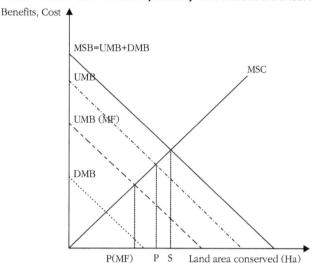

Source: Adapted by the authors.

Note: MSB = Marginal Social Benefit of Conservation; MSC = Marginal Social Cost of Conservation; UMB = Upstream Marginal Benefit (i.e., Private Benefit) of Conservation; UMB (MF) = Upstream Marginal Benefit of Conservation with Market Failures; DMB = Downstream Marginal Benefit; S = Social Optimal Extent of Cultivation; P = Private (marked) Optimum Level of Conservation.

Benefit (MSB) is the sum of DMB and UMB. Social value/welfare is maximised where MSC of watershed conservation is equal to the MSB of watershed conservation. With reference to Figure 13.1, the socially optimal level of conservation is S (as land area conserved).

In a market economic system, individuals take independent decision on production and consumption. Individual decision making is based on a comparison of private costs and benefits and not on comparison of social costs and benefits. Private decision making in a market economy is guided by market prices of resources and commodities. Where market prices do not encompass the social costs and/or benefits, private decisions will be not congruent with socially optimal resource use. Some of the benefits and costs of watershed conservation are not encompassed in market prices. Hence the socially optimal extent of land — S — will not be conserved as explained in the section below.

The upstream private benefits may be lower than UMB due to market failures. The market failures could be the lack of non-attenuated property rights for land to assure appropriation of benefits of soil conservation to the private farmers. Further as the benefits of conservation accrues over a long period of time (temporal externality) and with high rates of time preferences of particularly poor farmers, the level of benefits accrued at present will be low. The reduced upstream marginal benefits from conservation due to market failures are shown by UMB (MF) in Figure 13.1. Hence with market failures, the extent that would be conserved would be P (MF) where UMB (MF) is equal to MSC.

Social optimal level of conservation, S, will not be achieved due to spatial and temporal externalities and public goods nature of downstream benefits of watershed conservation. An externality is a situation where one person's activities (upstream watershed conservation) affect another person (downstream) and is not being compensated by the market. The spatial externality is that the upstream conservation activities undertaken at a cost by upstream farmers' leads to benefits of reduced erosion, and thereby sustained supplies of water for irrigation and hydroelectricity generation, which would be enjoyed by downstream populations. The downstream benefits also have a public goods nature. That is once upstream farmers undertake soil conservation, the downstream benefits that are not divisible, are available to all beneficiaries to benefit without rivalry. In such situations, the beneficiaries will avoid payment. The downstream users of water and electricity, who receive a public good, will not compensate the upstream farmers, who bear the cost of watershed conservation. The temporal externality is such that the cost of watershed conservation is undertaken by the present generation, whilst the benefits are enjoyed mostly by the future generation (as benefits from improved land productivity and reduced silting takes a long time to be realised).

Where the markets fail to compensate those who incur costs on watershed conservation, the level of investment by those who bear the cost would be low, as they would invest only up to where private benefits [UMB (MF)] equate private costs (MSC). Hence with market failures that reduce private benefits, the extent of conservation will be P(MF) and with corrections of market failure the extent conserved will be P. If the downstream users compensate the upstream users by the magnitude of the difference between MSB and UMB the social optimum of, S, will be achieved. In order to achieve the socially optimal area of conservation, upstream market failures should be corrected, and the downstream beneficiaries need to compensate the upstream farmers who bear the cost of conservation. Such compensation or PWS, referred to as a 'Coasian solution' in environmental economic theory, has been experimented, as mentioned earlier. As the literature suggest, wider application of PWS, particularly in large watersheds, would be limited due to high transaction costs.

The theoretical model suggests that if the market fails to coordinate between the downstream and upstream and make payments effective, the government could tax the downstream beneficiaries and finance the costs of conservation undertaken by upstream farmers, to conserve an area that is socially optimal. If the beneficiaries cannot be taxed directly due to high transaction costs or equity reasons, then the government has to finance watershed management through general taxes charged from other sectors.

Economic Significance of Watershed Conservation in Sri Lanka

The significance of watershed conservation to achieve sustainable development in Sri Lanka has been recognised by the government (Ministry of Lands and Land Development 1984, quoted in Flemming 1991). The economic contribution of the upper Mahaweli watershed is explained as a case of analysis. The upper watershed of Sri Lanka comprises the central hill region above the 300-meter contour. This region, which is about 25 per cent of the islands land area, receives approximately 37 per cent of the annual precipitation. All major rivers have their source of origin in this region. It is also a region hosting high endemic biodiversity and aesthetically appealing landscapes. The upper watershed provides a base for livelihood not only to those living in the watersheds (upstream) but also to others living outside of the watershed (downstream) who are dependent upon the flow of water for irrigated agriculture and hydropower, etc. For example the Kothmale, Victoria, Randenigala and Rantambe reservoirs fed by the Mahaweli river, provide irrigation for 300,000 ha and generate over 54 per cent of the country's hydropower (State of the Environment Report 2001).

Over the past half-century, forest in the upper watershed in particular has been subject to overexploitation, extensive and unplanned human settlement and erosive use of land for agriculture. The annual soil loss in areas ranging in elevation between 300 and 1,000 metres is more than 100 tonnes per hectare (Ministry of Forestry and Environment 2001a). This has lead to degradation of land fertility, reduced domestic water supplies, landslides, silting of streams reservoirs and depletion of the water table with the consequence of declining agricultural productivity and increasing economic costs in maintaining public utilities such as roads and reservoirs supplying irrigation water and generating electricity (Ministry of Forestry and Environment 2001b). Somarathne (2001) has estimated that the on-site cost due to soil erosion and off-site cost due to silting of reservoirs and loss of hydropower generation as LKR 953 and 15 million per year, respectively.

Several foreign financed projects have been undertaken to avert the degradation of the upper Mahaweli and other watersheds in Sri Lanka. Though these projects have achieved some desired results, given the discontinuity of financing, the sustainability of project benefits are at stake. Unless watershed conservation is undertaken with commitment of national financial allocations, degradation of watersheds in Sri Lanka would continue unabated.

Financing Watershed Management in Sri Lanka

A policy on watershed management should provide the governments intention on how to orderly acquire and allocate finances for watershed management. The financing policy should give guidance to formulate strategies on from whom, how much and how could finances be appropriated and who should, how much and how should expenditures be made on watershed management.

Level of Financing

Pilapitiya (1996) had conducted a study to determine the adequacy of public financing on environmental projects identified in the National Environmental Action Plan (NEAP) of Sri Lanka through the public investment programme of the government. According to estimates made, only 1 per cent of the Gross Domestic Product (GDP) has been spent on environmental related projects in 1995 in Sri Lanka. It is reported by Pilapitiya (ibid.) that the World Bank has recommended that a country should spend 2 to 3 per cent of the GDP on environmental investments to ensure sustainable development. The above analysis reveals that the level of financing for environmental conservation in Sri Lanka is suboptimal.

Theoretically, for development to be sustainable, one view is that investment should be equal to the sum of rents from natural resource depletion and the environmental damage (Hartwick and Olewiler 1986, quoted by Panayotou n.d.). The empirical testing of the above condition is fraught with difficulty due to problems associated with estimating rents on natural resource depletion and damage. However a 'back of the envelope' estimate based on available Sri Lankan data on investments on soil conservation and damage due to soil erosion is considered (disregarding rents of soil erosion). The investment on soil conservation is the sum of expenditures on capital (physical and human) and on soil fertility maintenance. The Natural Resources Management Center (NRMC) of the Department of Agriculture in Sri Lanka is mandated to implement the Soil Conservation Act. The annual expenditure of the NRMC from years 1995 to 1997 ranged from LKR 4.1 to 4.7 million (National Planning Department 1995–1997). The whole expenditure of the NRMC is considered as investment on soil conservation. Fertiliser is applied to replace the fertility lost due to crop cultivation. Hence expenses on fertiliser application could be considered as an expense on soil fertility maintenance. The fertiliser subsidy (which is part of the expense on fertiliser) provided to the whole country has been LKR 1,733 million in year 2000 (Central Bank of Sri Lanka 2000). Considering that the Upper Watershed is about 25 per cent of the land area of the country, the cost of fertiliser subsidy appropriable to the Upper Watershed is about LKR 425 million. Further, considering the fertiliser subsidy as 50 per cent of the total expenses (subsidy plus private cost), then total expenditure on fertiliser would be LKR 850 million (i.e., LKR 425 million each of public and private expenses). Hence the total annual investment on soil conservation is LKR 855 million (LKR 4.7 + LKR 850). The estimated loss of nutrients, which is due to erosion in the Upper Mahaweli Watershed, is LKR 953 million per annum (Somaratne 2001). It is apparent that the total investment (public and private) to maintain soil fertility (LKR 855 million) is less than the environmental damage/loss in soil fertility (LKR 953 million) in Upper Mahaweli Watersheds. Hence the level of investment is below the expected levels of investment for sustainable management of Upper Mahaweli Watershed.

Sources of Financing

In Sri Lanka between 1990 and 1999, approximately 33 projects on environmental conservation have been implemented with foreign financing, of which about 50 per cent has been on water and forest resource conservation (National Status Report on Land Degradation

2001). On the one hand, the availability of foreign financing for Sri Lanka is on the decline. With Sri Lanka's per capita income surpassing US $ 800, it is nearing the status of a middle income country, thus not warranting it to receive concessionaire foreign finances for its development investments. The global political changes have too diverted foreign finances more to East European, Russian and African nations. On the other hand, the Sri Lankan government is faced with increasing difficulty in financing public expenditures. As late as 1999, the domestic and foreign debt was LKR 1,051 billion or 95 per cent of GDP and took up more than 50 per cent of the Sri Lankan government revenue for debt servicing alone. By 2001, the debt servicing need (LKR 295 billion) has surpassed the government tax revenue (LKR 275 million). It is thus reasonable to expect lesser allocation of public finances for environmental conservation. This is so due to the following reasons. Allocation of public finances is done through a political and bureaucratic process. The politicians would favour allocating finances to sectors that yield benefits in the short term and in geographically clearly identified areas such as their own constituency. However, benefits from environmental conservation take a long time to realise and are geographically dispersed. It is apparent that financing of environmental conservation in Sri Lanka, which is at present suboptimal, would further decline in the future.

Financing Mechanisms
National Level

The national government obtains finances from domestic funds (government revenue) and from foreign loans and grants. The Ministry of Finance and Planning, via the National Treasury, allocates these funds to ministries and their line departments and agencies. The basis for the allocation is provided by the annual budgets prepared by the ministries and its departments and agencies. The budgets are examined at the ministry and forwarded to the treasury for approval. The necessary adjustments in the budgets are made at this level through a process of negotiation. Once the budgets are finalised, they are submitted to the parliament for formal approval. As it is the universal case, it is evident that allocation of public finances in Sri Lanka too is done through a political and bureaucratic process. The rationalising of financial allocation through such a process, if not guided by a policy, could be non-optimal. The environmental sector particularly would receive less than optimal allocations on reasons mentioned in an earlier section of the chapter.

Within the budget of a ministry, finances are allocated both for capital and recurrent expenditure. The capital budget allocations comprise government funds and foreign aid loans and grants while the recurrent budget allocation comprises entirely of government funds. The latter generally accounts for a relatively small proportion of the total budget of a ministry. In 1999 for example, the recurrent budgetary allocation for the Ministry of Forestry and Environment amounted to less than 30 per cent of the total budget. Experiences in the irrigation sector have clearly shown that inadequate allocation of finances for operational and maintenance has lead to poor performance of the irrigation sector (Small and Carruthers 1991). The same is applicable to the environmental sector too.

The capital budget consists of both government funds and foreign funds. The relative importance of the two sources of funds varies with each ministry and from year to year depending on the number of projects that are being implemented with donor assistance. The Ministry of Forestry and Environment implemented several foreign-funded projects pertaining to environmental management; hence approximately 75 per cent of the capital budget in 1999 and 2000 came from foreign aid. This can be observed in Table 13.1. As indicated in an earlier section of the chapter, continued dependence on foreign financing may not be sustainable. Hence national sources to finance environmental conservation are necessary.

Table 13.1: Per cent Share of Sources of Finance of the Capital Budget of the Ministry of Forestry and Environment (1999–2000)

Source	Year	
	1999	*2000*
Government allocation	26.6	22.2
Foreign aid	22.4	12.9
Reimbursable foreign aid	51.0	64.9

Source: Prepared by the authors.

Provincial

Some aspects of environmental conservation have been devolved to the jurisdiction of the provincial councils by the 13th amendment to the Sri Lankan constitution. Hence financing mechanisms of the provincial councils will have drastic influence on environmental conservation, particularly watershed management. It is known that provincial councils operate on budgetary deficiencies. In the year

2000, the total income of provincial councils have been LKR 7,534 million and the expenditure have been LKR 37,328 million. Nearly 77 per cent of the expenditure has been on salaries and other recurrent expenditures. Given the essentiality of spending on salaries, the finances available for investment on environmental conservation, which would only provide dispersed and long-term benefits, would be marginal. The central province had only identified 0.1 per cent of its total budget for environmental conservation in the year 2001 (Dissanayake 2002).

The funds at the disposal of the provincial council are from government transfers (about 77 per cent), and revenue generated by the province (about 23 per cent). In addition to the above sources, development and environmental-based activities within the province are also financed through the decentralised budget with finances allocated to members of parliament to enable them to undertake activities in their electorates. Finances are also channeled to different activities through non-devolved line agencies, statutory boards and corporations and special projects or donor-assisted projects.

The grants made by the central government to the provinces generally represent only a relatively small proportion of the finance available for investment in the provinces. In the North Central Province for example in 1998, of the LKR 2.3 billion available for investment 76 per cent came from projects, national line agencies and statuary boards and corporations. Grants to Provincial Council from the central government amounted to 18.65 per cent and the Local Authority and DCB funds to 1.96 per cent.Where income generation potential at provincial level is limited (due to existence of substantial central governmental taxes) and allocations from the central government too are restrictive, long-term investment on watershed management could be restricted at the provincial level.

Pradeshiya Sabha

The funds for Pradeshiya Sabha (regional councils) are obtained in two ways — funds provided by the provincial council for the payment of salaries of officials and funds generated by the Pradeshiya Sabha mainly through local taxes, fees and charges. These are utilised mainly for services such as maintenance of roads and disposal of wastes. The analysis and conclusion on the provincial level of declining financial allocations for environmental conservation are equally applicable to the pradeshiya sabahas, municipal councils, etc.

Non-governmental Organisations (NGO) and Community-based Organisations (CBO):

Most of the NGOs and CBOs are funded through foreign finances. Recorded information on levels of financing environmental conservation through this sector is not available. It is presumed that NGOs and CBOs substantially finance environmental conservation at the village level.

The ability to generate finances by provincial governments is very limited in Sri Lanka. The ability to acquire finance is highest at the national level through general taxes. Hence the transfer of finances from the central government to the provincial governments would be required until the provincial governments are enabled to acquire finance. The central government would need a policy guide to rationally allocate finances for environmental conservation to different ministries and provinces.

Financing Options

Financing of environmental conservation could be done through three approaches, namely reducing financial needs for environmental conservation, improving the efficiency of use of financial resource allocated to environmental conservation and through acquiring finances from environmental resource users.

Reducing Need

Financial need for watershed management could be reduced using policy options to reduce soil erosion, which would not incur financial costs. For example as experienced in 1997–1998, the increase of importation of potatoes, influenced through import tariff reduction, led to a reduction in potato cultivation in the upper watersheds (Kotagama et al. 1998). The reduction of potato cultivation would have reduced erosion as potato cultivation is a major factor causing soil erosion. Reducing soil erosion through such policy changes does not require public finances to achieve objectives of watershed conservation.

Improving Efficiency

Available financial resources could be used more effectively, thereby reducing the need for additional funds. In the year 2000, the fertiliser subsidy costs the government LKR 1,733 million. Considering that the Upper Mahaweli Watershed is about 25 per cent of the land area of the country, the cost of fertiliser subsidy applicable to the Upper Mahaweli Watershed is about LKR 425 million per year. This subsidy is a curative expenditure to partly replace the fertility lost due to soil erosion. Thus part of the fertiliser subsidy could be provided to finance soil conservation which will prevent loss of soil and fertility.

Acquiring Finances

Issues on who should pay for environmental conservation, how much should be paid and how should the payment be obtained are briefly discussed in the section that folllows, with respect to watershed management.

WHO SHOULD PAY?

Environmental conservation may be financed either by foreign or national sources and the national sources could be either charging natural and environmental resource users' or general taxes.

FOREIGN VERSUS NATIONAL

Environmental conservation may be financed by either foreign or national sources. The primary foreign sources are foreign aid and international borrowing. Some other foreign sources are the GEF under the biodiversity and greenhouse gas mitigation conventions, the CDM under the Kyoto Protocol on greenhouse gas mitigation and Nature for Debt Swap (NDS) under the United States Tropical Forest Protection Act. Finances through GEF could be obtained through clear demonstration of international benefits achievable through biodiversity conservation and mitigation of greenhouse gas emission. The incremental cost beyond meeting costs for nationally sustainable development could be obtained for specific projects. The mechanism of financing CDM is rapidly evolving. The principle is that developed countries that are required to reduce carbon emission through the Kyoto Protocol could finance projects that reduce emissions of carbon in developed countries. De Silva and Kotagama (1998) estimated the global benefit (carbon dioxide sequestration and maintenance of carbon stocks) of forest conservation in Sri Lanka as approximately LKR 91,000/hectare/year. The principle of NDS provides debt relief funding on conditions of conserving rainforests to countries that are facing difficulty in servicing international debt. Some other possibilities of generating finances for environmental conservation could be promotion of gene prospecting for the pharmaceuticals and/or agricultural breeding. Pushpakumara et al. (2002) have examined the prospects of pharmaceutical prospecting to finance biodiversity conservation in Sri Lanka and have concluded that there are 'reasonable' possibilities. Most benefits from watershed conservation are national. Hence continued financing of watershed conservation through foreign sources of financing, or by mechanisms explained above, is limited.

Foreign financing is generally sought for new project-based investments and not to cover recurrent costs. Recurrent costs must be financed from national sources. Foreign financing for environmental conservation may also be perceived as a threat to national sovereignty. Hence seeking to acquire finance nationally to conserve the environment is considered prudent in the long run. However, all efforts must be undertaken to acquire internationally available finances for environmental conservation that are provided as grants (such as GEF) and on commercial conditions if such are worthy for Sri Lanka (such as CDM). This requires not only careful scrutiny of economic worthiness but also other social and political implications.

GENERAL VERSUS SPECIFIC TAXES

Finances for public investments can be acquired from general taxes and from specific taxes imposed on resource users. The present major practice in Sri Lanka is to acquire finances through general taxation to finance environmental conservation. General taxation does not impose payments specifically for benefits derived from environmental conservation. These taxes do not influence the decision making of environmental users; therefore these does not influence better management of the environment. The current trend in Sri Lanka is towards acquiring finances through specific taxes such as resource user fees and pollution fees.

As Kallesoe and De Alwis (2005) mentions, the 'Environmental Action Plan: Caring for the Environment 2003–2007' of Sri Lanka, contains proposals for fiscal policy instruments such as taxes, charges and liability rules formulated on the polluter pay principle and subsidies to encourage environmentally less stress on environment and natural resources. The polluter pay principle is the most well known rule on deciding who should pay for use of environmental sink services. According to the Soil Conservation Act of Sri Lanka, the polluter pay principle has been accepted. Upstream farmers are required to adjust use of land to conserve watersheds. If such is not adhered to, the 'owners' of the land could be financially penalised. However, this has perhaps never been implemented. A major reason for non-implementation, other than the poverty of farmers, is the high transaction cost of implementation. The transaction cost of taxing non-point (geographically dispersed) resources users or polluters are high. Hence it is not reasonable to expect to generate finances for particularly watershed conservation from taxing the farmers who are the major users. In cases where the transaction costs of collecting specific taxes are high, general tax-based financing of environmental conservation would have to continue.

Some societies may allocate the rights to use the environment to the polluters, in which case the operative rule is the beneficiary pay principle. The beneficiary pay rule particularly applies where the polluter is considered poor, as most of the farmers upstream of watersheds, and the beneficiary of pollution abatement are richer, as electricity consumers (Panayotou n.d.). Although it is known that the benefits of watershed conservation are upstream farm productivity improvements and downstream sustenance of reservoir capacity enabling sustained supply of irrigation and hydropower, these beneficiaries (particularly the farmers), in practice, can neither be identified nor could the benefits be individually quantified. Further, where poor farmers upstream and downstream lack the ability to pay and also where it may not be equitable to tax the poor, the option that could be considered pragmatic is to tax hydropower users. However, the downstream benefit is relatively small compared to upstream benefits of soil conservation (Somarathne, 2001).

Not all farmers deriving benefits from on-site use of watersheds are poor, such as plantation companies in Sri Lanka. It may be argued that these companies can be taxed on the rent earned by using the fertility of land. These companies are paying product taxes at present and with improved productivity of plantations with watershed conservation, they may also be paying a higher quantum of product taxes. Hence it may not be appropriate to further tax these companies. Further, plantation companies may be privately financing watershed conservation activities through adoption of appropriate soil conservation, stream conservation, etc.

How much to charge?

The question of how much to charge is relevant only if there exist a potential for specific taxes. In the case of watershed conservation, a possibility is to tax the electricity users. In such case, it can be considered that the electricity user would be willing to pay to avoid the loss of electricity generation due to silting of reservoirs, which is LKR 15 million per year as estimated by Somarathne (2001).

How to Charge?

The method of acquiring finance would depend on the administrative costs of acquisition, equity and the influence on resource use efficiency. For example in the case of watershed conservation, it is easier to link a watershed conservation tax on the electricity fee, as the administrative cost will be less, given the current existence of an electricity fee. Electricity may also be having an inelastic demand paving the way for better possibilities for taxing. It also may be equitable to tax the electricity users, as they are generally richer than farmers. However, a tax on the electricity fee would not influence improved efficiency of watershed conservation, as it does not influence the decision making of the upstream watershed users.

Allocation of Finances

Who should Allocate?

The current procedure of allocation of finances is that the central government allocates finances to central ministries and to the provincial governments, and these in turn allocates to provincial implementing agencies. The sustainability and effectiveness of this system of financing environmental conservation can be improved through earmarking as specifically as possible the quantum of finances to be allocated for environmental conservation by the central government through provinces up to the level of implementing agencies. In the longrun, provincial governments must be allowed to generate the required finances, at least to finance environmental conservation needs at the provincial level.

How much should be allocated?

If the central government continues to allocate finances to provinces for environmental conservation through the central governmental budget, then allocations to provinces could be based on provincial contributions to generating environmental benefits. The provincial contribution to environmental benefits could be based on provincial income statements. For example, it has been proposed in the Concept

Paper on the Establishment of a Watershed Management Unit (Ministry of Forestry and Environment, 2000) that the proposed National Watershed Management Unit would establish an accounting system to value the benefits of watershed management. These estimates could be used to guide allocation of finances for watershed management in different provinces. The provincial budget could allocate finances to implementing departments based on project proposals for watershed management.

Conclusion

Watersheds are the foundation of the water sector. A policy to guide strategies to finance watershed management will ensure sustained water services to the society. This chapter through review of economic theory, international experience and current practices in Sri Lanka, related to financing the environment and water sector, suggests a conceptual framework to guide formulation of a policy to finance watershed management in Sri Lanka.

The present level of financing watershed management in Sri Lanka is suboptimal. On the one hand, the availability of concessionary foreign finances, upon which Sri Lanka largely depended for watershed management investments, is declining. The continued dependence on foreign finance is neither desirable nor sustainable. Foreign finances have been in the form of project investments and have not been consistent. Benefits of watershed management are largely consumed nationally; hence finances to manage watersheds should be secured nationally. On the other hand, the allocation of national public finances to watershed management may also decline given the difficulties faced in public financing in Sri Lanka. Yet, watershed management will have to continue to depend on public financing, given the 'public good' nature of the benefits of watershed management. Hence earmarking the allocation of public finance for watershed management from the central government to provincial governments and implementing agencies would be required. The earmarking of financial allocation could be guided by the value of watershed services sustained by provinces. Provincial governments must be encouraged and conditions must be created to generate finances for watershed management. Innovative market-based approaches that are being experimented internationally to generate finances for the water sector investments must be positively pursued.

References and Select Bibliography

Bandaragoda, D. J. 2006. 'Water–Land Linkages: A Relatively Neglected Issue in IWRM', in P. P. Molinga, A. Dixit and K. Athukorale (eds), *Integrated Water Resource Management: Global Theory, Emerging Practices and Local Needs*, pp. 21–37. Delhi: Sage Publications.

Barrett, A., J. Lawlor, and S. Scott. 1997. *The Fiscal System and the Polluter Pays Principle: A Case Study of Ireland*. Aldershot: Ashgate.

Biswas, A. K., O. Varis and C. Tortajada. eds. 2005. *Integrated Water Resource Management in South and South East Asia*. Delhi: Oxford University Press.

Camdessus, M. and J. Winpenny. 2003. *Financing Water for All: Report of the World Panel on Financing Water Infrastructure*. Marseille: World Water Council.

Central Bank of Sri Lanka. 2000. *Annual Report of the Central Bank of Sri Lanka*. Colombo: Government of Sri Lanka.

CPEWM. 2000. 'Concept Paper on the Establishment of a Watershed Management Unit, Upper Watershed Management Project', Ministry of Forestry and Environment, Colombo.

De Silva, S. and H. Kotagama. 1998. 'Value of Carbon Sequestration and Sink Services of Forests in Sri Lanka: Justification for International Resource Transfers for Forest Conservation', *Sri Lankan Journal of Agricultural Sciences*, 35: 1–9.

Dillaha, T., P. Ferraro, M. Huang, D. Southgate, S. Upadhyaya and Sven Wunder. 2008. 'Payments for Watershed Services Regional Synthesis', USAID PES Brief 7. http://www.oired.vt.edu/sanremcrsp/documents/PES.Sourcebook.Oct.2007/PESbrief7.Regional%20Synth.pdf (accessed 12 May 2011).

Dissanayake, D. M. C. 2002. 'An Analysis on Financial Allocation for Environmental Conservation: A Case Study Nuwareliya Municipal Council'. Unpublished thesis, Postgraduate Institute of Agriculture, University of Peradeniya.

Echavarria, M., J. Vogel, M. Albán and F. Meneses. 2004. 'The Impacts of Payments for Watershed Services in Ecuador: Emerging Lessons from Pimampiro and Cuenca'. http://pubs.iied.org/pdfs/9285IIED.pdf (accessed 12 May 2011).

Environment Foundation Ltd (EFL). 2001. 'Sri Lanka: EFL Position on Proposed Tropical Forest Agreement under US Tropical Forest Conservation Act'. http://www.elaw.org/node/1821 (accessed 12 May 2011).

Fleming, W. M. 1991. 'Watershed Management Policies of Sri Lanka', Irrigation Support Project for Asia and the Near-East Policy Paper, US Agency for International Development.

Kallesoe, M. F. and D. J. De Alwis. 2005. 'Financial Incentives for Ecosystem Conservation: A Review of Development of Markets for Environmental Services in Sri Lanka', Nature and Economics Technical Paper No. 4.

Gland: International Union for Conservation of Nature (IUCN)-The World Conservation Union.

Kotagama, H. 2001. 'Conceptual Framework for a National Policy on Financing Watershed Management', Report submitted to Upper Watershed Management Project, Ministry of Environment and Forestry, Sri Lanka.

Kotagama, H., S. Thrikawala and N. Gunawardena. 1998. 'Impact of Macro-Economic Policies on Soil Erosion: A Simulation Study', Report submitted to the Environment Division, Ministry of Forestry and Environment, Sri Lanka.

Ministry of Forestry and Environment. 2000. 'Concept Paper on the Establishment of a Watershed Management Unit, Upper Watershed Management Project'. Colombo: Government of Sri Lanka.

———. 2001a. 'National Land Use Plan (Draft)', Ministry of Forestry and Environment, Government of Sri Lanka.

———. 2001b. *State of the Environment Report*. Colombo: Government of Sri Lanka.

National Planning Department. 1995–1997. *Annual Reports of the National Planning Department*. Colombo: Ministry of National Planning, Government of Sri Lanka.

Norwegian Institute for Water Research (NIVA). 2007. 'Feasibility of Payments for Watershed Services: Part 1: A Methodological Review and Survey of Experiences in India and Asia', Report SNO 5393-2007. Oslo: NIVA.

Panayotou, T. 1993. *Financing Mechanisms for Agenda 21*. Cambridge: Harvard Institute for International Development.

———. n.d. 'Instruments of Change', International Environment Programme Report, Harvard Institute for International Development, Cambridge.

Pilapitiya, S. 1996. 'Report on Investments on NEAP Projects', Ministry of Forestry and Environment, Colombo.

Pushpakumara, D. K. N. G., H. B. Kotagama, B. Marambe, G. Gamage, K. A. I. D. Silva, L. H. P. Gunaratne and S. S. D. K. Karaluvinne. 2002. 'Prospects of Pharmaceutical Prospecting to Finance Biodiversity Conservation in Sri Lanka', *Sri Lankan Journal of Agricultural Economics*, 3 (1): 39–71.

Rees, J. A., J. Winpenny and A. W. Hall. 2008. 'Water Financing and Governance', TEC Background Paper No. 12. Stockholm: Global Water Partnership.

Small, L. E. and I. Carruthers. 1991. *Farmer-financed Irrigation: The Economics of Reform*. Cambridge: Cambridge University Press.

Somaratne W. G. 2001. 'Policy Liberalisation and the Environment in Sri Lanka: A Computable General Equilibrium Analysis' in Ministry of Forestry and Environment (ed.), *National Status Report on Land degradation Implementation of the Convention to Combat Desertification in Sri Lanka*. Colombo: Natural Resources Management Division, Ministry of Forestry and Environment.

Tecco, N. 2008. 'Financially Sustainable Investments in Developing Countries Water Sectors: What Conditions Could Promote Private Sector Involvement?', *International Environment Agreements*, 8 (2): 129–42.

Upper Watershed Management Project (UWMP). 1997. *The Feasibility Report*. Colombo: Ministry of Forestry and Environment, Government of Sri Lanka.

Van Howegen, P. 2006. 'Enhancing Access to Finance for Local Governments, Financing Water for Agriculture', Report from Task Force on Financing Water for All Chaired by Angel Gurria, World Water Council.

Water Resources Council and Secretariat. 2000a. *National Water Resources Policy and Institutional Arrangements*. Colombo: Ministry of Irrigation, Government of Sri Lanka.

———. 2000b. *Institutional Development and Capacity Building for Integrated Water Resource Management: Final Report*. Colombo: Ministry of Irrigation, Government of Sri Lanka.

Water Resources Secretariat. 2001. 'Proposed Revisions for the National Water Resources Policy', Water Resources Council, Government of Sri Lanka.

About the Editors

E. R. N. Gunawardena is Professor of Agricultural Engineering at the University of Peradeniya, Sri Lanka. Previously, he was country coordinator of the CapNet, a United Nations Development Programme (UNDP) project on capacity building in Integrated Resources Management (IWRM), the Executive Director of the South Asia Consortium for Interdisciplinary Water Resources Studies (SaciWATERS) in Hyderabad, India and the Project Director of the Crossing Boundaries Project. He has been a consultant to the Food and Agricultural Organisation (FOA), the United Nations Economic and Social Commission for Asia and the Pacific (UN-ESCAP) and the United Nations Office for Project Services (UNOPS). He has published extensively in the areas of hydrological simulation, irrigation, drainage, watershed management and has contributed to the formulation of protected areas of Sri Lanka and the forestry sector master plan. His current research interest lies in the area of Integrated Water Resources Management.

Brij Gopal is a Member of the Board of Directors of the International Society for River Science, and the Executive Vice President (Developing Countries) of the International Society of Limnology (SIL). Previously, he was Professor of Environmental Science at the Jawaharlal Nehru University, New Delhi and Visiting Professor at the Swiss Federal Institute of Technology, Zurich. He has been a member of many committees of the Ministry of Environment and Forests, the Ministry of Water Resources, Government of India, and several state governments, for their programmes on river, lake and wetland conservation. He was also one of the lead authors of the chapter on Ecosystems for the Fourth Assessment Report of the Intergovernmental Panel on Climate Change (IPCC) for which this Panel was awarded the Nobel Peace Prize in 2007.

Hemesiri Kotagama is Assistant Professor at the Department of Natural Resource Economics, Sultan Qaboos University, Sultanate of Oman. His research has been in the applied areas of Environmental and Natural resource Economics and Management. He has served in several national committees guiding environmental policy and in national technical committees evaluating major investment projects in Sri Lanka. He has been extensively involved in training professionals in Environmental Impact Assessment and Environmental Economic Valuation in the South Asian region.

Notes on Contributors

Ajith de Alwis is Professor in the Department of Chemical and Process Engineering, University of Moratuwa Sri Lanka.

Kusum Athukorala is a pioneer researcher in gender and water in Sri Lanka. She is associated with SaciWATERs, Hyderabad; Network of Women Water Professionals (NetWwater), Sri Lanka; Global Water Partnership (GWP); and Women for Water Partnership (WfWP).

Jeb A. Barzen represents the International Crane Foundation (ICF), and has worked on management issues at the Tram Chim National Park since 1988.

Shymal Chandra Bhadra is a Research Officer at the Bangladesh Centre for Advanced Studies (BCAS), Dhaka.

Jayati Chourey is Senior Fellow (Education and Networking) with SaciWATERs, Hyderabad. She also acts as the Network Manager of SaciWATERs-CapNet Network (SCaN).

Bishnu Prasad Das is Vice-chairman of the Expert Appraisal Committee of Ministry of Environment & Forests, Delhi.

Missaka Hettiarachchi is a Ph.D. researcher at the School of Geography Planning and Environmental Management of University of Queensland, Australia.

M. Anisul Islam is a senior Monitoring and Evaluation Expert and Director, Center for Natural Resource Studies (CNRS), Dhaka, Bangladesh.

Peter-John Meynell works as an independent environmental consultant based out of Lao PDR. He is a member of the IUCN Commission on Ecosystem Management.

Subhadarshi Mishra leads the Software Services and R&D divisions of the Spatial Planning and Analysis Research Centre (SPARC).

Monirul Mirza is Physical Scientist with the Adaptation and Impacts Research Section (AIRS), Environment, Canada.

N. C. Narayanan is Associate Professor at CTARA, Indian Institute of Technology, Mumbai.

Prakash Nelliyat is a Research Coordinator at the Centre for Water Resources, Anna University India.

Duong Van Ni is leading and supervising research on areas such as rural development, natural resources conservation, wetland treatments, decision support systems and integrated environmental education.

Dhruba Pant is Professor at Nepal Engineering College, Pokhara University, Nepal.

Ravi Peris is a freelance researcher currently focusing on social issues in agricultural settlement in Sri Lanka.

Kriteshwar Prasad is a senior and distinguished Professor of Geology at Patna University, Patna.

Md. Golam Rabbani is a Research Fellow of the Bangladesh Centre for Advanced Studies (BCAS), Dhaka.

Atiq Rahman is the Executive Director of the Bangladesh Centre for Advanced Studies (BCAS), Dhaka.

Mokhlesur Rahman is doing Ph.D. on climate change adaptation at Curtin University, Australia.

Rezaur Rahman is Professor at the Institute of Water and Flood Management (IWFM) of Bangladesh University of Engineering and Technology (BUET).

Martin van der Schans is working as a Water Management Consultant with Grontmij. He is a hydrogeologist by training with expertise in wetland management and the ecosystem approach.

Gill Shepherd is a Visiting Senior Fellow in the International Development Department at the London School of Economics.

Deanne Shulman is a Fire and Emergency Management Specialist with the Office of International Programmes, U.S. Department of Agriculture, Forest Service.

K. A. I. D. Silva is Director, Policy and Planning, Ministry of Environment, Government of Sri Lanka.

Ravindra Kumar Sinha is Professor and Head of Environmental Science at Central University of Bihar, Patna.

Nguyen Huu Thien is a conservationist by training with 21 years of experience in wetland conservation and rural development works. He is also acting as a wetland conservation advisor to the Plain of Reeds Wetlands Restoration Project.

Julian Thompson leads the Wetland Research Unit of the UCL Department of Geography, UK.

Tran Triet is Director, Center for Wetland Studies, Vietnam National University, Ho Chi Minh City and Coordinator of the Southeast Asia Programme of the International Crane Foundation.

Index